W0037346

# INVESTIGATION AND EXPLOITATION OF ANTIBODY COMBINING SITES

# METHODOLOGICAL SURVEYS IN BIOCHEMISTRY AND ANALYSIS

Series Editor: Eric Reid

Guildford Academic Associates
72 The Chase
Guildford GU2 5UL, United Kingdom

The series is divided into Subseries A: Analysis, and B: Biochemistry
Enquiries concerning Volumes 1–11 should be sent to the above address.

Volumes 1–10 Edited by Eric Reid

Volume 1   (B):   Separations with Zonal Rotors

Volume 2   (B):   Preparative Techniques

Volume 3   (B):   Advances with Zonal Rotors

Volume 4   (B):   Subcelullar Studies

Volume 5   (A):   Assays of Drugs and Other Trace Compounds in Biological Fluids

Volume 6   (B):   Membranous Elements and Movement of Molecules

Volume 7   (A):   Blood Drugs and Other Analytical Challenges

Volume 8   (B):   Cell Populations

Volume 9   (B):   Plant Organelles

Volume 10 (A):   Trace-Organic Sample Handling

Volume 11 (B):   Cancer-Cell Organelles
                         Edited by Eric Reid, G. M. W. Cook, and D. J. Morré

Volume 12 (A):   Drug Metabolite Isolation and Determination
                         Edited by Eric Reid and J. P. Leppard
                         (includes a cumulative compound-type index)

Volume 13 (B):   Investigation of Membrane-Located Receptors
                         Edited by Eric Reid, G. M. W. Cook, and D. J. Morré

Volume 14 (A):   Drug Determination in Therapeutic and Forensic Contexts
                         Edited by Eric Reid and Ian D. Wilson

Volume 15 (B):   Investigation and Exploitation of Antibody Combining Sites
                         Edited by Eric Reid, G. M. W. Cook, and D. J. Morré

A Continuation Order Plan is available for this series. A continuation order will bring delivery of each new volume immediately upon publication. Volumes are billed only upon actual shipment. For further information please contact the publisher.

# INVESTIGATION AND EXPLOITATION OF ANTIBODY COMBINING SITES

Edited by

## Eric Reid
Guildford Academic Associates
Guildford, United Kingdom

## G. M. W. Cook
University of Cambridge
Cambridge, United Kingdom

and

## D. J. Morré
Cancer Center
Purdue University
West Lafayette, Indiana

PLENUM PRESS • NEW YORK AND LONDON

Library of Congress Cataloging in Publication Data

International Subcellular Methodology Forum (9th: 1984: Guildford, Surrey)
    Investigation and exploitation of antibody combining sites.

    (Methodological surveys in biochemistry and analysis; v. 15 (B))
    "Based on the Ninth International Subcellular Methodology Forum entitled Antibody
combining sites, held September 3–6, 1984, in Guildford, United Kingdom"–T.p. verso.
    Includes bibliographies and index.
    1. Immunoglobulins–Congresses. 2. Binding sites (Biochemistry)–Congresses. I.
Reid, Eric, 1922–    . II. Cook, G. M. W. (Geoffrey Malcolm Weston), 1938–    . III.
Morré, D. James, 1935–    . IV. Title. V. Series: Methodological surveys in
biochemistry and analysis; v. 15. [DNLM: 1. Antibody Formation–congresses. 2. Bind-
ing Sites, Antibody–congresses. W1 ME9612NT v.15 / QW 575 I615 1985i]
QR186.7.I56   1984                            574.2'93                          85-12471
ISBN-13: 978-1-4684-5008-8      e-ISBN-13: 978-1-4684-5006-4
DOI:   10.1007/978-1-4684-5006-4

Based on the Ninth International Subcellular Methodology Forum entitled
Antibody Combining Sites, held September 3–6, 1984, in Guildford,
United Kingdom, and supported by the U.S. Army Medical Research
and Development Command

©1985 Plenum Press, New York
Softcover reprint of the hardcover 1st edition 1985
A Division of Plenum Publishing Corporation
233 Spring Street, New York, N.Y. 10013

All rights reserved

No part of this book may be reproduced, stored in a retrieval system, or transmitted,
in any form or by any means, electronic, mechanical, photocopying, microfilming,
recording, or otherwise, without written permission from the Publisher

## Senior Editor's Preface

Immunochemists as well as 'immuno users' may benefit from this techniques-oriented book. It is the outcome of a gathering of eminent investigators in September 1984, but is not from the type of 'Conference-Proceedings stable' whose horses run on a loose rein. In respect of publication texts, the individual horses have been subjected to the tight rein that is traditional in this 15 year-old book series, for the sake of a smooth and rewarding ride. A minority of authors have felt the editing to be rather fierce or fussy. (Two pet aversions, usually left unaltered, are 'extremely', connoting merely 'very', and 'incubation at 0°' rather than 'kept at 0°'.) Other authors, however, have indicated contentment that bodes well for public approbation such as past Volumes have earned. Most Forum contributors did in fact find time to prepare a publication text, maybe late but nevertheless appreciated, in the cause of disseminating 'know-how' that is rapidly gaining in power. Especially welcome were the narrative and reflections (#A-1) of the pioneer investigator, Elvin A. Kabat, whose 70th birthday coincided with his journey to the Forum. The Forum was a notable climax to D. James Morré's long and effective tenure of the Chairmanship of the Advisory Board for the Forum series. He gained a gratifying support award (to Purdue University), crucial to the calibre of the Forum and of this book, from the U.S. Army Medical Research and Development Command (Fort Detrick, Md.; ref. 2132040 375-8119).

*Other acknowledgements.* Glaxo Research and ICI Pharmaceuticals (U.K.) gave valuable support, reinforcing the USAMRDC award. Permission to reproduce published material was kindly given by Pergamon Press [art. #NC(B)-5] and, as acknowledged in the texts concerned, by other sources including *Journal of Neurochemistry*.

*Some concepts, abbreviations & conventions.* Perspective on the contents of this book that N.A. Mitchison gives in an unorthodox 'Foreword' (complementing #A-1) is rightly on a 'square-two' basis. Any reader who is understandably somewhat bewildered by 'genetic boutiques', monoclonal antibodies (MAb's) and earlier developments in the Ab field may win some help from the 'Guidance' at the end of the book, supplemented by the following explanation of anti-idiotypic (anti-Id) Ab's (from V.P. Whittaker's Forum account in *ISN News*, Oct. 1984). "Ab's generate other Ab's some of which are anti-Id, that is, they are directed against the region of the first Ab that is complementary to the original antigen and may therefore closely resemble in configuration the haptenic group in the antigen. This can be of great value to the neuropharmacologist: a receptor agonist or antagonist may generate an anti-Id Ab which resembles the agonist/antagonist and can be used to isolate, or at least identify the receptor for it." He continues: "Surface antigens are being identified which are specific for certain cell lines: this is proving extremely useful in cancer research and in neurobiology"; the contributions by Whittaker, Richardson and Raff exemplify this (Sect. #C).

Abbreviations as given within each article include those given above and on p. 346, and the following.-

ACh(E), acetylcholine(sterase)
BSA, bovine serum albumin
cDNA, complementary DNA
CCDR - *see* #NC(A)-1
EIA, enzyme immunoassay
ELISA, enzyme-linked immunosor-
        bent assay
e.m., electron microscope
Fab, antigen-binding fragment
  ('Fab fragment' a tautology?)
HLA, human leukocyte antigen
HTLV, " T-cell leukaemia virus
IF, immunofluorescence
Ig, immunoglobulin

i.p., intraperitoneal
i.v., intravenous
MHC, major histocompatibility
        complex
PAGE, polyacrylamide gel electro-
        phoresis
PBS, phosphate-buffered saline
p.m., plasma (cell) membrane
RIA, radioimmunoassay
s.c., subcutaneous
SDS, sodium dodecyl sulphate
Th, T (cell) helper
TLC, thin-layer chromatography
UV, ultraviolet (absorbance)

Throughout the book, '°' connotes °C, and 'mol. wt.' is used (a policy which may be changed for future Vols.; but '$m_r$' is a typographical nuisance) even where the author had virtuously used an SI convention. Where a previous Vol. in this series is cited, the list facing the title page may be consulted for its title   (ed. E. Reid or E. Reid et al.; acquisition of a retrospective shelf is still feasible). The 'Comments' within 'NC' do not purport to be a complete and faithful record of the lively debates at the Forum.

*Guildford Academic Associates*                    ERIC REID
*72 The Chase, Guildford GU2 5UL*
*Surrey, U.K.*                              29 April 1985

# CONTENTS

The NOTES & COMMENTS ('NC' items) at the end of each section include supplementary short articles, discussion points from the Forum which led to the book [those from the ensuing Workshop are in #NC(E)-2], and some additional references. The concluding 'Guidance' notes are an optional lead-in to the initial wide-ranging surveys (Foreword and #A-1).

----

* *Sublisted here: comments on particular articles in the section; likewise on pp. 123, 215, 275 & 333.*

x                              Contents

# List of authors

| *Primary author* | *Co-authors, with relevant name to be consulted in left column* |
|---|---|
| P. Alexander - pp. 293-298<br>Univ. of Southampton, U.K. | B. Baumann - Köhler<br>C. Berek - Griffiths |
| R.W. Baldwin - pp. 299-305<br>Univ. of Nottingham, U.K. | T.N. Bhat - Davies<br>C.R. Birkett - Gull<br>A. Bond - Roitt |
| M.E. Bardsley - pp. 219-222<br>Univ. of Southampton, U.K. | W. Born - Marrack |
| O.J. Bjerrum - pp. 231-246<br>Univ. of Copenhagen, Denmark | E. Borroni - Whittaker<br>M. Boyer - Fotedar<br>S.P.J. Brennan - Lane |
| C. Bona - p. 125<br>Mount Sinai Sch. of Medicine,<br>New York | A.K. Campbell - Luzio<br>G. Carpenter - Headon<br>S. Chamat - Hoebeke<br>Y. Chien - Davis |
| M.J. Crumpton - (i) pp. 255-265<br>(ii) pp. 279-280<br>ICRF, London | O.P. Chilson - Crumpton (ii)<br>W.L. Cleveland - Erlanger<br>G.H. Cohen - Davies |
| D.R. Davies - pp. 51-60<br>NIH, Bethesda, MD, U.S.A. | M. Cope - Headon<br>I.R. Cottingham - Bardsley<br>M.J. Darsley - Sutton |
| M.M. Davis - pp. 281-282<br>Stanford Univ., U.S.A. | A.A. Davies - Crumpton (i)<br>C.J. Dean - Hobbs<br>P. de la Paz - Sutton |
| A.B. Edmundson - pp. 33-50<br>Univ. of Utah, U.S.A. | K.R. Ely - Edmundson |
| B.F. Erlanger - pp. 91-107<br>Columbia Univ., NY, U.S.A. | T. Feizi - Hounsell |
| A. Fotedar - pp. 283-286<br>Univ. of Alberta, Canada | P. Ferretti - Whittaker<br>C. Ferroli - Morré<br>R.F. Foster - Kaufman<br>W.W. Franke - Quinlan |
| M.J. Glennie - pp. 307-315<br>Southampton Genl. Hosp., U.K. | |
| E.S. Golub - pp. 277-278<br>Purdue Univ., IN, U.S.A. | C.J. Gallagher - Crumpton (i)<br>J. Gannon - Lane |
| G.M. Griffiths - pp. 141-142<br>MRC Lab. Mol. Biol., Cambridge | N.R.J. Gascoigne - Davis<br>A.L. Gibson - Edmundson |
| K. Gull - pp. 153-166<br>Univ. of Kent, U.K. | H.C. Gooi - Hounsell<br>J.-G. Guillet - Hoebeke |

| *Primary author* | *Co-authors, with relevant name to be consulted in left column* |
|---|---|
| D.R. Headon - pp. 83-90<br>Univ. Coll., Galway, Ireland | K. Haskins - Marrack<br>F.C. Hay - Roitt |
| S.M. Hobbs - p. 143<br>Inst. of Cancer Res., Sutton,<br>Surrey, U.K. | S.M. Hedrick - Davis<br>J.N. Herron - Edmundson<br>M. Hexham - Crumpton (i)<br>B.L. Hill - Erlanger |
| J. Hoebeke - pp. 115-122<br>Inst. Jacques Monod, Paris | |
| E.F. Hounsell - 317-322<br>Clin. Res. Centre, Harrow, U.K. | |
| G.E. Isom - pp. 109-114<br>Purdue Univ., IN, U.S.A. | A. Iglesias - Köhler<br>P. Jackson - Luzio |
| E.A. Kabat - pp. 3-22<br>Columbia Univ., NY, U.S.A. | M. Kaartinen - Griffiths<br>J.M. Kanellopoulos - Crumpton (ii)<br>J.W. Kappler - Marrack |
| S.J. Kaufman - pp. 167-176<br>Univ. of Illinois, Urbana, U.S.A. | H.H.Ku - Erlanger |
| K.L. Knight - pp. 129-130<br>Univ. of Illinois, Chicago | |
| G. Köhler - pp. 133-138<br>Basel Inst. for Immunology,<br>Switzerland | |
| D.P. Lane - pp. 75-82<br>Imperial Coll., London | F.S. Larsen - Bjerrum<br>J. McCubrey - Köhler |
| J.P. Luzio - pp. 247-253<br>Univ. of Cambridge, U.K. | A. Manheimer - Bona<br>G.R. Matyas - Morré<br>C. Milstein - Griffiths |
| P. Marrack - pp. 267-274<br>Nat. Jewish Hosp., Denver, U.S.A. | S.E. Mole - Lane<br>S.E. Moore - Walsh |
| N.A. Mitchison - (i) pp. xv-xxi<br>(ii) p. 335<br>Univ. Coll. London., U.K. | B.P. Morgan - Luzio<br>S.L. Morrison - Oi |
| D.J. Morré - pp. 323-332<br>Purdue Univ., IN, U.S.A. | S. Naaby-Hansen - Bjerrum<br>R. Nayak - Walsh<br>M. Neuberger - Winter<br>B.W. O'Malley - Headon |
| V.T. Oi - pp. 131-132<br>Beckton-Dickinson Monoclonal<br>Center, Mountainview, CA, U.S.A. | R.J. Owens - Crumpton (i)<br>E.A. Padlan - Davies<br>E. Palmer - Marrack<br>A.S. Penn - Erlanger |
| R.A. Quinlan - pp. 223-226<br>German Cancer Research Center,<br>Heidelberg, FRG | K.R. Pennington - Morré<br>D.C. Phillips - Sutton<br>D. Pikaard - Morré |

| *Primary author* | *Co-authors, with relevant name to be consulted in left column* |
|---|---|
| | A.R. Rees - Sutton |
| M.C. Raff - pp. 217-218 | G.P. Richardson - Whittaker |
| Univ. Coll. London, U.K. | P.J. Roberts - Bardsley |
| | N. Roehm - Marrack |
| P.J. Richardson - pp. 207-214 | R. Sarangarajan - Erlanger |
| Univ. of Cambridge, U.K. | T.Z. Schulz - Headon |
| | J.C. Selmer - Bjerrum |
| I.M. Roitt - pp. 127-128 | B.L. Sibanda - Thornton |
| Middlesex Hosp. Med. Sch., | K. Siddle - Luzio |
| London | F.A. Simmen - Headon |
| | B. Singh - Fotedar |
| | W. Smart - Fotedar |
| | K.K. Stanley - Luzio |
| | G.T. Stevenson - Glennie |
| B.J. Sutton - pp. 63-68 | A.D. Strosberg - Hoebeke |
| Univ. of Oxford, U.K. | J.M. Styles - Hobbs |
| | |
| J.M. Thornton - pp. 23-31 | W.R. Taylor - Thornton |
| Birkbeck Coll., London | R. Tedder - Roitt |
| | Y.M. Thanavala - Roitt |
| F.S. Walsh - pp. 177-188 | A. Traunecker - Köhler |
| Inst. of Neurology, London | C. Victor-Kobrin - Bona |
| | V.P. Walter-Doelling - Morre |
| V.P. Whittaker - pp. 189-206 | N.H. Wasserman - Erlanger |
| Max-Planck-Inst. für Biophysika- | J. White - Marrack |
| lische Chemie, Göttingen, FRG | D.A. Wright - Headon |
| | J. Yagüe - Marrack |
| G. Winter - pp. 139-140 | D. Zhu - Köhler |
| MRC Lab. Mol. Biol., Cambridge | |

*Introductory Foreword on*

# EPITOPES AND RECOGNITION, ESPECIALLY T–CELL

**N.A. Mitchison**

Imperial Cancer Research Fund
Tumour Immunology Unit
University College London
London WC1E 6BT, U.K.

It was at a good moment in the history of immunology that the Forum which led to this book took place. Our subject is the combining site of the antigen-receptors of the immune system, and the epitopes which they recognize. We deal not only with the combining sites of immunoglobulin (Ig) molecules, but also with those of the T-cell receptor. This represents significant progress from what would have been possible just a short time ago, for only lately has it become possible to discuss the latter aspect in any molecular detail (as in section #D; the former aspect falls mainly in #A). It is particularly satisfactory to note that our consideration embraces both of the two major threads which led to the molecular elucidation of the T-cell receptor. The contributions of serology as in the article by Marrack et al. (#D-4), and of molecular genetics as outlined by Davis et al. [#NC(D)-3], were made essentially independently of each other. Science worked out in the way that it is supposed to do, and it was the concordance between the conclusions reached from such very different kinds of evidence that has led us all to accept so readily the present synthesis.

The presentations and discussion points in #D highlight not only the validity of our present understanding of the T-cell receptor [also appraised by Golub, in #NC(D)-1], but also the limitations of that understanding in comparison with what is known of Ig's. For both types of receptor the molecular genetics and protein structure are now understood in some detail. What is lacking for the T-cell receptor is still any system in which to examine the antigen-binding properties of isolated molecules and, therefore, of information about the configuration of their binding sites. Those who are engaged in work on the receptor are well aware of these gaps, and their presentations below illustrate the difficulties which are holding up further progress.

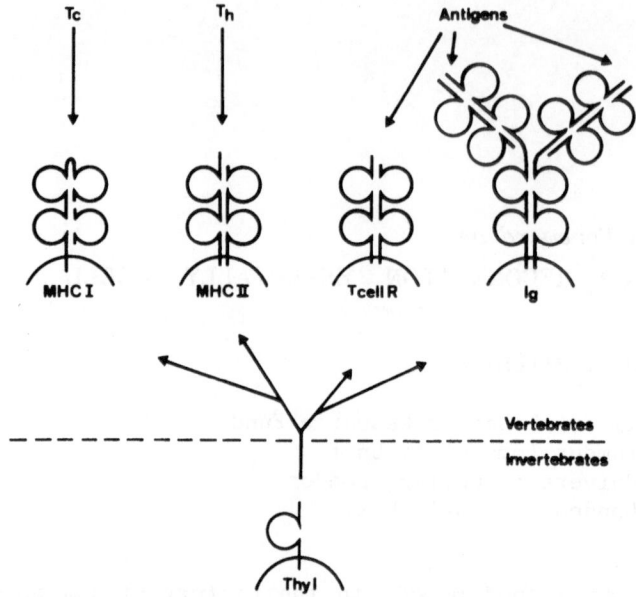

**Fig. 1.** The family of membrane glycoproteins responsible for the specificity within the immune system: multi-domain structures (Thy-1) evolve from a single-domain ancestral molecule by gene duplication.

In introducing these topics it is worth starting from an evolutionary standpoint which emphasizes the common evolution of both types of receptor. Fig. 1 outlines the evolution of the gene family to which the two types of receptor belong. It includes all the molecules which give the immune system its unique property of specificity. The family has its origin among invertebrates in the Thy-1-like molecule isolated from squid brain [1]. It has a single-domain structure, such as is represented in man by the Thy-1 molecule itself and by β2-microglobulin. The other members of what may therefore properly be termed the Thy-1 gene family have diversified through the process of gene duplication (Fig. 1). Their domain structure has been identified through tertiary structure (two layers of β-pleated sheet, usually but not always tied together by a di-sulphide bridge) and to a lesser extent through sequence homology. The family divides into two main branches, one encompassing the Ig's and the T-cell receptor, the other the major histocompatibility molecules which are themselves divided into two classes.

Note that we have very little information about the evolutionary stages in this diversification. So far as is at present known, it occurred during the early evolution of the vertebrates from their invertebrate ancestors some $400 \times 10^6$ years ago, and no group of primitive vertebrates is known in which any of these immunolo-

gical molecules are missing.    Meanwhile we may also note that
these molecules are by now scattered around the genome, more-or-less
at random, as evidenced by the Thy-1 gene family's chromosomal locations [2]:- IgH: human - 14, *mouse* - *12*; IgK: 2, *6*; Ig: 22, *16*;
Antigen R α: 14, *14*; Antigen R β: 7, *6*; Antigen R γ (an alternative to the α chain, expressed early in T-cell ontogeny): 7, *13*;
MHC Class I: 6, *17*; MHC Class II: 6, *17*; Thy-1: non-ascertainable, *9*.
This scatter is just what would be expected of such an ancient gene
family.    The only exceptions are   the Class I and II major histocompatibility. genes, which remain linked.    The reason for this
linkage is not understood:    perhaps they are subject to common
control, although the information at present available suggests
that this is not the rule.    One further evolutionary point:    the
location of Thy-1-like molecules in the squid, where they occur in
the brain, lends no support to the much-touted view that the family
traces back to molecules involved in self-recognition by our colonial
invertebrate ancestors.    It suggests rather an origin from molecules
involved in cell-cell recognition *within* a multicellular body.

The recognition of antigen by T-cells is shown in greater
detail in Fig. 2 [2] and Fig. 3 [3].    They illustrate the two most
remarkable features of this process - dual recognition, and the
involvement of subsidiary glycoproteins on the T-cell surface.    Dual
recognition, the great discovery of Zinkernagel & Doherty, receives
much attention later.    Its key element is that the T-cell receptor
does not recognize antigen alone, but rather antigen plus a major
histocompatibility molecule.    This ensures that T-cells interact with
antigen only when docked at the surface of another cell, something
which is entirely appropriate for a cell which can produce only a
local effect.    In this respect there is a marked contrast with
B-cells, which do not employ dual recognition and which release
antibodies (Ab's) as long-range effector molecules.

In contrast to any imaginable mechanism in which the antigen-
recognizing and cell-cell docking functions are separate from one
another, dual recognition also prevents T-cells from getting blocked
by antigen occurring free in body fluids, before they have reached
their proper destination.    Furthermore, dual recognition enables
TR-cells to specialize, as shown in Fig. 3.    The differentiation of
these cells into effector (Tc) and regulatory (Th and Ts) types is
accompanied and enabled by a differentiation of major histocompatibility molecules into types I and II;    by allocating expression of
these molecules to particular stimulatory and target cells, the
various specialized T-cell types are permitted to dock only at
appropriate locations.

Because of the lack of a cell-free system in which the interaction between antigen and receptor can be examined directly, the
molecular mechanism of dual recognition remains a mystery.    New

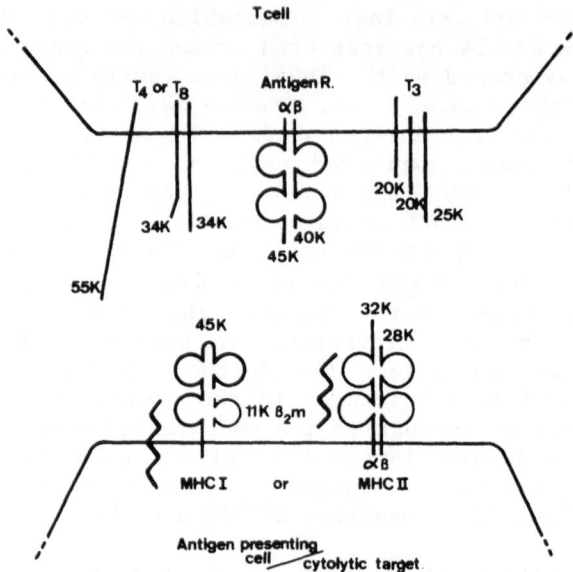

**Fig. 2.** T cells recognize antigen by a mechanism which is unique in two features: dual recognition – the antigen receptor binding to an MHC molecule and also to an antigen (shown as a zig-zag) – and subsidiary molecules ($T_3$, $T_4$, $T_8$) participate in the binding. Cytolytic cells utilize MHC I, $T_3$ and $T_8$; regulatory cells utilize MHC II, $T_3$ and $T_4$.- The 'T-cell recognition complex'.

ideas about the mechanism do emerge below, but they are as yet tenuous and supported only by indirect evidence.

The second feature of recognition, the involvement of subsidiary glycoproteins, is exemplified by the $T_3$, $T_4$ and $T_8$ molecules depicted in Fig. 2. Just how these molecules are involved in binding is unknown. They are believed to join with the receptor to form a recognition complex, on the basis of inhibition studies with monoclonal antibodies (MAb's), co-precipitation by Ab, and coordinated expression of $T_3$ and the β-chain of the receptor during thymic ontogeny [2]. These other molecules probably contribute to recognition in an optional manner, which is important only when weak binding of antigen by the T-cell receptor needs strengthening. These molecules seem to have come on in the T-cell recognition act at a fairly late stage of its evolution.

What are the implications of dual recognition in terms of epitopes recognized by T-cells? Now that attention is focusing on the second generation of vaccines based on molecular tailoring, there is an urgent need for more information about T-cell sites on vaccine molecules. This question has been the subject of several

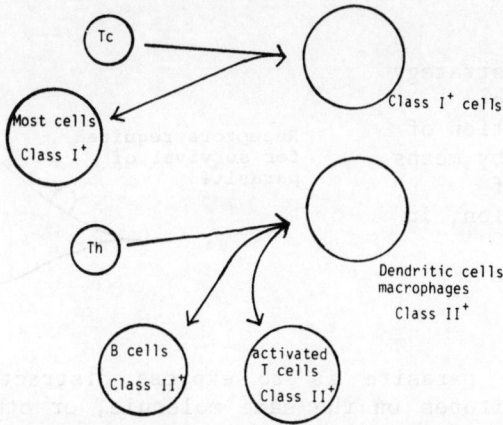

**Fig. 3.**   The differentiation of MHC molecules into class I and class II enables Tc (cytolytic cells) and Th (helper cells) to dock at appropriate stimulator and target cells (→ in-contact triggering).

recent reviews [4].   Perhaps we should begin by asking what benefits molecular tailoring has to offer, apart from cheapness and safety. I believe that there are at least two, concerned with (a) suppressor epitopes, and (b) antigenic competition.  Of these the possibility of amputating suppressor epitopes is the more obviously important, as it involves only the assumption that suppressor epitopes do in fact exist and can be identified.   This is a case which my colleagues and I have argued strongly in recent publications, following in the footsteps of Sercarz [5, 6].  We believe that the receptor repertoire for such epitopes is generated as a direct consequence of repertoire-purging by self-macromolecules   occurring within the body at intermediate concentrations.

Yet however reassuring it may be to establish that suppressor determinants do actually exist, what the vaccine developers need is something more, namely practical advice on how to spot them.  What they would like is a set of structural rules, analogous to the rules about Ab-defined epitopes which begin to emerge from consideration reflected in this book.   At the present stage of immunology they are unlikely to get them.  All that is at present on offer is, I fear, merely advice to avoid variable epitopes (used by the parasite to escape helper and effector responses, but presumably not suppressor ones), coupled with a distant hope that parasite molecular genetics (scanning  gt 11 libraries and so on [4]) will eventually, if carried out on a sufficient scale, generate such rules  empirically.

Antigenic competition is our other source of potential benefits.  Consider a parasite which needs to protect one of its molecules, e.g. a receptor, which is vital for its survival.  A logical

DISTRACTORS

**Fig. 4.** A logical strategy for parasites is to distract the attention of the immune system by means of the mechanism of antigenic competition, in the context of the 'Cloner's nightmare'.

Receptors required for survival of parasite

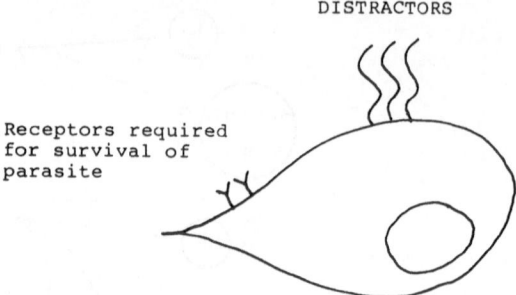

strategy for the parasite is to express distractors  - either immunodominant epitopes on the same molecule, or other molecules - able to divert competitively the attention of the immune system (Fig. 4).   While no parasite is yet known to operate this strategy, there are many examples of competition from model systems such as alloantigens [7].   Furthermore, we know that the T-cell system is more prone to this kind of competition than are B-cells as a consequence of dual recognition [8, 9] and also because T-cells are relatively more prone to homeostatic control of proliferation [10]. These distractors are at the same time the cloner's nightmare and his greatest hope.   A nightmare because gene-cloning of vaccine antigens based on screening with Ab's or T-cells is likely initially to pick up mainly distractors; and a hope because if and when the vital molecule is cloned, it should prove a more effective vaccine on its own than when used in combination with other  molecules on the rest of the parasite as in first-generation vaccines.

     Consider, in this context, Fig. 5 which illustrates the type of results that are currently being obtained with overlapping sets of peptides. We compare the results of immunization performed either with intact protein or with peptide presented one-by-one on a neutral carrier.   Epitope (1) proves active both with the intact protein and with the corresponding peptide, so no problem arises.  Epitope (2) is active with the peptide but not with the protein; this could be because intrastructural competition, say from epitope (1), inhibits the response [7], although another possibility is that the epitope is an interior structure not exposed during antigen presentation. Epitope (3) in Fig. 5 is active with the protein but not with the peptide, so perhaps it needs help such as could be furnished by Th-cells recognizing epitope [1].   Clearly  epitope  analysis is  a fascinating and worth-while enterprise, and equally clearly  it  re- quires sophisticated interpretation.   It is against  this  background that the various examples of epitope analysis that follow need to be judged.   Possible benefit to 'exploitation' studies from taking cog- nizance of epitope analyses (which feature  largely  in  the  earlier part of the book) is exemplified by subtle investigations  (late  in art. #C-1) of tubulin polypeptides with MAb's.

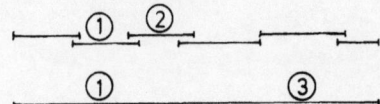

**Fig. 5.** Problems encountered with overlapping peptide sets in epitope analysis. It reveals structures of types (2) and (3) in the example, which are indicative respectively of competitive and helper interactions.

## References

1. Williams, A.F. & Gagnon, J. (1982) *Science 216*, 696-703.
2. Owen, M. & Collins, M.K.L. (1985) *Immunol. Lett. 9*, 175-183.
3. Czitrom, A.A., Sunshine, G.H., Reme, T., Ceredig, R., Glasbrook, A.L., Kelso, A. & MacDonald, H.R. (1983) *J. Immunol. 130*, 546-550.
4. Bell, R. & Torrigiani, G., eds. (1983) *New Approaches to Vaccine Development*, Schwabe, Basel, 515 pp.
5. Lukic, M.L. & Mitchison, N.A. (1984) *Eur. J. Immunol. 14*, 766-768.
6. Mitchison, N.A. (1985) *Clin. Immunol. Newsletter 6*, 12-14.
7. Clark, E.A., Lake, P. & Favila-Castillo, L. (1981) *J. Immunol. 127*, 2135-2140.
8. Mitchison, N.A. (1980) *Regulation of the immune response to cell surface antigens*, in *Regulatory T Lymphocytes* (Pernis, B. Vogel, H.J., eds.), Academic Press, New York, pp. 147-157.
9. Rock, K.L. & Benaccerraf, B. (1983) *J. Exp. Med. 157*, 1618-1634.
10. Mitchison, N.A. & Pettersson, S. (1983) *Annales d'Immunol. Inst. Pasteur 134D*, 37-43.

Fig (5). Microemulcombined with a simulated micelle pair in spatial analysis. The micelle structures of types (A) and (D) in the examples which are behaving respectively at comparative and major intersections.

References.

1. Williams, R.P. & Baramori, (1982) ...
2. Uoa, M. & Miller, M.A. ...
3. Salaron, M.A., Sunshine, C.A. ...
4. Salbo, F. & Matheson, R.B. (1981) ...

5. Hall, E. & Morthland ... (1984) New Approaches to Colloidal Recognition. Solenty, South, Sks, pp.

6. Lin, M.L. & Mitchelson, R.A. (1980) ... Francols for Inhiand

7. Mitchison, M.A. (1985) ...

8. Mccollister, M.H. (1980) Regulation of the Immune Response to ... major antigens. In Regulation Lymphocyte Response, b. Vol. 2, (L.L. ...), Academic Press, New York, pp. ...

9. Koch, K.R. & Mitchelson ... (1983) ...

10. Mitchelson ... Petersson, J. (1983) ...

Section #A

ANTIBODY COMBINING SITES, ESPECIALLY STRUCTURE

#A-1

# ANTIBODY COMBINING SITES — PAST, PRESENT AND FUTURE

**Elvin A. Kabat**

Departments of Microbiology,* Human Genetics and
Development, and Neurology, and the Cancer Center
/Institute for Cancer Research, Columbia University
College of Physicians and Surgeons
701 West 168th Street, New York, NY 10032
and
National Institute of Allergy and Infectious Diseases
National Institutes of Health
Bethesda, MD 20205, U.S.A.

*Editorial note (G.M.W.C.).— During the 9th International Sub-
cellular Methodology Forum, Dr. Kabat, as well as delivering the
keynote address, celebrated his 70th birthday. Forum participants
were privileged to listen to a retrospective look at the way in
which our concepts of antibody combining site structure have
developed over the past 50 years, from an investigator who has been
intimately involved in much of this work. In this article, based on
the keynote address, the reader is guided through the ways which
have ultimately led to our present understanding of the sizes and
shapes of antibody molecules, their specific combining sites and the
genetic control of the antibody response. The author concludes with
a summary of the limitations of our present understanding and the
way in which future studies are likely to develop.*

I shall begin by trying to present a picture of immunochemistry
when I entered the field having just finished college in September
of 1932. I hope you will forgive my omissions and recognize that
this is necessarily a somewhat partisan view, since one's excitement
and enthusiasm is influenced largely by proposing and testing new
hypotheses of one's own and by examining critically the hypotheses
and assumptions of others.

---

* This Department is the address for correspondence.

In January 1933, when I began working with Michael Heidelberger as a laboratory assistant and as his first graduate student, the chemical nature of antibodies was still controversial.  Although Felton [1] had been able to precipitate and concentrate antibodies by addition of horse antipneumococcal serum to 20 vol. of distilled water (pseudoglobulin fraction), to produce a purified and concentrated material used clinically for the treatment of lobar pneumonia, and although Heidelberger & Kendall [2] were analyzing washed antigen-antibody precipitates of nitrogen-free type III pneumococcal polysaccharide and antibody and determining the antibody as protein nitrogen by the micro-Kjeldahl method, there were papers claiming that antibody activity was demonstrable in solutions which did not contain detectable nitrogen.  Analogous types of experiments were being reported by Rohdewald & Willstatter [3] on "nitrogen-free" solutions with enzymatic activity.

Although Sumner had crystallized urease and Northrop & Kunitz were crystallizing pepsin, trypsin, etc., it was not always appreciated that tests for serological activity were much more sensitive than the then available chemical methods of detecting proteins.  The Heidelberger laboratory [4] had begun use of the quantitative precipitin reaction to study protein-antiprotein systems.  The trend towards abandoning immunization with complex mixtures of antigens was just beginning [see 5], and the powerful gel diffusion [6-8] and immunoelectrophoretic [9] methods for detecting mixtures of antibodies and antigens were more than one and two decades away.  The Svedberg ultracentrifuge had only recently come into use [10] to characterize proteins, and Tiselius [11] had not yet perfected the moving-boundary mode of electrophoresis.  There was no high-resolution X-ray crystallography of globular proteins.  At that moment in time an understanding of antibody specificity and complementarity in terms of antibody combining site structure could be said neither to have a past nor a present, and it was doubtful as to whether it had a future.

Nevertheless the concept of immunological specificity was well advanced, notably by the studies of Landsteiner, Obermayer & Pick and Breinl & Haurowitz.  In 1933 Landsteiner had published in German his first book on the specificity of serological reactions [12]; John Marrack's *The Chemistry of Antigens and Antibodies* would not appear till 1934.  Blood group antigens A, B, M, N and P on erythrocytes had been recognized [refs. in 5].  The role of the pneumococcal polysaccharides in conferring type specificity on pneumococci was clear.  Retrospectively, one can now see that the dogma of the times which insisted that only proteins were antigenic retarded the discovery of glycolipid antigens for many years.  Although Schiemann & Casper [13] had shown that immunization with pneumococcal polysaccharides protected mice against infection with many lethal doses of virulent pneumococci, and Francis & Tillett [14] had shown that humans injected with purified pneumococcal polysaccharides developed a positive wheal and erythema, this indication of antigenicity was considered to be due to protein attached to the type-specific poly-saccharide, and the failure to obtain antibody in rabbits on immuni-zation with polysaccharides essentially led many immunologists to discount or ignore the mouse studies [see 15]. During World War II Heidelberger & coworkers [16] demonstrated unequivocally that pneumo-coccal polysaccharides were antigenic in man. MacLeod, Hodges, Heidelberger & Bernhard [17] in a large field trial for the U.S. Army conclusively established the value of prophylactic immunization with these purified type-specific pneumococcal polysaccharides against lobar pneumonia of the corresponding types.

Yet even in 1951, when allergic reactions were encountered on infusion of dextran as a plasma expander, they again were ascribed to contamination with protein. It was not until I injected myself with a milligram of dextran and demonstrated at a meeting of the Subcommittee on Shock of the National Research Council [18] that I had developed wheal and erythema type skin sensitivity as well as precipitating antibodies to dextran [19], and shortly thereafter showed that the antibodies precipitated [20] $^{14}$C-biosynthetically labelled dextran, that the antigenicity of polysaccharides in man became generally accepted.

Another tremendous conceptual obstacle to developing a model of antibody site structure was the heterogeneity of antibodies [21]. Even to single antigens the antibody response to protein and even to polysaccharide antigens was extremely complex. Cross-reactions were seen the intensity of which correlated with the evolutionary rela-tionship of the various species being studied, a phenomenon first described with mixtures of antigens by Nuttall [22].

By successive absorptions of an antiserum to a single protein one could remove that portion of the antibodies cross-reacting with a distantly related species, leaving antibodies reacting with more closely related species as well as a considerable proportion which reacted only with the antigen used for immunization. This was very difficult to understand until studies by Lapresle [23, 24] and by Porter [25] showed by enzymatic digestion into fragments that serum albumin was made up of a number of antigenic determinants and that certain of these would cross-react whereas others would not. This led to extensive studies on mapping the antigenic determinants of various proteins, synthetic peptides, etc. [26]. With synthetic polypeptides and especially with azoproteins and other chemically modified proteins, additional heterogeneity of antigenic determin-ants was created since each introduced group might be attached to a different sequence in the protein and the introduced group was often sufficiently small to comprise only a portion of the antigenic determinant; thus on immunization numerous populations of antibody molecules might be formed, each of which had a site only a portion of which was complementary to the introduced group. This heterogeneity continues to make life complicated even today with respect to the interpretation of the sequence differences found among such antibodies.

During this period, physicochemical studies were being carried out on various preparations of purified antibodies although these preparations contained populations of molecules with combining sites specific for various antigenic determinants.   Tiselius and I [27] showed that in potent anti-egg albumin  and antipneumococcal sera there was a large increase in the gamma-globulin fraction which could comprise as much as 40% of the total protein;  on removal of the antibody with antigen, the decrease in gamma-globulin correspon- ded precisely to the quantity of antibody determined by the quanti- tative precipitin reaction.   Various species of antibody were found initially by the ultracentrifuge to fall into two mol. wt. classes, ~160,000 and ~900,000 [28], now termed IgG and IgM.   With the development of immunoelectrophoresis the presence of two additional classes, IgA and IgD, was manifest.  Immunoelectrophoresis demonstra- ted that these antibodies were heterogeneous in that, unlike a single protein, they showed bands migrating over a wide range of mobility, indicating that there were mixtures of proteins with dif- ferent isoelectric points.    These are a consequence of the poly- clonal response to antigen.

Equilibrium dialysis [cf. 29] had first been used by Marrack & Smith [30] to demonstrate that purified serum globulins containing anti-hapten antibodies bound the low mol. wt. dialyzable hapten reversibly.   Eisen & Karush [31] subsequently used this method to establish a valence of two for purified rabbit IgG antibody to azo- proteins.  IgM antibodies were shown to have valences of 10 [32, 33]. These findings became beautifully clear from the electron micro- graphs of Valentine & Green [34] and Svehag [35].  Dan Campbell [36] searched intensively but unsuccessfully for what he called hetero- ligating antibody, namely a bivalent molecule with antibody com- bining sites directed toward two different antigenic determinants. Karl Landsteiner had earlier shown that antibody to an azoprotein antigen with two different determinant groups attached to the same aromatic ring manifested specificity for one or another but not both [37].

## DEXTRAN–ANTIDEXTRAN SYSTEM

Important insights leading to measurements of the sizes and shapes of antigenic determinants came from quantitative precipitin and inhibition studies of homopolysaccharides, notably with the dextran-antidextran system using the sera of individuals, including myself, who had been immunized with dextrans having predominantly α1→6 linkages with small proportions of non-α1→6 linkages.  This system of dextran, antidextran and the isomaltose (α1→6 linked) oligosaccharides from the di- to the hepta- [38, 39, & see 21] and

more recently the octa- and nona-saccharides [40, 41] provided a
molecular ruler for mapping the size and fine structure of antidextran combining sites.  Essentially it was established that humans
immunized with dextran still produced heterogeneous populations of
antibody combining sites which could vary in size from those complementary to between 1 and 2 glucoses to those complementary to 6 and
7 glucoses with α1→6 linkages.   Oligosaccharides in this size range
but with one or more non-α1→6 linkages or with side-chains of glucose residues exerted lesser inhibitions.  Attempts to fractionate
these mixtures to obtain homogeneous populations of antibody combining sites, using dextran as an insoluble adsorbent (Sephadex), were
successful in separating the antidextrans into antibodies with
smaller and larger combining sites but did not yield materials
sufficiently uniform for amino acid sequencing [42, 43].   The pre-
dominant antibody formed in humans to dextran was IgG2, but IgM, IgA
and the other IgG subclasses were also produced, albeit usually in
lesser amounts [44].

Numerous other studies [see 39, 45] on polysaccharide, polypeptide, nucleotide and protein determinants were shown to have combining sites falling between the lower and upper limits for the
dextran-antidextran system.   It should be borne in mind that the
measurements of the upper and lower size limits are arbitrarily
expressed in the most extended conformation, and that many kinds of
determinants including polysaccharides may exist in more compact or
folded conformations as do proteins, in which helices, β-sheets and
disulphide bonds produce more stable 3-dimensional structures.

## FINE STRUCTURE OF ANTIBODY COMBINING SITES

These insights into antibody combining sites led to numerous
studies aimed at elucidating the fine structures [46, & see 47] of
various other antibody combining sites, using quantitative precipitin inhibition assays or, more recently, RIA and ELISA assays. These
studies were often carried out in conjunction with the isolation of
fragments of the antigens involved, as in the mapping of the combining sites of the antigenic determinants of blood groups A, B, H,
Le[a], Le[b], I and i [48, & see 39] and of various protein antigens
[see 49], notably myoglobin [50, 51] and lysozyme [52], entailing
the basic assumption that the best inhibitor of the antigen-antibody
interaction on a molar basis fitted most closely into the combining
site.  Two types of antigenic determinants were defined − sequential
and conformational, the former being those to linear chains of polysaccharides or amino acid residues whereas the latter involved
determinants formed by non-contiguous amino acid residues brought
into close proximity by 3-dimensional folding [53].

Landsteiner [54, 55] and Avery & Goebel [56] had shown with
amino acids and with disaccharides which were coupled to protein
that the end amino acid and the non-reducing sugar moiety were

immunodominant.   An additional important parameter in classifying
antigenic determinants was the recognition, among polysaccharide
antigens, that antibodies might be directed not only against the
non-reducing ends of chains with the one or two sugars at the non-
reducing ends contributing a substantial proportion of the total
binding energy, but that other antibodies reacted with interior
linear segments of sugar chains.   Although they were first recognized by their differences in association constants for methyl α-D-
glucoside and isomaltose, relative to the larger isomaltose oligosaccharides [57], this distinction could clearly be made in the
α1→6 dextran-antidextran system [57] using a completely linear dextran of ~200 sugar units [58].   Since it had only one non-reducing
end it could not precipitate with the antibodies specific for the
non-reducing ends but was multivalent towards and hence precipitated
with antibodies specific for interior chains.

We have termed the sites specific for the non-reducing end
*cavity-type* and those specific for the internal chains, and resemb-
ling the lysozyme site, *groove-like*.   The evidence to date only imp-
licates the two non-reducing residues [57, 59, 60] as being held in
three dimensions, and the remaining sugars may well be held in a
depression or groove.   The cavity-type could be a deeper portion of
the site holding the non-reducing end more firmly.   Studies in other
systems have given comparable results [61].

More recent inhibition studies of blood-group antigenic determinants, notably anti-I [48] and anti-B [62], using sets of synthetic oligosaccharides and monoclonal antibodies (MAb's), have made
possible more extensive definition of the fine structure of these
combining sites, and further established the importance of hydrophobic interactions [39, 63] in polysaccharide-antipolysaccharide
systems.

The hybridoma technique of Köhler & Milstein [64] provided
additional insights into the nature of the repertoire of MAb's to the
α1→6 antidextrans:   12 BALB/c α1→6 antidextrans [65, 66] were shown
to precipitate with a synthetic linear dextran, thus having groove-
type sites. Some of the sites were complementary to 6 and others to 7
α1→6 linked glucoses.  However, their association constants by affinity gel electrophoresis were different, as were the ratios of relative inhibiting power of the isomaltose oligosaccharides to one
another.   Thus their combining sites all differed in fine structure
indicating that, even to a single antigenic determinant, populations
of groove-type antibody combining sites were formed in addition to
the distinction between groove- and cavity-type sites made earlier
[57].

Using hybridoma antibodies, this heterogeneity of sites with
respect to single antigenic determinants of protein antigens has
become very clear, necessitating extensive revision [49] of the

inferences of Atassi [51, 52] that single antigenic determinants would always give rise to antibodies of the same specificity. The conclusion of Benjamin et al. [49] is reasonable that "The surface of a protein antigen consists of a complex array of overlapping potential antigenic determinants; in aggregate these approach a continuum. Most determinants depend upon the conformational integrity of the native molecule". The data on antipolysaccharide MAb's had led to the same conclusion, save that numerous conformations besides the preferred conformation might exist in solution. It is not known whether or not antibodies to more than the most stable conformations can be formed.

## STRUCTURAL CHEMISTRY OF ANTIBODY MOLECULES

The studies outlined above provided a degree of insight into the fine structure of antibody combining sites. The indication was that there would be many distinct sites even to a single antigenic determinant, but a detailed knowledge of the structural chemistry of antibody molecules was required beyond what was known about classes and subclasses.

Parfentiev [67, 68] had obtained a patent on the digestion of diphtheria horse antitoxic sera with pepsin at a pH close to 4; this treatment produced a fragment with intact antitoxic activity but which had only $\frac{2}{3}$rds the mol. wt. This material came into widespread use, supplanting whole diphtheria horse antitoxin. Porter in 1958 had shown that, on papain digestion, rabbit IgG and rabbit IgG antibodies would split into three fragments separable on carboxymethyl-cellulose, the first two of which, Fab fragments, each possessed one intact antibody combining site whereas the third, termed Fc, crystallized and had no antibody activity [69, 70]. Pepsin digestion was subsequently found by Nisonoff to give a bivalent molecule which on reduction split into two monovalent fragments [71].

Isoelectric focussing showed whole IgG to give a large number of bands, whereas fewer bands were obtained with purified antibodies or monoclonal myeloma proteins and Waldenstrom macroglobulins.

When whole IgG or monoclonal immunoglobulins were reduced with cysteine or mercaptoethanol, alkylated and subjected to urea starch gel electrophoresis at acid pH, two very diffuse bands were obtained with whole IgG, but with the monoclonals two sharp bands differing in mol. wt. were obtained, the faster-moving being the light and the slower the heavy chain. Bence Jones proteins from a given individual when reduced and alkylated were shown by Edelman & Gally [72] to have the same light-chain bands as found with the intact monoclonal myeloma protein but the light chains from different patients varied. In urea starch gel at pH 7.0, reduced and alkylated whole IgG of various species gave a series of ~10 sharp bands whereas myeloma proteins gave single bands [73]. These differences were not surprising since earlier studies had shown the Bence Jones proteins of different individuals to differ in electrophoretic mobility and in antigenic specificity.

The stage was now set for amino acid sequencing of immunoglobu- lin chains, the technique having been introduced by Sanger in the 1950's. The monoclonal Bence Jones proteins which were obtainable in large amounts were the ideal choice, and sequencing was begun by various groups in the early 1960's. Fortunately a large number of mouse monoclonal proteins had also been produced in BALB/c mice by injection of paraffin oil, and many of these were later found to show antibody activity. These mice developed plasmacytomas and many excreted Bence Jones proteins in the urine [74, 75]. Hilschmann & Craig [76] sequenced two human Bence Jones proteins and found many differences in the amino-terminal half, the first 107 residues, whereas the carboxy-terminal half was essentially constant; these were termed the variable or V- and the constant or C-regions respec- tively. These studies were rapidly substantiated and it became clear that antibody chains were built upon entirely different prin- ciples from other proteins.

As more sequences accumulated several points became evident from examination of homologized V- and C-region sequences. The C- regions were essentially not different from other proteins, in that they showed relatively limited sequence variation, whereas the sequence variability in the two V-regions was extraordinary. When the first mouse myeloma protein was sequenced it showed a lack of species-associated residues, the two human chains differing at more positions than did the mouse chain from one of the human chains [77]. Several studies [78-80] in 1967 classified human $\kappa$ light chains into three subgroups, but certain positions of the V-region did not con- form to the classification in that they had many more substitutions. Of especial significance was the finding that 5 of 15 $V_\kappa I$ and 3 of 9 $V_\kappa III$ chains were identical in the sequence of amino acids 1 to 23. It was clear that this segment of the V-region could not be making contact with antigen in the site since specificity differences would be minimal. Thus other portions of the $V_L$ regions were involved in site complementarity. Similar findings were noted for the first 33 residues of the $V_H$ region [81, 82].

As I analyzed this finding in 1968 [82] the subgroups of human $\kappa$ Bence Jones protein recognized by sequences of residues 1 to 23 (the position of the first cysteine), although they might be under the control of three genes, would not be involved in the genetic control of antibody complementarity except possibly insofar as this terminal chain may have a stabilizing effect on the antibody combin- ing site or on the 3-dimensional folding of the molecule.

## ANTIBODY COMPLEMENTARITY

In the late-1960's there were thought to be two possible ways of forming antibody combining sites. One was that different portions of the V-region could fold differently to form various sites and the second was that the site would always be in the same place as in other proteins with receptor sites, e.g. enzymes. When complete and partial sequence data on 77 Bence Jones proteins and immunoglobulin light chains had been published, Dr. Wu and I [83] undertook to examine these two alternatives.

We defined the parameter:

$$\text{Variability} = \frac{\text{Number of different amino acids at a given position}}{\text{Frequency of the most common amino acid at that position}}$$

When the variability at each of the positions in the V-region was plotted against residue number, three regions of hypervariability were noted, and it was predicted that the chain would fold so that these hypervariable regions formed the walls of the antibody combining site with the rest of the $V_L$ region constituting a framework. Thus all antibody combining sites would be formed by the same segments of the V-region. A model was suggested in which the genes coding for hypervariable regions would be inserted into genes coding for the framework; this could contribute to the generation of diversity [83].

Shortly thereafter Weigert et al. [84] showed that mouse $V_\lambda$ chains were very highly restricted in their framework segments, about 12 chains now having the identical sequence which was termed the $\lambda_o$ germline gene, and that all amino acid substitutions were associated with the hypervariable regions. Since all substitutions from the germline sequence involved single base changes, this variation was ascribed to somatic mutation. When sufficient sequences of $V_H$ chains became available, they too were found to have three corresponding hypervariable segments [85]. Affinity labelling studies [see 39, 86] showed that residues in the hypervariable regions were labelled and when high-resolution X-ray crystallography studies on Fab fragments and $V_L$ dimers became available, the hypervariable residues as defined by the variability plots were seen to form the walls of the antibody combining site [see 39, 87-89]. Accordingly we have termed them complementarity-determining segments (CDR) with the remainder of the chain constituting the framework (FR).

Variability plots have recently been used to localize the polymorphisms in class I [90] and class II [91] antigens of the major histocompatibility complex. They were not found on examination of 67 cytochromes $c$ [see 92].

The discovery of idiotypes[*] by Oudin [93] and by Kunkel [94] introduced another important and puzzling parameter into the understanding of antibody complementarity. Antibodies specific for idiotypic determinants were formed when purified antibodies were injected into animals of the same species (homo anti-idiotypic) or of different species (hetero anti-idiotypic). A great simplification was effected when MAb's were used to produce the anti-idiotypic sera. Idiotypic determinants are antigenic determinants (idiotypes) occurring on the Fv fragment of an antibody molecule; many idiotype-anti-idiotype reactions are inhibited by the antigen used to induce the antibody possessing the idiotype.

Idiotypic antibodies are an important regulatory element of the immune response and are basic to the Jerne Network Theory. Indeed, either spontaneously or on prolonged immunization the antibody initially produced by injection of antigen may induce the formation of anti-idiotypic antibody in the same animal. A unique feature of the induction of anti-idiotypic antibodies was shown by Oudin & Casenave [95, 96] in that the same idiotype may be present on antibody specific for different determinants on the antigen used for immunization (ovalbumin and serum albumin) as well as on antibody molecules with no detectable antibody specificity to the antigen; these idiotypes were not present in the pre-immune serum.

These findings established that there were two distinct but interrelated universes associated with Fv fragments, a universe of antibody specificities and a universe of idiotypic specificities. Boyd & Bernard [97] showed convincingly 47 years ago that immunization with a given antigen also produced a rise in non-antibody- as well as in antibody-globulin, and the idiotypic data provide an explanation for this.

From the structural point of view one may consider idiotypes which are inhibitable by antigen to be directed towards the amino acid side chains on the outer surface of the antibody combining site [39, 98; cf. 99]. This would readily account for the same idiotype on different antigenic determinants and for the non-antibody induced by immunization, since the exterior amino acid side chains forming the idiotype might be similar or identical whereas the internal side chains forming the various antibody sites could be different. This hypothesis will ultimately be verified by X-ray crystallography. The widely used term 'internal image' would appear to be a non-structural expression of this [100].

## SEQUENCE DATA AND THE USE OF THE COMPUTER

It became evident by 1973 or 1974 that further progress towards explaining antibody site complementarity and specificity would

---

[*]The 'idiotype' concept here outlined is mentioned in the Preface too. - *Ed*.

depend upon ready accessibility to the rapidly increasing body of sequence data.    While spending a sabbatical year, 1974-75, as a Fogarty scholar at NIH, I was introduced to the PROPHET computer network supported by NIH and to its originator, Dr. William F. Raub. It proved extraordinarily valuable in keeping track of variable region sequences and it was decided, when my term as a Fogarty Scholar ended, that I would continue to keep track of all V-region sequences spending two days a week at NIH and continuing the collaboration with Tai Te Wu and with Howard Bilofsky and later with Margaret Reid-Miller and Harold Perry of Bolt, Beranek & Newman who were responsible for the PROPHET network.    This has continued up to the present time, and three compilations have been published, in 1970, 1979 and 1983.

The scope of the data has increased from only variable regions. The second edition included constant regions and $\beta_2$ microglobulins, and the third has a self-explanatory title, *Sequences of Proteins of Immunological Interest.  Tabulation and analysis of amino acid and nucleic acid sequences of precursors, V-regions, C-regions, J-chain, $\beta_2$-microglobulins, major histocompatibility antigens, Thy-1, complement, C-reactive protein, thymopoietin, post-gammaglobulin and $\alpha_2$-microglobulin.*  Future editions will include the now explosively developing data on the T cell receptor for antigen, interleukins, etc.

It has always been rather amusing to me to realize that the usefulness of the book depended in large part on the decision to list each V-region amino acid sequence vertically so that each column represents a given domain with FR and CDR segments separated by lines.  This made all of the pages equivalent and when nucleotide sequences  were added this domain format was preserved.  It is thus readily possible to compare codons and amino acid sequences, most simply by photocopying one of the two sets of data placing it alongside the other. We have continuously devoted much time to designing simple ways of organizing the data so as to make it most useful especially to individuals who do not have access to a computer.   I personally find studying data in the book provides the best leads about new problems and suggests questions to ask the computer rather than using the computer directly.

Planning the structure of the book had some extraordinary rewards.  For the first edition, in order to make the existing data rapidly available, sequences of chains of various species were entered essentially as they were located by searching the literature. For the second edition we attempted to rearrange them so that sequences with identical FR1 were placed together, and then attempted to order CDR1,  FR2, etc.  We discovered that the FR segments appeared to assort independently [99], viz. sequences identical in a given FR1 could be associated with different FR2, etc.  This suggested that the genetic units might be the FR segments and by implication the CDR segments.  Although at the time too few identical CDR's

were available to demonstrate assortment, we had reported that a
$V_\lambda II$ and a $V_\lambda V$ which had 21 differences in V-region nevertheless had
an identical CDR1 of 14 residues [101]; Klapper & Capra [102] subsequently reported a $V_H I$ and a $V_H III$ with 30 amino acid differences
to have an identical CDR3 of 9 residues consistent with FR segments
also assorting.   We called these assorting FR and CDR segments
minigenes [101, & see 103].

At that time a $V_\lambda$ clone from a 12-day old mouse embryo had been
sequenced by Tonegawa and was found to contain a leader sequence, an
intervening sequence   and then coded for $V_\lambda$ from residues 1 to 95
[104].    Our assortment data had indicated that FR4 was missing.
Since the adult myeloma V-region had 108 residues, we suggested [99]
a process of somatic assembly by which FR4 was added between the
twelfth day of embryonic life and the time the adult myeloma developed.   Shortly thereafter the Tonegawa group isolated a second clone
from a 12-day old embryo containing the segment coding for amino
acids 96 through 108, which they termed the J segment followed by a
long intervening sequence and the exon for the C-region [105].   As
only FR's had been assorted we could not know that two residues of
CDR3 plus all of FR4 were involved.   Since the J segment was preceded and followed by intervening sequences it was a minigene as we
had defined it.   Subsequent studies demonstrated three other J segments in mouse $\lambda$  [106, 107], 5 J segments (one, J 3, a pseudogene)
in mouse $\kappa$ chains [108, 109], 5 J in human $\kappa$  [110] and 4  in rat  $\kappa$
[111, 112] and 4 J in mouse [113-115], and 6 J [116] in human H
chains all with introns before and after their coding segments.

When $V_H$ regions were cloned it was  found that the $V_H$ and $J_H$
coding segments were not contiguous, $V_H$  ending after FR3 and $J_H$
just beginning close to FR4 (nucleotides coding for 5 [113] and 14
amino acids being missing) [114]   and it was postulated that a D
(diversity) segment existed which comprised most of CDR3.

Subsequently a D exon was located — a second minigene;  4 human
[116] and 12 mouse [117-119] D minigenes have been sequenced.   The
functional $V_H$ region involves VDJ joining; CDR3 may vary substantially in length largely based on differences in codons involved at
the DJ junction.   Moreover at the DJ junction additional nucleotides
termed the N region may be inserted [120];   their role is not under-
stood.   The recently described β-chain of the human T cell receptor
for antigen also has D and J minigenes, the D minigene being inferred
from the non-contiguous nucleotide sequences of $V_{T\beta}$ and $J_{T\beta}$, the
former encoding residues 1 to 94 and the latter 101-115 [121-124];
this suggests the presence of a $D_T$ minigene.   These minigenes have
now been found and sequenced [125, 126];   of seven D regions
sequenced, all differ [127].

The Tonegawa group have recently cloned what may be the α chain
of the T-cell receptor for antigen [128].   One can anticipate that

studies of the diversity of T-cell clones will contribute many additional sequences to our data bank and hopefully will throw much
light on T-cell helper asnd suppressor functions and T cell subpopulations and their interactions with β cells, macrophages, etc.
Of especial interest are the findings that the T-cell receptor gene
in the mouse is on the same chromosome (chromosome 6) as the κ chain
gene [129], but in the human it is on a chromosome (chromosome 7)
[130] not known to carry immunoglobulin genes or those of the major
histocompatibility complex. This area of study is definitely a wave
of the future.

In the germ line each V-region minus J in the light chain and
minus D and J in the heavy chain occurs as a contiguous nucleotide
sequence;   individual V-regions are separated by considerable distances and families of $V_H$ genes occur in groups [see 131].   These
data could appear to be in conflict with the assortment data, but
Egel [132] and Baltimore [133] have proposed that a gene conversion
mechanism would account for the discrepancy, and Krawinkel et al.
[134] have found a double recombinant in which a segment containing
CDR1 and CDR2 of one gene was joined somatically to one containing
CDR3 of another gene presumably by gene conversion.

Substantial additional data supporting assortment [92, 99, 135,
136] and gene conversions have been presented by Komoromy & Wall
[137] at the nucleotide level using probes of mouse FR1, FR2 and FR3
and by the Zachau laboratory [138-140] for human $V_k$I chains using
additional amino acid data from sequenced $V_K$ clones.   The latter
group found substantial increased diversity among the CDR's compared
with the FR's;   substantial diversity was also seen in the introns
and flanking sequences. Their data, therefore, accord with the idea
that the sequences would result from recombinations or gene conversions of assembled minigenes. However, they note that stretches of
identical sequences tend to span several subregions.

A region between the 5' flanking sequence and the intron preceding the signal sequence also showed low variability comparable to
the FR segments;   it is termed L (for low copy number).   Clarke &
Rudikoff [141] also report gene conversions involving phosphoryl-
choline-binding myelomas.

There are other aspects of the data which require analysis. One
of these is the finding that segments of the human D2 and D1 minigenes code for portions of CDR2 rather than in the expected CDR3
[142];   14, 13, 12 [142-146] and 14 [K. Bernstein & R.G. Mage, pers.
comm.] nucleotides of human D2 matched human $V_H$III and $V_H$II genomic
clones and mouse and rabbit cDNA sequences respectively and 8 nucleotides of the human D1 minigenes matched a rabbit cDNA sequence
[145].   A clone with almost all of CDR2 deleted has also been described [146].

Another is the remarkable preservation of FR2 segments in $V_K$ and $V_H$ over evolutionary time both at the amino acid and at the nucleotide level. Thus one human, 21 mouse and 13 rabbit $V_\kappa$ chains have the identical amino acid sequence for FR2 residues 35 through 49 [135]. In the two instances sequenced, a human $V_H$III germ line gene [147] and a rabbit cDNA [145] have identical nucleotides coding for residues 36 through 47 with one and two nucleotide differences in codons for residues 48 and 49. The other segments of the V-region differ considerably. Moreover the FR2 of $V_\kappa$ is not the only sequence found, there being many alternate forms of FR2 in both the mouse and the rabbit chains although these occur infrequently [135]. If all of the $V_\kappa$ genes (minus J) occur sequentially in the genome it becomes important to explain why one FR2 segment should occur in such high frequency over 80 million years.

Yet another seminal finding is that two MAb's against different antigens, anti-NP and anti-GAT, had the same germ-line V gene (minus D and J) with only one amino acid difference and one change in a third base in residues 1 to 94, and expressed the $NP^b$ or the CGAT idiotype when joined with $\lambda$ or with $\kappa$ light chains respectively [148]. Differences in the D and J segments would also be contributing to the specificity differences.

## THE FUTURE DIRECTION OF RESEARCH ON ANTIBODY COMBINING SITES

This survey essentially brings us to the present with respect to antibody combining sites. The history has been comprehensively surveyed from a somewhat different point of view by Kindt & Capra [149] in a book most appropriately titled *The Antibody Enigma*. As to the future, we have a long way to go. It seemed impossible a few years back to imagine how we could ever account for the enormous repertoire of antibody specificities. At the rate that the T-cell receptor work is going, we should be hard pressed to keep up with the new $V_T$, $D_T$ and $J_T$ data for our book of sequences. We are now facing what may prove to be an even more complicated problem. The genetic findings have established so much diversity — many different $V_L$ and $V_H$ genes, $D_H$, $J_H$ and $J_L$ minigenes; additional junctional diversity created by $V_H D_H$, $V_L J_L$ and $D_H J_H$ joining; the mysterious extra nucleotides inserted at the DJ junction (the N region) and diversity created by assortment of assembled $V_H$ and $V_L$ chains — that now one must not only question but also study how much of this diversity is affecting antibody complementarity and how much is noise generated by these recombinational mechanisms.

Such investigation will require additional high-resolution X-ray crystallographic structures, hopefully for systems whose combining sites have been or can be mapped by immunochemical methods. We sorely need a precise understanding of contacting residues in combining sites in relation to binding and to idiotypic specificity: whether the idiotypic determinants are on the outer surface of the site and the extent to which framework procedures can influence or be part of idiotypic determinants. We also need more information as to the extent of variation in sites, the precise delineation of anticarbohydrate combining sites termed groove-type and cavity-type, and the shapes of sites of antigenic determinants of proteins in which the determinant is held in a fairly rigid as compared with a more mobile 3-dimensional orientation [150].

It would also be important to evaluate the effects of differences in lengths of the CDR's and how they affect the shapes and structures of the sites as well as the extent to which framework differences can affect site complementarity. There are very important aspects of site complementarity and idiotypic specificity which can be learned by $V_H$-$V_L$ recombination among antibodies of similar specificities as well as among antibodies of unrelated specificities.

Also of unusual interest are the findings that the $J_T\beta$ segments, but not the proposed $J_T\alpha$ [128], generally contain Gly-X-Gly at positions 108 and 110 like those in immunoglobulin $J_L$ at residues 99 and 101 and $J_H$ at positions 104 and 106, which are essentially invariant. In 1967 I suggested that these would function as a pivot, providing flexibility for the CDR's to adjust optimally in binding antigen [151].

The minigene hypothesis is acquiring more significance from the recent findings of Reth & Alt [152] that an actual mini-protein consisting of $D_H J_H C_H$ is synthesized by B cells and that the D minigene has its own 5' transcriptional promoter. T cells also synthesize a $D_T\beta J_T\beta C_T$ mini-protein and $D_T\beta$ also has its own 5' promoter [153]. The role of these mini-proteins in regulation might prove startling.

## Acknowledgements

Work in the laboratories named on the opening page is supported by National Science Foundation Grant NSF PCM-81-02321, National Institutes of Health Grant AI 19042 to E.A.K. and CA 13696 to the Cancer Center, Columbia University. Work with the PROPHET computer system is supported by the National Cancer Institute, National Institute of Allergy and Infectious Diseases, the National Institute of Arthritis, Diabetes, and Digestive and Kidney Diseases, the National Institute of General Medical Sciences, and Division of Research Resources (contract NO12-RR8-2118) of the National Institutes of Health.

## References

1. Felton, L.D. (1928) *J. Inf. Dis. 42*, 248-255.
2. Heidelberger, M. & Kendall, F.E. (1929) *J. Exp. Med. 50*, 809-823.
3. Rohdewald, M. & Willstatter, R. (1934) *Hoppe-Seyler's Z. Physiol. Chem. 225*, 103-124, & *229*, 241-254.

4.  Heidelberger, M. (1956) *Lectures in Immunochemistry*, Academic Press, New York, 150 pp.
5.  Landsteiner, K. (1945) *The Specificity of Serological Reactions*, 2nd edn., Harvard Univ. Press, Cambridge, Mass., 297 pp.
6.  Oudin, J. (1946) *C.R. Acad. Sci. Paris 222*, 115–116.
7.  Ouchterlony, O. (1949) *Acta Path. Microbiol. Scand. 26*, 507–515.
8.  Elek, S.D. (1949) *Br. J. Exp. Path. 30*, 484–500.
9.  Grabar, P. & Williams, C.A. Jr. (1955) *Biochim. Biophys. Acta 17*, 67–74.
10. Svedberg, T. & Pedersen, K.O. (1940) *The Ultracentrifuge*, Clarendon Press, Oxford, 478 pp.
11. Tiselius, A. (1937) *Trans. Farad. Soc. 33*, 524–531.
12. Landsteiner, K. (1933) *Die Spezifizitat der serologischen Reaktionen*, Springer, Berlin, 123 pp. [cf. ref. 5].
13. Schiemann, G. & Casper, W. (1927) *Z. Hyg. Infektionskr. 108*, 220–257.
14. Francis, T. Jr. & Tillett, W.S. (1930) *J. Exp. Med. 52*, 573–585.
15. White, B. (1938) *The Biology of the Pneumococcus*, Commonwealth Fund, New York, 769 pp.
16. Heidelberger, M., Macleod C.M., Kaiser, S.J. & Robinson, B. (1946) *J. Exp. Med. 83*, 303–320.
17. Macleod, C.M., Hodges, R.G., Heidelberger, M. & Bernhard, W.G. (1945) *J. Exp. Med. 82*, 445–465.
18. Kabat, E.A. (1983) *Ann. Rev. Immunol. 1*, 1–32.
19. Kabat, E.A. & Berg, D. (1953) *J. Immunol. 70*, 514–532.
20. Kabat, E.A., Berg, D., Rittenberg, D., Pontecorvo, L., Eidinoff, M.L. & Hellman, L. (1954) *J. Am. Chem. Soc. 76*, 564–566.
21. Kabat, E.A. (1961) Kabat and Mayer's *Experimental Immunochemistry*, 2nd edn., Chas. C. Thomas, Chicago, 905 pp.
22. Nuttall, G.H.F. (1904) *Blood Immunity and Relationship*, Cambridge University Press, Cambridge, U.K., 444 pp.
23. Lapresle, C. (1955) *Ann. Inst. Pasteur 89*, 654–665.
24. Lapresle, C. & Durieux, J. (1957) *Ann. Inst. Pasteur 92*, 62–73.
25. Porter, R.R. (1957) *Biochem. J. 66*, 677–686.
26. Berzofsky, J.A. & Berkower, I.J. (1984) in *Fundamental Immunology* (Paul, W.E., ed.), Raven Press, New York, pp. 595–644.
27. Tiselius, A. & Kabat, E.A. (1939) *J. Exp. Med. 69*, 119–131.
28. Kabat, E.A. (1939) *J. Exp. Med. 69*, 103–118.
29. Klotz, I.M. (1953) in *The Proteins* (Neurath, H. & Bailey, K., eds.), *Vol. 1B*, Academic Press, New York, pp. 727–806.
30. Marrack, J.R. & Smith, F.C. (1932) *J. Exp. Path. 13*, 394–402.
31. Eisen, H. & Karush, F. (1949) *J. Am. Chem. Soc. 71*, 363–364.
32. Merler, E., Karlin, L. & Matsumoto, S. (1968) *J. Biol. Chem. 243*, 386–390.
33. Ashman, R.F. & Metzger, H. (1968) *J. Biol. Chem. 244*, 3405–3414.
34. Valentine, R.C. & Green, N.M. (1967) *J. Mol. Biol. 27*, 615–617.
35. Svehag, S.E. (1973) in *3rd Int. Convocation on Immunol.* (Pressman, D., Tomasi, T.B., Grossberg, A.L. & Rose, N.R., eds.), S. Karger, Basel, pp. 80–91.

36. Campbell, D.H. & Bulman, N. (1952) *Fortschr. Chem. Org. Naturstoffe 9*, 443-484.
37. Landsteiner, K. & van der Scheer, J. (1938) *J. Exp. Med. 67*, 709-723.
38. Kabat, E.A. (1956) *J. Immunol. 77*, 377-385; (1960) *84*, 82-85.
39. Kabat, E.A. (1976) *Structural Concepts in Immunology and Immunochemistry*, 2nd edn., Holt Rinehart & Winston, N.Y., 547 pp.
40. Newman, B.A. (1984) PhD. Dissertation, Columbia University.*
41. Lai, E. (1984) PhD. Dissertation, Columbia University.*
42. Schlossman, S.F. & Kabat, E.A. (1962) *J. Exp. Med. 116*, 535-552.
43. Gelzer, J. & Kabat, E.A. (1964) *Immunochemistry 1*, 303-316.
44. Yount, W.J., Dorner, M.M., Kunkel, H.G. & Kabat, E.A. (1968) *J. Exp. Med. 127*, 633-646.
45. Goodman, J.W. (1975) in *The Antigens, Vol. 3* (Sela, M., ed.), Academic Press, New York, pp. 127-187.
46. Rao, A.S., Liao, J., Kabat, E.A., Osserman, E.F., Harboe, M. & Nimmich, W. (1984) *J. Biol. Chem. 259*, 1018-1026.
47. Kabat, E.A. (1983) in *Fifth Int. Congr. Immunol. - Progress in Immunology, Vol. 5* (Yamamura, Y. & Tada, T., eds.), Academic Press, New York & Tokyo, pp. 67-85.
48. Lemieux, R.V., Wong, T.C., Liao, J. & Kabat, E.A. (1984) *Mol. Immunol. 21*, 751-759.
49. Benjamin, D.C., Berzofsky, J.A., East, I.J., Gurd, F.R.N., Hannum, C., Leach, S.J., Margoliash, E., Michael, J.G., Miller, A., Prager, E.M., Reichlin, M., Sercarz, E.E., Smith-Gill, S.J., Todd, P.E. & Wilson, A.C. (1984) *Ann. Rev. Immunol. 2*, 67-101.
50. Crumpton, M.J. (1974) in *The Antigens. Vol. 2* (Sela, M., ed.), Academic Press, New York, pp. 1-78.
51. Atassi, M.Z. (1975) *Immunochemistry 12*, 423-438.
52. Atassi, M.Z. & Lee, C.-L. (1978) *Biochem. J. 171*, 429-434.
53. Hurrell, J.G.R., Smith, J.A., Todd, P.E. & Leach, S.J. (1980) *Immunochemistry 14*, 283-288.
54. Landsteiner, K. & van der Scheer, J. (1932) *J. Exp. Med. 55*, 781-796.
55. Landsteiner, K. & van der Scheer, J. (1934) *J. Exp. Med. 60*, 769-780.
56. Goebel, W.F., Avery, O.T. & Babers, F.H. (1934) *J. Exp. Med. 60*, 599-617.
57. Cisar, J.O., Kabat, E.A., Dorner, M.M. & Liao, J. (1975) *J. Exp. Med. 142*, 435-459.
58. Ruckel, E.R. & Schuerch, C. (1967) *Biopolymers 5*, 515-523.
59. Bennett, L.G. & Glaudemans, C.P.J. (1979) *Carbohydrate Res. 72*, 315-319.
60. Bhattacharjee, A.K., Das, M.K., Roy, A. & Glaudemans, C.P.J. (1981) *Mol. Immunol. 18*, 277-280.
61. Roy, A., Manjula, B.N. & Glaudemans, C.P.J. (1981) *Mol. Immunol. 18*, 79-84.
62. Lemieux, R.U., Venot, A.P., Spohr, U., Bird, P., Mandal, G., Morischima, N. & Hindsgaul, O. (1985), in preparation.

_____

* & (with Kabat) submitted for publication.

63. Lemieux, R.U. (1982) in *The Binding of Carbohydrate Structures with Antibodies and Lectin*, IUPAC Frontiers of Chemistry (Laidler, K.J., ed.), Pergamon, New York, pp. 3-24.
64. Köhler, G. & Milstein, C. (1976) *Eur. J. Immunol. 6*, 389-397.
65. Sharon, J., Kabat, E.A. & Morrison, S. (1982) *Mol. Immunol. 19*, 375-388.
66. Sharon, J., Kabat, E.A. & Morrison, S. (1962) *Mol. Immunol. 19*, 389-397.
67. Parfentiev, I.A. (1936) U.S. Patent 2,065,196.
68. Parfentiev, I.A. (1938) U.S. Patent 2,123,198.
69. Porter, R.R. (1958) *Nature 182*, 670-671.
70. Porter, R.R. (1959) *Biochem. J. 73*, 119-127.
71. Nisonoff, A., Wissler, F.C. & Woernley, D.L. (1959) *Biochem. Biophys. Res. Comm. 1*, 318-322.
72. Edelman, G.M. & Gally, J.A. (1962) *J. Exp. Med. 116*, 207-227.
73. Cohen, S. (1966) *Proc. Roy. Soc. B 166*, 114-123.
74. Potter, M. (1972) *Physiol. Rev. 52*, 631-719.
75. Potter, M. (1977) *Adv. Immunol. 25*, 141-211.
76. Hilschmann, N. & Craig, L.C. (1965) *Proc. Nat. Acad. Sci. 53*, 1403-1409.
77. Kabat, E.A. (1967) *Proc. Nat. Acad. Sci. 57*, 1345-1349.
78. Milstein, C. (1967) *Nature 216*, 330-332.
79. Niall, H.D. & Edman, P. (1967) *Nature 216*, 262-263.
80. Hood, L., Gray, W.R., Sanders, B.G. & Dreyer, W.J. (1967) *Cold Spr. Harb. Symp. Quant. Biol. 32*, 133-146.
81. Kabat, E.A. (1968) *Proc. Nat. Acad. Sci. 59*, 613-619.
82. Kabat, E.A. (1968) *Landsteiner Centennial. Ann. N.Y. Acad. Sci. 169*, 43-54.
83. Wu, T.T. & Kabat, E.A. (1970) *J. Exp. Med. 132*, 211-250.
84. Weigert, M., Cesari, I.M., Yonkevich, S. & Cohn, M. (1970) *Nature 228*, 1045-1047.
85. Kabat, E.A. & Wu, T.T. (1971) *Ann.N.Y. Acad. Sci. 190*, 382-393.
86. Givol, D. (1974) *Essays in Biochem. 10*, 1-31.
87. Davies, D.R. & Metzger, H. (1983) *Ann. Rev. Immunol. 1*, 87-117.
88. Novotny,J., Bruccoleri, R., Newell, J., Murphy, D., Haber, E. & Karplus, M. (1983) *J. Biol. Chem. 258*, 14433-14437.
89. Marquart, M. & Deisenhofer, J. (1982) *Immunol. Today 3*, 160-166.
90. Lopez de Castro, J.A., Strominger, J.L., Strong, D.M. & Orr, H.T. (1982) *Proc. Nat. Acad. Sci. 79*, 3813-3817.
91. Benoist, C.O.,Mathis, D.J., Kanter, M.R., Williams, V.E. & McDevitt, H.O. (1983) *Cell 34*, 169-177.
92. Kabat, E.A., Wu, T.T., Bilofsky, H., Reid-Miller, M. & Perry, H. (1983) *Sequences of Proteins of Immunological Interest*, U.S. Dept. of Health & Human Sciences, Public Health Service, National Institutes of Health, Bethesda, MD, 323 pp.
93. Oudin, J. & Michel, M. (1963) *C.R. Soc. Biol. Paris 257*, 805-808.
94. Kunkel, H.G., Mannik, M. & Williams, R.C. (1963) *Science 140*, 1218-1219.
95. Oudin, J. & Casenave, P.-A. (1971) *Proc. Nat. Acad. Sci. 68*, 2616-2620.

96. Casenave, P.-A., Ternynck, T. & Avrameas, S. (1974) *Proc. Nat. Acad. Sci. 71*, 4500-4502.
97. Boyd, W.C. & Bernard, H. (1937) *J. Immunol. 33*, 111-122.
98. Kabat, E.A. (1984) *Idiotypic determinants, minigenes and the antibody combining site*, in *The Biology of Idiotypes* (Greene, M.I. & Nisonoff, A., eds.), Plenum Press, New York, pp. 3-17.
99. Kabat, E.A., Wu, T.T. & Bilofsky, H. (1978) *Proc. Nat. Acad. Sci. 75*, 2429-2433.
100. Rajewsky, K. & Takemori, T. (1983) *Ann. Rev. Immunol. 1*, 569-607.
101. Wu, T.T., Kabat, E.A. & Bilofsky, H. (1975) *Proc. Nat. Acad. Sci. 72*, 5107-5110.
102. Klapper, D.G. & Capra, J.D. (1976) *Ann. Immunol. (Paris) 127c*, 233-235.
103. Kabat, E.A. (1980) *J. Immunol. 125*, 961-969.
104. Tonegawa, S., Maxam, A.M., Tizard, R., Bernard, O. & Gilbert, W. (1978) *Proc. Nat. Acad. Sci. 75*, 1485-1489.
105. Bernard, O., Hozumi, N. & Tonegawa, S. (1978) *Cell 15*, 1133-1144.
106. Blomberg, B. & Tonegawa, S. (1982) *Proc. Nat. Acad. Sci. 79*, 530-533.
107. Blomberg, B., Traunecker, A., Eisen, H.N. & Tonegawa, S. (1981) *Proc. Nat. Acad. Sci. 78*, 3765-3769.
108. Max, E.E., Seidman, J.G. & Leder, P. (1979) *Proc. Nat. Acad. Sci. 76*, 3450-3454.
109. Sakano, H., Huppi, K., Heinrich, G. & Tonegawa, S. (1979) *Nature 280*, 288-294.
110. Hieter, P.A., Maizel, J.V. & Leder, P. (1982) *J. Biol. Chem. 257*, 1516-1522.
111. Sheppard, H.W. & Gutman, G.A. (1982) *Cell 29*, 121-127.
112. Burstein, Y., Breiner, A.V., Brandt, C.R., Milcarek, C., Sweet, R.W., Warszawski, D., Ziv, E. & Schechter, I. (1982) *Proc. Nat. Acad. Sci. 79*, 5993-5997.
113. Early, P.W., Huang, H., Davis, M., Calame, K. & Hood, L. (1980) *Cell 19*, 981-992.
114. Sakano, H., Maki, R., Kurosawa, Y., Roeder, W. & Tonegawa, S. (1980) *Nature 286*, 676-683.
115. Gough, N.M. & Bernard, O. (1981) *Proc. Nat. Acad. Sci. 78*, 509-513.
116. Siebenlist, V., Ravetch, J.V., Korsmeyer, S., Waldman, T. & Leder, P. (1981) *Nature 294*, 631-635.
117. Sakano, H., Kurosawa, Y., Weigert, M. & Tonegawa, S. (1981) *Nature 290*, 562-565.
118. Kurosawa, Y. & Tonegawa, S. (1982) *J. Exp. Med. 155*, 201-218.
119. Wood, C. & Tonegawa, S. (1983) *Proc. Nat. Acad. Sci. 80*, 3030-3034.
120. Alt, F.W. & Baltimore, D. (1982) *Proc. Nat. Acad. Sci. 79*, 4118-4122.
121. Yanagi, Y., Yoshikai, Y., Leggett, K., Clark, S.P., Aleksander, I. & Mak, T.W. (1984) *Nature 308*, 145-149.
122. Hedrick, S.M., Nielsen, E.A., Kavale, J., Cohen, D.I. & Davis, M.M. (1984) *Nature 308*, 154-158.
123. Hedrick, S.M., Cohen, D.I., Nielsen, E.A. & Davis, M.M. (1984) *Nature 308*, 149-153.
124. Chien, Y.-H., Gascoigne, N.R.J., Kavaler, J., Lee, N.E. & Davis, M.M. (1984) *Nature 309*, 322-326.

125. Kavaler, J., Davis, M.M. & Chien, Y.H. (1984) *Nature 310*, 421-423.
126. Malissen, M., Minard, K., Mjolsness, S., Kronenberg, M., Governman, J., Hunkapillar, T., Prystowsky, M.B., Yoshikai, Y., Fitch, F., Mak, T.W. & Hood, L. (1984) *Cell 38*, 1101-1110.
127. Patten, P., Yokota, Y., Rothbard, J., Arai, K.-I. & Davis, M.M. (1984) *Nature 312*, 40-46.
128. Saito, H., Kranz, D.M., Tagaki, Y., Hayday, A.C., Eisen, H.N. & Tonegawa, S. (1984) *Nature 309*, 757-762.
129. Lee, N.E., D'Eustachio, P., Pravtcheva, D., Ruddle, F.H., Hedrick, S.M. & Davis, M.M. (1984) *J. Exp. Med. 160*, 905-913.
130. Caccia, N., Kronenberg, M., Saxe, D., Haars, R., Bruns, G.A.P., Goverman, J., Malissen, M., Hunt, W., Yoshikai, Y., Simon, M., Hood, L. & Mak, T.W. (1984) *Cell 37*, 1091-1099.
131. Dildrop, R. (1984) *Immunology Today 5*, 85-86 & *insert*.
132. Egel, R. (1981) *Nature 290*, 191-192.
133. Baltimore, D. (1981) *Cell 24*, 592-594.
134. Krawinkel, U., Zoebelein, G., Bruggemann, M., Radbruch, A., Rajewsky, K. & Bayreuther, K. (1983) *Proc. Nat. Acad. Sci. 80*, 4997-5001.
135. Kabat, E.A., Wu, T.T. & Bilofsky, H. (1979) *J. Exp. Med. 149*, 1299-1313.
136. Kabat, E.A., Wu, T.T. & Bilofsky, H. (1980) *J. Exp. Med. 152*, 72-84.
137. Komoromy, M. & Wall,R. (1981) in *ICN-UCLA Symposium - Immunoglobulin Idiotypes* (Saneway, C., et al., eds.), pp. 59-64.
138. Jaenichen, H.R., Pech, M., Lindemaier, W., Wildgruber, N. & Zachau, H.G. (1984) *Nucleic Acids Res. 12*, 5249-5263.
139. Straubinger, B., Pech, M., Muhlebach, K., Jaenichen, H.R., Bauer, H.G. & Zachau, H.G. (1984) *Nucleic Acids Res. 12*, 5265-5275.
140. Pech, M., Jaenichen, H.R., Pohlenz, H.D., Neumaier, P.S., Klobeck, H.G. & Zachau, H.G. (1984) *J. Mol. Biol. 176*, 189-204.
141. Clarke, S.H. & Rudikoff, S. (1984) *J. Exp. Med. 159*, 773-782.
142. Wu, T.T. & Kabat, E.A. (1982) *Proc. Nat. Acad. Sci. 79*, 5031-5032.
143. Rechavi, G.,Ram, D., Glazer, L., Zakut, R. & Givol, D. (1983) *Proc. Nat. Acad. Sci. 80*, 855-859.
144. Loh, D., Bothwell, A.L.M., Whit-Scharf, M.E., Imanish-Kari, T. & Baltimore, D. (1983) *Cell 33*, 85-93.
145. Bernstein, K.E., Reddy, E.P., Alexander, C.B. & Mage, R. (1982) *Nature 300*, 74-76.
146. Takahashi, N., Noma, T. & Honjo, T. (1984) *Proc. Nat. Acad. Sci. 81*, 5194-5198.
147. Matthyssens, G. & Rabbits, T.H. (1980) *Proc. Nat. Acad. Sci. 77*, 6561-6565.
148. Rocca-Serra, J., Tonelle, C. & Fougereau, M. (1983) *Nature 304*, 353-355.
149. Kindt, T.J. & Capra, J.D. (1984) *The Antibody Enigma*, Plenum Press, New York, 270 pp.
150. Tainer, J.A., Getzoff, E.D., Alexander, H., Houghten, R.A., Olsen, A.J., Lerner, R.A. & Hendrickson, W. (1984) *Nature 312*, 127-134.
151. Kabat, E.A. (1967) *Proc. Nat. Acad. Sci. 58*, 229-232.
152. Reth, M.G. & Alt, F.W. (1984) *Nature 312*, 418-423.
153. Siu, G., Kronenberg, M., Strauss, E., Haars, R., Mak, T.W. & Hood, L. (1984) *Nature 311*, 344-350.

#A-2

# VARIABILITY, MODELLING AND PREDICTION OF β-HAIRPINS WITH REFERENCE TO THE IMMUNOGLOBULIN FOLD

J.M. Thornton, B.L. Sibanda and W.R. Taylor

Laboratory of Molecular Biology
Crystallography Department
Birkbeck College
Malet Street, London WC1E 7HX, U.K.

*In the immunoglobulin (Ig) β-sandwich domain there are two types of connections between sequentially adjacent β-strands. One links adjacent hydrogen-bonded strands - the β-hairpin; the other links strands on opposite sides of the sandwich forming a β-arch. The antibody (Ab) combining site is built from these loop regions, and insertions and deletions found in related sequences commonly occur here. To understand more about the relationship between structure and sequence, we have studied β-hairpins and β-arches in many proteins of known structure, using coordinates taken from the Protein Data Bank. We find that the 'tight' hairpins - classified by the lengths and conformations of the loop regions - form distinct families with characteristic sequences. We are using the results of this study to help define template or consensus sequences. These have been incorporated into a supersecondary structure prediction algorithm and are found to improve prediction accuracy. In addition this information can be used in model-building homologous proteins.*

Although Ig's are very large complex molecules, high-resolution X-ray crystallography has shown that, as with all large proteins, they are built from much smaller domains [1-4]. These domains are all very similar and comprise the familiar β-sandwich structure in which two β-sheets stack together, stabilized by an inter-sheet disulphide bridge. The β-sandwich structure is by no means specific to the Ig's. It is found with different topologies and connectivities in a variety of proteins which have no apparent sequence or functional similarity (Fig. 1).

Many workers have studied the packing together of β-sheets [5,6] and have shown that when stacking occurs, the strands of the two

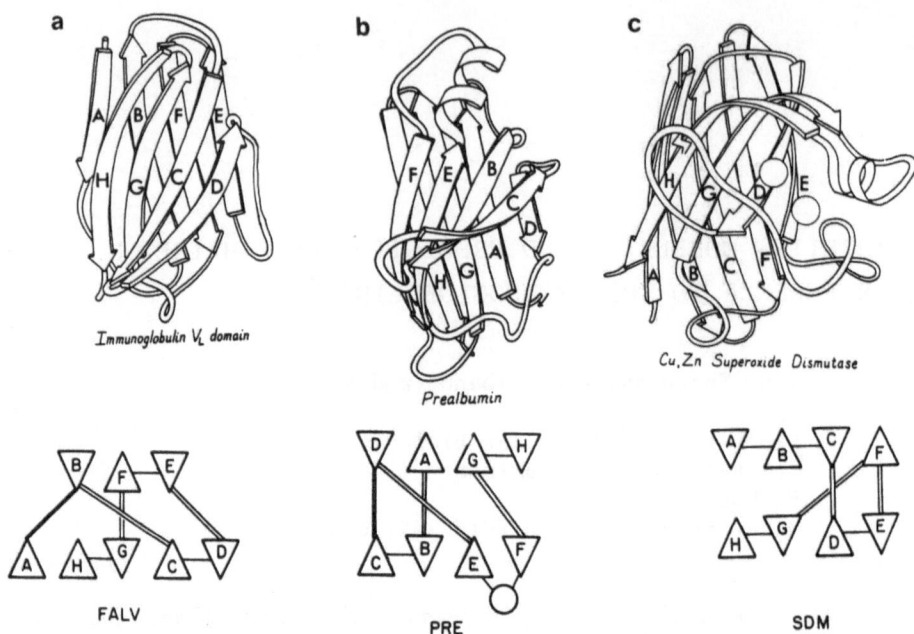

**Fig. 1.** Various β–sandwich proteins. They have non–homologous sequences and are not functionally related. For the triangle–and–circle diagrams, each β–sheet is viewed along the strand direction. Each β–strand is represented by a triangle whose apex points up or down according to whether the strand is viewed from the N or the C end. β–Hairpin connections between strands are shown in single lines, β–arch connections in double lines.
**(a)** Ig – light chain: variable domain from Fab'(New) [1].
**(b)** Prealbumin [8]. **(c)** Copper, zinc dismutase.
*'Ribbon' diagrams taken*, with some modifications, from Richardson [7].*

sheets do not lie parallel but are twisted anticlockwise about an axis perpendicular to the plane of the two sheets at an angle of between 20° and 50°. This is observed within the Ig domains, as can be seen in Fig. 1a. Detailed comparisons of the constant and variable regions show that sequence and structure are more highly conserved in the strands and that insertions and deletions occur in the connecting loop regions [9]. From a structural viewpoint the connecting loop regions have been largely ignored, reflecting the idea that bends are purely passive in the folding process. This extreme viewpoint suggests that the structures of the loop regions are essentially 'random', and any sequence can be accommodated. To study the crystallographic evidence we have initiated detailed analyses of the sequences and 3-dimensional structures of the loop regions in proteins. These regions are important for the following reasons.-
(1) In the Ig's the Ab combining site is essentially 'built' from 6 loops each connecting a pair of β–strands. The light and the

---

* by kind permission of Dr. Jane S. Richardson as copyright-holder.

**Fig. 2.** Definition of loop
connections in β-sandwich
structures:-    β-hairpin, with
hydrogen bonds between two
sequential strands, and β-arch
- with no hydrogen bonds between
the sequential strands, which
usually lie on opposite sides of
the sandwich.

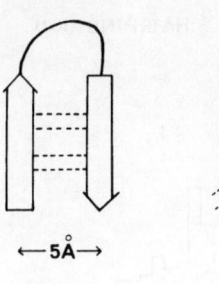

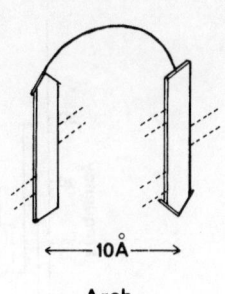

←—5Å—→

**Hairpin**

←——10Å——→

**Arch**

heavy chains each contribute three of these loops – the hypervariable regions, where changes in sequence allow recognition
of different antigens. Can we 'predict' the structures of the
loops?

(2) Since insertions/deletions in sequence occur mainly in the loop
regions of proteins [10], model-building an unknown structure by
homology with a related structure essentially involves model-
building the loop regions.

(3) Secondary structure prediction commonly goes wrong in the loop
regions, incorrectly predicting sheet or helix. A thorough
investigation of loop sequences may improve their identification.

(4) For tertiary structure prediction the location of loop regions
is critical.

The prediction of structure of the all-β proteins [11] is of
particular importance in the Ig's. Having identified a protein as
all-β    (using secondary structure prediction methods [12] or spectroscopic measurements), it is quite probable that the polypeptide
chain will fold into a β-sandwich structure. The question now
becomes "which of the possible β-sandwich topologies is the most
likely?" - e.g. the Ig fold, a jelly roll (e.g. Tomato Bushy Stunt
virus structure [13]), the prealbumin fold [8], etc. Considering
the β-sandwich structures from the 'connections' point of view
(see Fig. 1), it becomes apparent that there are two distinct categories of connections (Fig. 2), the β-hairpins and the β-arches. If
the sequences/structures of these loop regions can in any way be
used to distinguish the type of connection, this would be a tremendous help for predicting the chain fold. Greek key strand patterns
[14] appear to form the 'nucleus' of these structures, and if these
could also be located from the sequence, more reliable predictions
may be possible.

This article summarizes the results of an analysis of all the
β-hairpin loops derived from 62 proteins of known structure, including REI  [2] and Fab'(New) [1]. Some preliminary results for the
β-arch connections will also be given, to indicate that distinction
between these two categories of loops may be possible. Finally a
short consideration of the loops in the Ig's, their structures and
sequences will be presented.

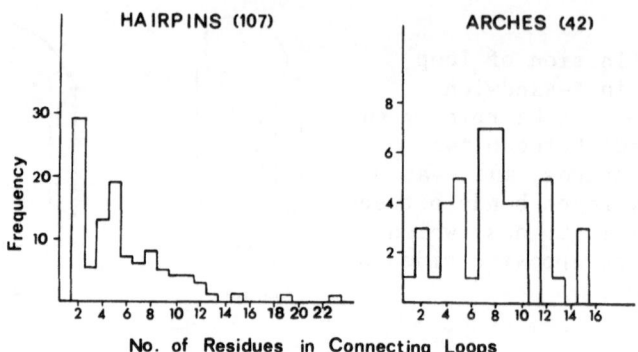

**Fig. 3.** Length of loop connections for β-hairpins *(left)* and β-arches *(right)*. The data are taken from Kabsch & Sander's secondary structure dictionary [16] for 62 proteins. A residue is assigned to the loop region if it is not in the β-ladder which forms the hairpin [16] or if it is not assigned as β-sheet for the arch.

## DATA

The coordinate data for the analyses came from the Cambridge Protein Data Bank [15]. The first step in our classification was to consider the number of residues in the loop region, necessitating a non-subjective definition of β-strands and coil regions. The assignments given in the 62 proteins included in Kabsch & Sander's secondary structures dictionary [16] were used, as defined using their powerful Pascal programme (now available on the Protein Data Bank [15]). The results presented, especially for the length of the loop regions, depend critically on these definitions, but visual inspection of many hairpins using computer graphics has shown these assignments to be remarkably consistent with a visual interpretation of the structure. The 62 proteins include several homologous families (e.g. 5 serine proteases), and to avoid biassing the results, whilst including all the available data, only one example of a homologous family of hairpins was included if the hairpin sequences were more than 50% homologous. Amongst the proteins of the original data set, 39 had β-hairpins, giving a total of 107 non-homologous hairpins.

## LENGTH OF CONNECTION LOOPS

Fig. 3 shows the number of residues in the loop region for 107 β-hairpins and 42 β-arches. The difference in the two distributions is striking, with β-hairpins clearly dominated by short loops (≤ 5 residues) whilst the connections for β-arches are usually much longer, 7-8 residues being the most common loop lengths. In the Ig structure there are several short loop regions [e.g. between strands (E & F) in Fig. 1a] and several long loop arches [e.g. between strands F & G in Fig. 1a] (see below).

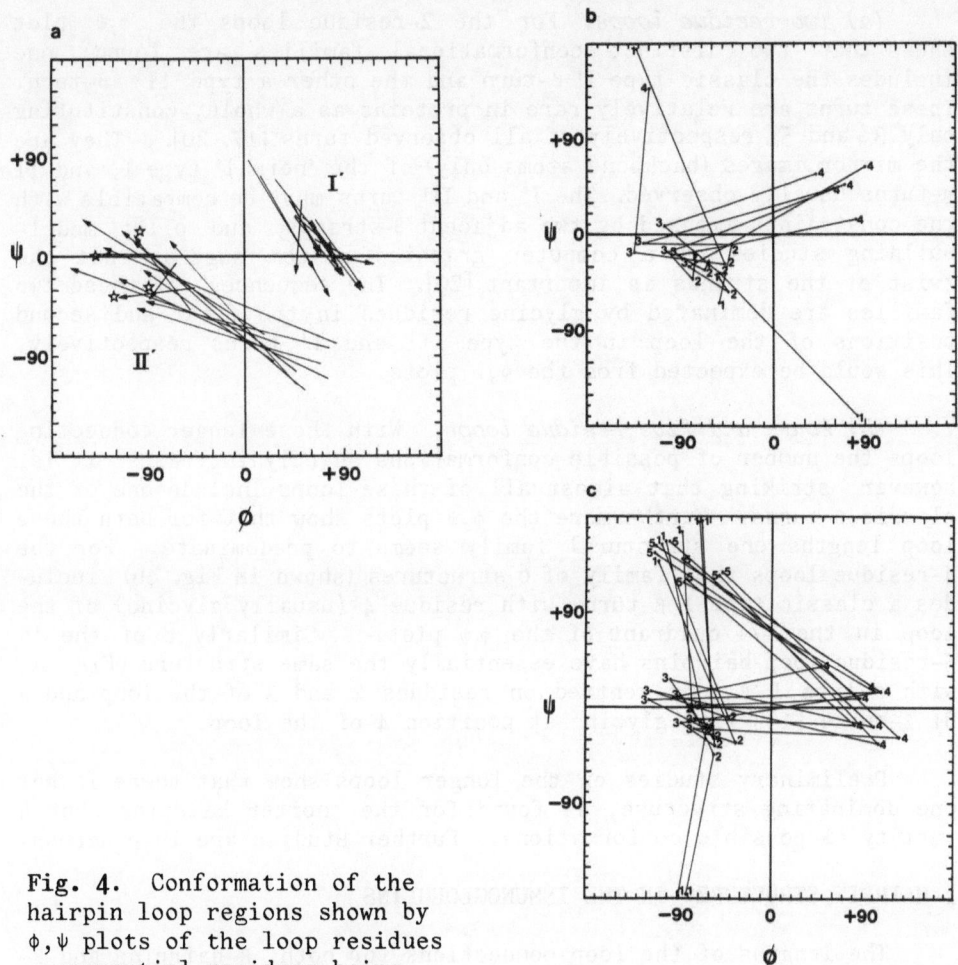

**Fig. 4.** Conformation of the
hairpin loop regions shown by
$\phi,\psi$ plots of the loop residues
– sequential residues being
joined by a line.
**a:** Plot for all the 2-residue hairpin loops.  From 29 hairpins, 15
include a type I' turn, 10 include a type II' turn and 4 (marked ☆)
have a type I turn [17, 18].  *Arrowheads* denote  the second residue.
**b:** Plot for 6 of the 13 4-residue loop hairpins, which include a
type I β-turn.  Sequential residues are numbered.
**c:** Plot for the 8 5-residue loop hairpins, which form a structural
family.  Note the type I turn at residues 2 & 3 and the +$\phi$ value
for residue 4, which is part of a G1 β-bulge [19].

## CONFORMATIONS AND SEQUENCES OF β-HAIRPIN LOOPS

The conformations of the hairpin loops were studied by plotting
the $\phi,\psi$ angles of the loop residues on a $\phi,\psi$ plot, and  joining consecutive residues (see Fig. 4, a–c).  The plots can then be used to
search for common structures and patterns.

*(a) Two-residue loops.* For the 2-residue loops the $\phi,\psi$ plot shows that two distinct conformational families are found; one includes the classic type I' β-turn and the other a type II' β-turn. These turns are relatively rare in proteins as a whole, constituting only 3% and 5% respectively of all observed turns [17, 20]. They are the mirror images (backbone atoms only) of the 'normal' type I and II β-turns usually observed. The I' and II' turns must be compatible with the constraints imposed by two adjacent β-strands, and pilot model-building studies on a computer graphics system suggest that the twist of the strands is important [20]. The sequences of these two families are dominated by glycine residues in the first and second positions of the loop in the type II' and I' turns respectively. This would be expected from the $\phi,\psi$ plots.

*(b) Four- and five-residue loops.* With these longer connecting loops the number of possible conformations sharply increases. It is, however, striking that almost all of these loops include one of the classic β-turns. Furthermore the $\phi,\psi$ plots show that for both these loop lengths one structural family seems to predominate. For the 4-residue loops this family of 6 structures (shown in Fig. 4b) includes a classic type I β-turn, with residue 4 (usually glycine) of the loop in the +/+ quadrant of the $\phi,\psi$ plot. Similarly 8 of the 19 5-residue loop hairpins have essentially the same structure (Fig. 4c) with a type I β-turn centred on residues 2 and 3 of the loop and a G1 β-bulge [19] with glycine at position 4 of the loop.

Preliminary studies on the longer loops show that there is not one dominating structure, as found for the shorter hairpins, but a variety of possible conformations. Further studies are in progress.

## β-HAIRPIN STRUCTURES IN THE IMMUNOGLOBULINS

The lengths of the loop connections for both β-hairpins and β-strands in the REI and Fab'(New) structures are shown in Fig. 5. The following observations may be made.-

(1) There are several short-loop 'standard' β-hairpins. Two-residue loops are found between strands E and F in FALV (Fig. 5a) and the homologous hairpin in REI (see Fig. 5b). The sequence of these type II' turns is Gly Ser and Gly Thr respectively. There is one other 2-residue loop between strands H and G in FALV. This is part of the Ab combining site and has the unusual sequence Arg Ser, but adopts a type II' conformation. One 4-residue loop (between strands F and G in FAHC) has the standard conformation shown in Fig. 4b but neither the sequence nor the structure is conserved in FALC. Similarly the CDR loop between strands D and E in FAHV adopts the standard 5-residue loop conformation, but this is not conserved in the other variable domains.

(2) The short β-arch connection between strands A and B in REI is unusual. It has the sequence Pro-Ser which cannot adopt a type I' or II' turn and may therefore effectively prohibit hairpin formation.

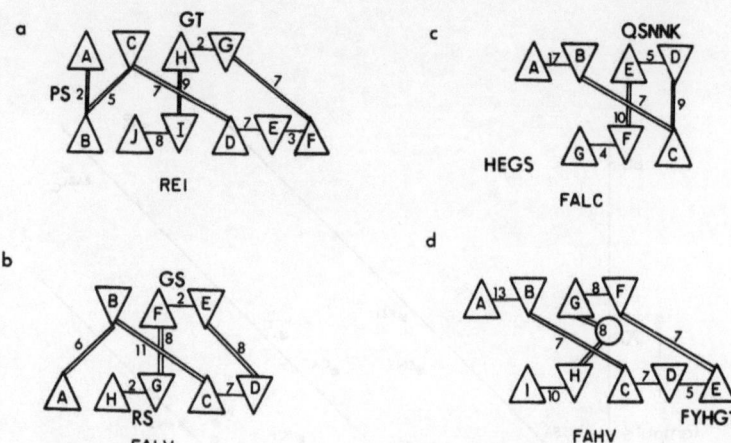

**Fig. 5.** Length of loop regions in the Ig structures REI [2] and Fab'(New) [1]. Definitions of loops taken from Kabsch & Sander [16]. **a**: Bence-Jones protein variable dimer REI [2]. **b-d**: Fab' fragment IgG($\lambda$)New: **b**, light chain variable domain - FALV; **c**, light chain constant domain - FALC; **d**, heavy chain variable domain.

(3) Although there is some conservation of loop length, wide variations can occur.

## STRUCTURE PREDICTION AND MODEL-BUILDING

The results of the β-hairpin analysis are being incorporated into our prediction algorithm [21, 12] which seeks to locate supersecondary structure - such as the βαβ unit or the β-hairpin. The sequences for the 2-residue loops in hairpins are highly selective, being dominated by glycines, and we have constructed templates to search for these specific sequence patterns. Currently for all the all-β proteins we obtain an 8% improvement over conventional secondary structure prediction accuracy (see Fig. 6) similar to our results for the βαβ unit [12].

The data are also useful for model-building. In this context it is of particular interest for predicting the structure of the Ab combining site which includes 4 β-hairpin loops and 2 β-arch loops. The sequence conservation and structural homology of these loop regions clearly needs further study. Work is in progress both at our own College and in Oxford [B.L. Sutton and collaborators; see #NC(A)-1, this vol.] to search for guidelines to direct model-building of loop regions.

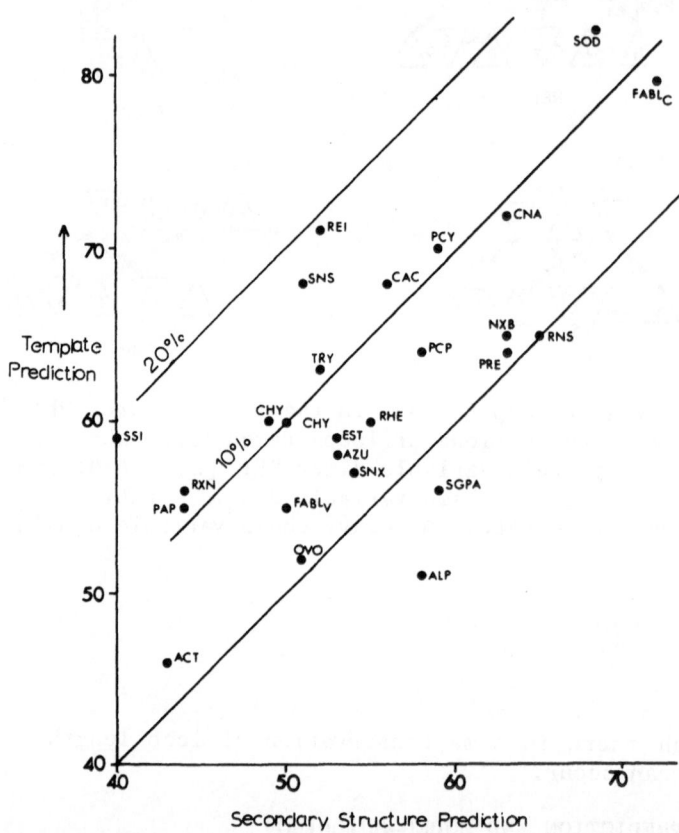

**Fig. 6.** Improvement in prediction accuracy for the all-β proteins obtained by searching for β-hairpins supersecondary structures [21]. Secondary structure prediction performed using standard Robson method [22]. The protein names are taken from the Protein Data Bank files, and the abbreviations denote:-
SOD, superoxide dismutase; FABL_C, Ig Fab' light chain - constant domain; CNA, concanavalin A; REI, Bence-Jones variable dimer REI; PCY, plastocyanin; SNS, staphylococcal nuclease; CAC, carbonic anhydrase; TRY, trypsin; PCP, penicillopepsin; NXB, neurotoxin B; RNS, ribonuclease S; PRE, prealbumin; CHY, α and γ chymotrypsin; EST, elastase; RHE, Ig λ variable light-chain dimer; SSI, strepto-myces subtilisin inhibitor; AZU, azurin; SNX, scorpion neurotoxin; RXN, rubredoxin; PAP, papain; FABL_V, Ig Fab' light-chain variable domain; SGPA, *Strep. Griseus* protease A; OVO, ovomucoid; ALP, α-lytic protease; ACT, actinidin.

## References

1.  Poljak, R.J., Amzel, L.M., Chen, B.L., Phizackerly, R.P. & Saul, F. (1974) *Proc. Nat. Acad. Sci. 71*, 3440-3444.
2.  Epp, O., Colman, P., Fehlhammer, H., Bode, W., Schiffer, M. & Huber, R. (1974) *Eur. J. Biochem. 45*, 513-524.
3.  Furey, W.J.R., Wang, B.C., Yoo, C.S. & Sax, M. (1983)
4.  Marquart, M., Deisenhofer, J., Huber, R. & Palm, W. (1980) *J. Mol. Biol. 141*, 369-391.
5.  Cohen, F.E., Sternberg, M.J.E. & Taylor, W.R. (1981) *J. Mol. Biol. 148*, 253-272.
6.  Chothia, C & Janin, J. (1981) *Proc. Nat. Acad. Sci. 78*, 4146-4152.
7.  Richardson, J.S. (1981) *Adv. Protein Chem. 34*, 167-339.
8.  Blake, C.C.F., Geisow, M.J., Oatley, S.J., Rerat, B. & Rerat, C. (1978) *J. Mol. Biol. 121*, 339-356.
9.  Chothia, C. & Lesk, A.M. (1982) *J. Mol. Biol. 160*, 285-308.
10. Greer, J. (1981) *J. Mol. Biol. 153*, 1027-1042.
11. Levitt, M. & Chothia, C. (1976) *Nature 261*, 552-558.
12. Taylor, W.R. & Thornton, J.M. (1984) *J. Mol. Biol. 173*, 487-514.
13. Harrison, S.C., Olson, A.J., Schutt, C.E., Winkler, F.K. & Bricogne, G. (1978) *Nature 276*, 368-373.
14. Richardson, J.S. (1977) *Nature 268*, 495-500.
15. Bernstein, F.C., Koetzle, T.F., Williams, G.J.B., Meyer, E.F. Jr., Brice, M.D., Rodgers, J.R., Kennard, O., Shimanouri, T. & Tasumi, M. (1977) *J. Mol. Biol. 112*, 535-542.
16. Kabsch, W. & Sander, C. (1983) *Biopolymers 22*, 2577-2637.
17. Chou, P.Y. & Fasman, G.D. (1977) *J. Mol. Biol. 115*, 135-175.
18. Lewis, P.N., Momany, F.A. & Scheraga, H.A. (1973) *Biochim. Biophys. Acta 303*, 211-229.
19. Richardson, J.S., Getzoff, E.D. & Richardson, D.C. (1978) *Proc. Nat. Acad. Sci. 75*, 2574-2578.
20. Chothia, C. (1973) *J. Mol. Biol. 75*, 295-302.
21. Taylor, W.R. & Thornton, J.M. (1983) *Nature 301*, 540-542.
22. Garnier, J., Osguthorp, D.J. & Robson, B. (1978) *J. Mol. Biol. 120*, 97-120.

#A-3

# PROBING THE BINDING SITES IN CRYSTALS OF IMMUNOGLOBULINS

Allen B. Edmundson,   Kathryn R. Ely,
James N. Herron and Amy L. Gibson

Department of Biology
University of Utah
Salt Lake City, UT 84112, U.S.A.

*In myeloma proteins from patient Mcg, 5 distinct 3-D structures assumed by one λ-type gene product have been defined by X-ray analysis: 2 conformational isomers in the Mcg Bence-Jones dimer crystallized in ammonium sulphate, 2 in the Mcg dimer crystallized in water (orthorhombic), and one in the Mcg IgG1 immunoglobulin (Ig). The 3 crystal systems differ in binding-cavity geometry, due to the conformational flexibility of the light and heavy chains. The Bence-Jones dimer binds 2 molecules of bis(DNP)lysine in solution and one in trigonal crystals. The IgG1 does not bind this ligand, nor does a λ-type Bence-Jones dimer from the patient Weir; but a hybrid dimer of the Weir and Mcg light chains binds one ligand molecule with the same affinity as the Mcg dimer. An explanation in 3-D terms is given.*

*With 2.7 Å (rather than 6.5 Å) resolution, observations of crystalline Mcg dimer binding with fluorescein, lucigenin and rhodamine derivatives have led to 3-D criteria for site-filling ligands. The binding cavity could distinguish between m- and p-isomers of carboxytetramethylrhodamine (CTMR). This cavity also provided a model for the binding of N-formylated chemotactic peptides. Extensive conformational changes to improve the complementarity of protein and ligand constituents were consistent with a neo-instructive theory of binding in this primitive cavity.*

In probing the ligand binding sites of crystalline Ig's and their fragments, we generally diffuse small-molecular compounds into crystals and determine their orientations and relative occupancies by difference Fourier analyses [1, 2]. The success of such studies has depended on the prior analysis of the crystal structure of the non-liganded protein molecule to intermediate or relatively high resolution (e.g. 2.7 Å [2]). Recently, an antigen-binding fragment (Fab) from an anti-fluorescyl monoclonal antibody (MAb), 4-

4-20[3], co-crystallized with its hapten in the active site [4]. Other groups have also obtained crystals of liganded fragments [5-7], and it appears that co-crystallization is becoming a viable alternative to diffusion methods for investigations of ligand binding.

Here we discuss preliminary results with murine MAb's of known specificity and more advanced studies of ligand binding in human myeloma proteins. The latter include the serum IgG1 and the urinary Bence-Jones (light chain) dimer from the patient Mcg, and heterologous dimers produced by hybridization of the Mcg light chain with Bence-Jones proteins from other sources.

## PRELIMINARY STRUCTURAL STUDIES OF MURINE MAb Fab's

Two MAb's from the laboratory of Edward Voss, Jr., of the University of Illinois, proved to be favourable for producing crystalline Fab's. The 4-4-20 anti-fluorescyl Ab was supplied by David Kranz [3], and the BVO4-01 anti-single-stranded-DNA protein by Dean Ballard [8]. Both Ab's were purified from ascites fluid by affinity chromatography. The 4-4-20 Ab was eluted from its column with fluorescein while the BVO4-01 protein was removed in non-liganded form with 0.5 M NaCl. The Ab's were hydrolyzed with papain to give Fab's, which were purified by isoelectric focusing for crystallization [4].

The 4-4-20 Fab held the fluorescent ligand very firmly (association constant of the parent IgG ~$10^{11}$ M$^{-1}$). The complex could therefore be subjected to crystallization trials without fear of dissociation of the ligand. Both ammonium sulphate and polyethyleneglycol (PEG) were used as crystallization media. The PEG crystals with a monoclinic ($P2_1$) space group were suitable for X-ray diffraction studies. These crystals were exceptionally resistant to radiation damage (Cu Kα X-rays), and data to 2.5 Å resolution were taken on one crystal. Efforts are now being made to solve the 3-D structure of the liganded protein by the multiple isomorphous replacement (MIR) method. It will be interesting to compare the high-affinity binding site with the fluorescein-binding sub-sites of the Mcg Bence-Jones dimer [2].

Triclinic ($P1$) crystals of the non-liganded BVO4-01 fragment were obtained in ammonium sulphate, and are excellent candidates for diffraction experiments; X-ray data to 2.0 Å resolution have been collected. Preliminary attempts to solve the 3-D structure have indicated that the Fab molecules are tightly packed in the crystal lattice (J.N. Herron & A.L. Gibson, unpublished work). As emphasized later, ligand-binding studies are strongly influenced by the availability of solvent channels through which added compounds can diffuse in the crystal lattice. It may be necessary to mount a major effort to co-crystallize oligonucleotides with the Fab fragments.

## X-RAY ANALYSIS OF THE INTACT IgG1 FROM PATIENT Mcg

The Mcg IgG1 ($\lambda$) protein was the first Ig recognized to have a hinge deletion [9], and moreover it was easy to crystallize. The Mcg Ig was a euglobulin (H.F. Deutsch, pers. comm., 1969), unlike the Dob, Kol and Zie cryoglobulins [10-13]. We exploited this to crystallize the Mcg Ig in water [14]. Diffraction data have been acquired to $d$ spacings of 2.7 Å and the crystal structure solved to 6.5 Å resolution by the MIR method [15]. The structural analysis is currently being extended to higher resolution by rigid body refinement with the CORELS program [16], in combination with interactive computer graphics for model-building.

Dinitrophenyl (DNP) ligands are bound by a significant number of Ig's, even myeloma proteins of unknown specificity. Bis(DNP)-lysine is among the test compounds which bind to the Mcg Bence-Jones dimer both in crystals and in solution [1, 17]. However, this compound failed to bind when added to the Mcg IgG1 protein in solution, whose binding site was thus structurally and functionally different from that of the dimer although the respective binding sites have identical light-chain amino acid sequences. IgG1 binding-site geometry will be examined at higher resolution with a view to selecting ligands most appropriate for binding.

## LIGAND BINDING IN HYBRIDS OF LIGHT CHAINS

The influence of the Mcg light chain on bis(DNP)lysine binding is substantially greater in heterologous dimers with other light chains than in the Fab region of the Mcg IgG1 protein. A method for obligatory and quantitative hybridization of light chains [18] allows any combination of interest to be examined; initially we studied Mcg hybrids which bound bis(DNP)lysine. As Kabat et al. [19] pointed out, the amino acid sequence of the first hypervariable region of the Vil $\lambda$-chain [20] is identical with that of the Mcg light chain [21]. Both the Vil dimer and the Mcg x Vil hybrid bound bis(DNP)lysine in solution in a manner indistinguishable from that of the Mcg dimer [18]. Moreover, a $\kappa$-chain dimer from a patient (Tew) with amyloidosis [22, 23] showed similar binding activity, as did the Mcg x Tew hybrid. The Weir $\lambda$-chain dimer [24] was a nonbinder, but the Mcg x Weir hybrid bound the ligand with the same affinity as the Mcg dimer.

In ammonium sulphate solutions the Mcg x Weir hybrid produced crystals suitable for X-ray diffraction [25, 26]. Surprisingly, the main form appearing early in the crystallization had the same space group ($P3_121$) as the trigonal crystals of the Mcg dimer. The $a$ cell dimensions (72.3 Å) were identical and those of $c$ (188.1 $vs.$ 185.9 Å) differed by only 1% in the two crystal systems. Thus, the two molecules could be treated as if they were nearly isomorphous. A 6.5 Å electron density map was calculated, and the resulting model of the hybrid was compared with that of the Mcg dimer (Fig. 1). At low resolution it is apparent that the two models are indistinguishable.

These marked similarities in 3-D structure partially explain why the hybrid binds bis(DNP)lysine like the Mcg dimer. However, there are 36 differences in the first 113 residues (V domains) of the Mcg and Weir light chains [21, 24]. Five of these substitutions involve contact residues for the DNP ligand in the Mcg dimer. Replacement of a key tyrosine residue (No. 34 in Mcg) by serine (Weir) would adversely affect the affinity for a DNP ring; the replacement of tyrosine 38, valine 48, glutamic acid 52 and serine 91 (Mcg) by phenylalanine, leucine, alanine and methionine (Weir) would tend to make a Weir dimer's binding cavity more hydrophobic and probably shallower than its counterpart in the Mcg dimer (see Fig. 2). From these considerations, we would not expect the Weir dimer to bind bis(DNP)lysine. In the hybrid, the Weir component participates in the formation of a different binding cavity but obviously does not interfere with the accessibility of Mcg contact residues for DNP ligands. The X-ray analysis of the hybrid is currently being extended to 3.5 Å resolution to clarify these complex structural relationships between such seemingly disparate light chains.

## THE BINDING REGION OF THE Mcg BENCE-JONES DIMER IN TWO CRYSTAL FORMS

The similarities in the 3-D structures of the hybrid and the Mcg dimer contrast sharply with the differences found in the Mcg dimer crystallized in water and in ammonium sulphate [27, 28]. The crystal structure of the water (orthorhombic) form is at present being analyzed and refined at 2.7 Å resolution, and being compared with the atomic model fitted to a 2.3 Å electron-density map of the ammonium sulphate (trigonal) form [29, 30].

Both monomers in the orthorhombic form are conformationally distinct from their counterparts in the trigonal system (Fig. 3). The dimers are as different from one another as the trigonal form is from an Fab: e.g. the 'elbow bend' angles between pseudodiads relating V- and C-domain pairs is 132° in the orthorhombic form and only 115° in the trigonal form. Bend angles in Fab regions have ranged from ~130° to 170°, with most near the value for the orthorhombic form [5, 11, 12, 31].

Major conformational differences in the two crystal forms were reflected in the binding cavities (Fig. 4). Significant changes in the geometry and the space available for ligand binding resulted when water was substituted for ammonium sulphate as the crystallizing medium. We have shown that bis(DNP)lysine can be bound to the Mcg dimer in orthorhombic crystals, but have deferred more extensive studies with other compounds in favour of ligand-binding studies with trigonal crystals at high resolution.

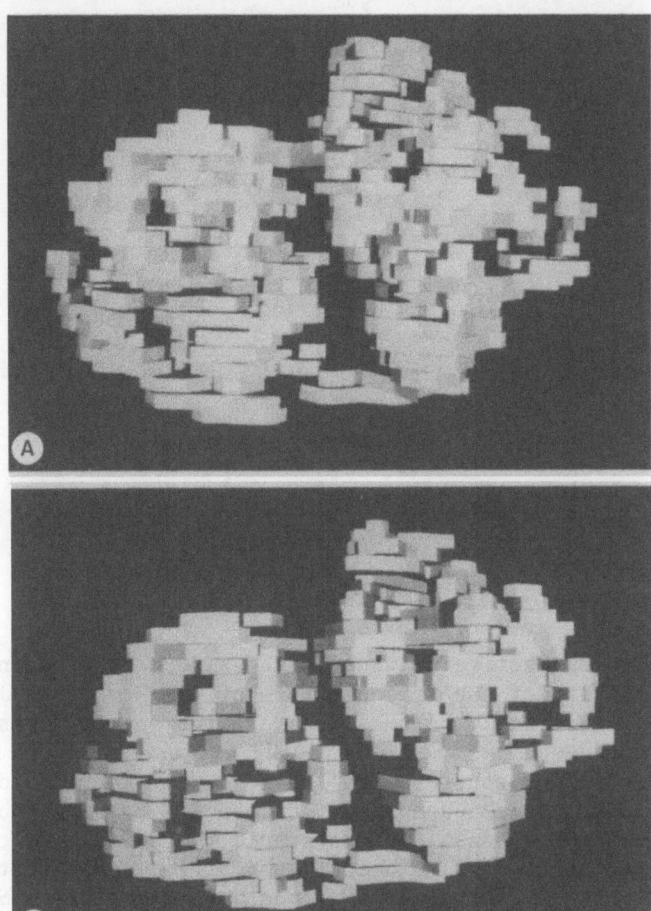

**Fig. 1.** Solid image models of the Mcg Bence-Jones dimer (panel **A**) and Mcg x Weir hybrid dimer (panel **B**) at 6.5 Å resolution. Programs for generating these models were written and implemented by Russell J. Athay. Contours in an electron-density map for each protein were assigned to an individual dimer in the unit cell. Each contour was given a solid image by offsetting it and constructing polygons for the top, bottom and side faces. The polygons were processed with a hidden surface algorithm, shaded, and displayed on a frame buffer. The larger module of electron density in each model corresponds to the pair of 'variable' (V) domains and the smaller module to the 'constant' (C) domain pair. Note the striking similarities in the structure of the Mcg dimer and the hybrid.

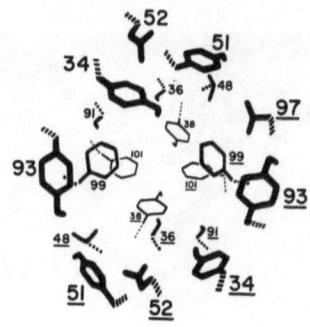

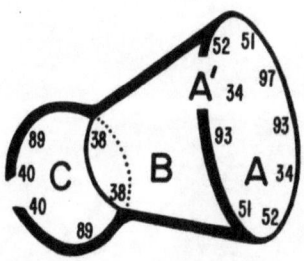

Fig. 2. *Top.*- Perspective drawing of the side chains lining the
main binding cavity of the trigonal form of the Mcg Bence-Jones
dimer (crystallized in ammonium sulphate). The sequence was determined by Fett & Deutsch [21]. This drawing (taken from [1]) helps
discussion of binding studies with the Mcg X Weir hybrid and the Mcg
dimer. *Underlined:* sequence nos. of monomer 2; ⋯⋯, bonds between
side-chain α and β carbons.  In the hybrid the Weir light chain
probably plays the role of monomer 1 [26]; if it were substituted
for monomer 1, Tyr 34 in the drawing would be replaced by Ser, and
Tyr 38, Val 48, Glu 52 and Ser 91 by Phe, Leu, Ala, Met [24]. Changes
in the overall geometry of the hybrid's binding cavity cannot as yet
be assessed (with 6.5 Å resolution), but clearly the substitutions
accentuate the hydrophobicity of the binding cavity.  The Mcg but
not the Weir dimer binds bis(DNP)Lys; but a Weir light chain present
in the hybrid dimer does not prevent the binding of a DNP ligand.
*Bottom.*- Schematic side-view of the conical main cavity and the
deep binding pocket of the Mcg dimer's trigonal form.  Positions of
key amino acid residues are *numbered* and sub-sites for the binding
of various ligands are *lettered*; e.g. bis(DNP)lysine bridges sites
A and B, and menadione occupies site C.  Site A', homologous with
site A in the dimer, is partially blocked by interactions with
another protein molecule in the crystal lattice.

**Fig. 3.** Comparison of αC skeletal
models of the Mcg Bence-Jones
dimer in two crystal systems.
*Top.* - Orthorhombic form crystallized in deionized water.
*Bottom.* - Trigonal form crystallized in ammonium sulphate.
In each model the V domains are
on *right* and the C domains on *left*.
Pseudo-two-fold axes of rotation
between pairs of homologous
domains are superimposed on the
models. The $V_1$-$V_2$ and $C_1$-$C_2$
pseudodiads intersect at angles of
$132°$ in the orthorhombic form and
$115°$ in the trigonal form [27, 28].
Comparable angles for Fab's are
closer to that of the orthorhombic
form [5, 11, 12, 31].

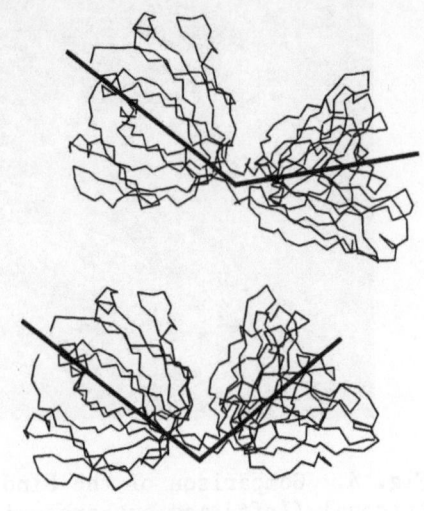

## LIGAND BINDING IN TRIGONAL CRYSTALS OF THE Mcg BENCE-JONES DIMER

*General description of the binding regions.* - The main binding
cavity is shaped like a truncated cone which is 15 Å across at the
cavity entrance and is 17 Å deep (Fig. 2). There are 3 negatively
charged side-chains around the rim of the cavity, but the interior
consists of hydrophobic side-chains and uncharged hydroxyl groups.
Twelve of the 21 side-chains are aromatic, with 8 Tyr and 4 Phe
residues. Below the floor of the main cavity is a hydrophobic binding pocket. At low resolution we had used aliphatic and aromatic
compounds to define 3 general sub-sites, A and B in the main cavity
and C in the pocket [1] (Fig. 2). Bis(DNP)lysine bridged sites A
and B, with one DNP ring in each sub-site. Phenanthroline, dansyl
derivatives and opioid compounds preferred site A, while purine
derivatives lodged in B; menadione and pyrimidine derivatives migrated to C.

The trigonal form is favourable for such binding studies for
several reasons. As we have seen, the dimer has a large, deep cavity
relative to other Ig's, and no other systems are known with an even
deeper pocket. Access to the binding cavity is not seriously inhibited by interactions between proteins in the crystal lattice,
although a potential sub-site (A') opposite site A is blocked (Fig.
2). Finally, a sizable solvent channel suitable for ligand diffusion
permeates the lattice and leads directly to the binding cavity's
entrance (Fig. 5).

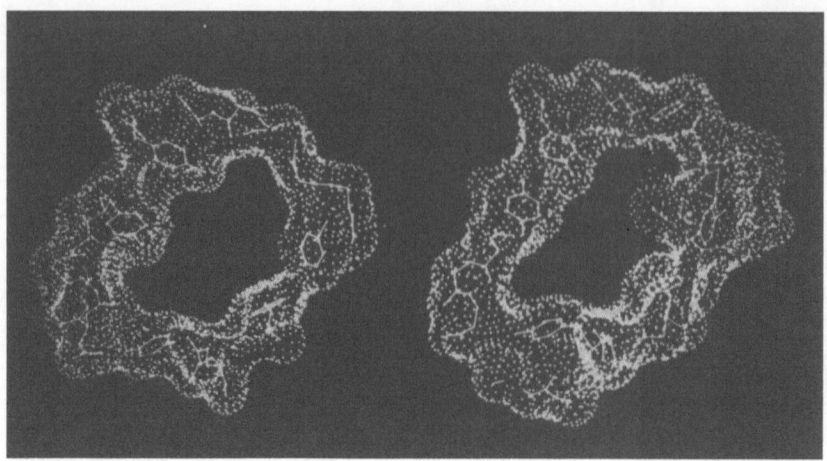

**Fig. 4.** Comparison of the binding cavities of the Mcg dimer in the trigonal *(left)* and orthorhombic *(right)* forms. Van der Waals surfaces were calculated [32] for the atoms of the polypeptide backbone and side chains lining the main binding cavity, and were represented by dots and co-displayed with skeletal models on an Evans & Sutherland computer graphics system. Note the marked differences in the sizes and shapes of the two cavities. *For the production of this picture we gratefully acknowledge the support of Judith McLarin, Robert Langridge and Thomas Ferrin at the NIH computer graphics resource at the University of California at San Francisco.*

*Search with fluorescent compounds for site-filling ligands.–* For a given series of compounds, it is generally assumed that the affinity increases with the number of contacts with binding-site components. This number should be maximized with a site-filling ligand. When we began working with anti-fluorescyl Ab's, we surmised that the Mcg dimer's main cavity was well suited for the binding of fluorescent compounds. To search for site-filling ligands among these compounds, we studied the modes of binding of a series progressively increasing in size: 1-anilinonaphthalene-8-sulphonate (ANS), fluorescein, bis($N$-methyl)acridine (lucigenin) and dimers of carboxytetramethylrhodamine (CTMR) [2]. Crystallographic data to 2.7 Å resolution were compared with data for the non-liganded protein by difference Fourier analyses. The results were interpreted with interactive computer graphics (R.J. Athay wrote and implemented the programs).

ANS proved to be a special case. It was first bound in the main cavity. In crystals soaked with ANS for periods longer than about a month, however, the ligand appeared exclusively in the deep pocket. Although the apparent migration of ANS was not accompanied by overt damage to the crystal or even conformational changes detectable in

**Fig. 5.** A solvent channel
through the crystal lattice
of the Mcg Bence-Jones dimer's
trigonal form. Parts of two
symmetry-related protein
molecules are shown as van der
Waals surface dot representations. Ligands can diffuse
through the channel directly
to the entrance of the main
binding cavity *(lower part of
Fig.)*. Some segments of the
upper molecule closely approach the lower molecule's
binding cavity and prevent
the ligands from entering
site A' (see Fig. 2).

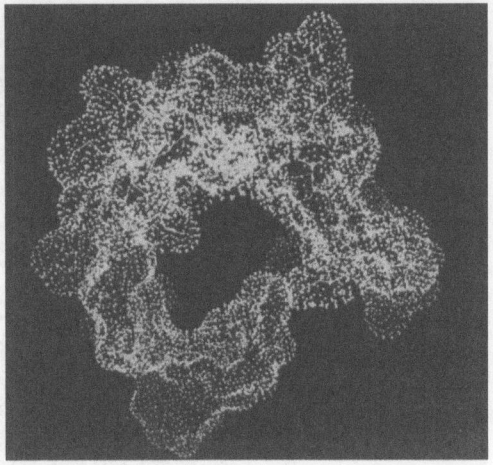

the difference Fourier map, the main cavity obviously lost its capability to bind the ligand after site C was occupied.

The larger fluorescent compounds were restricted to the main
cavity. There were two mutually exclusive (i.e. overlapping) binding
sites for fluorescein, with the major site favoured 2:1 in terms of
relative occupancy and similar in binding mode to that expected in
an antigen-Ab reaction (Fig. 6A). The xanthonyl moiety was embedded
in a deeply situated hydrophobic region, while the more polar carb-
oxyphenyl faced outward towards the solvent. The angle between the
two ring systems (which have some freedom of rotation in solution)
was influenced by interactions with protein aromatic side chains.

While lucigenin occupied more space than fluorescein, it still
could not be regarded as site-filling (Fig. 6B). The structure of
the ligand was clearly indicated in the difference electron-density
map (Fig. 7). Again the angle between the two acridine rings was
influenced by binding-site components. With two formal positive
charges on its rings, lucigenin was not expected to penetrate so far
into the hydrophobic cavity. The explanation for this penetration
became clear when extra electron density was found near the *N*-methyl
group of the more deeply situated acridine ring. This electron density was of the size and shape associated with a sulphate ion, which
accompanied the ligand into the cavity and neutralized its charge.

A commercial sample of CTMR did fulfil the requirements for a
a site-filling ligand (Fig. 6C). It proved to be a mixture of the

Fig. 6.
*For legend
see
opposite.*

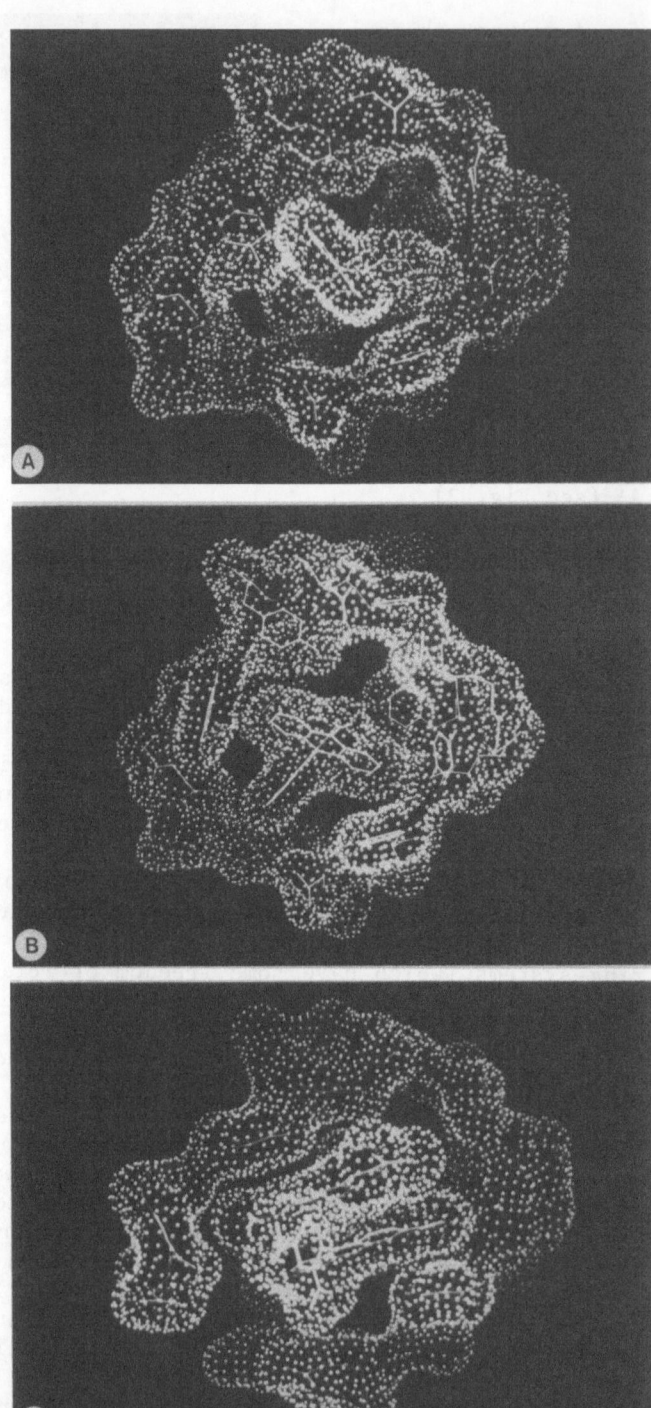

Fig. 6 *(opposite)*. Van der Waals surface dot representations of
fluorescent compounds as they are bound in the main cavity of the
trigonal form of the Mcg Bence-Jones dimer.
**A**, fluorescein;  **B**, lucigenin;  **C**, dimer of CTMR.
Note that the ligands occupy progressively more space in the binding
cavity.  There were extensive conformational changes in the binding
cavity to accommodate the ligands, particularly the rhodamine dimer.

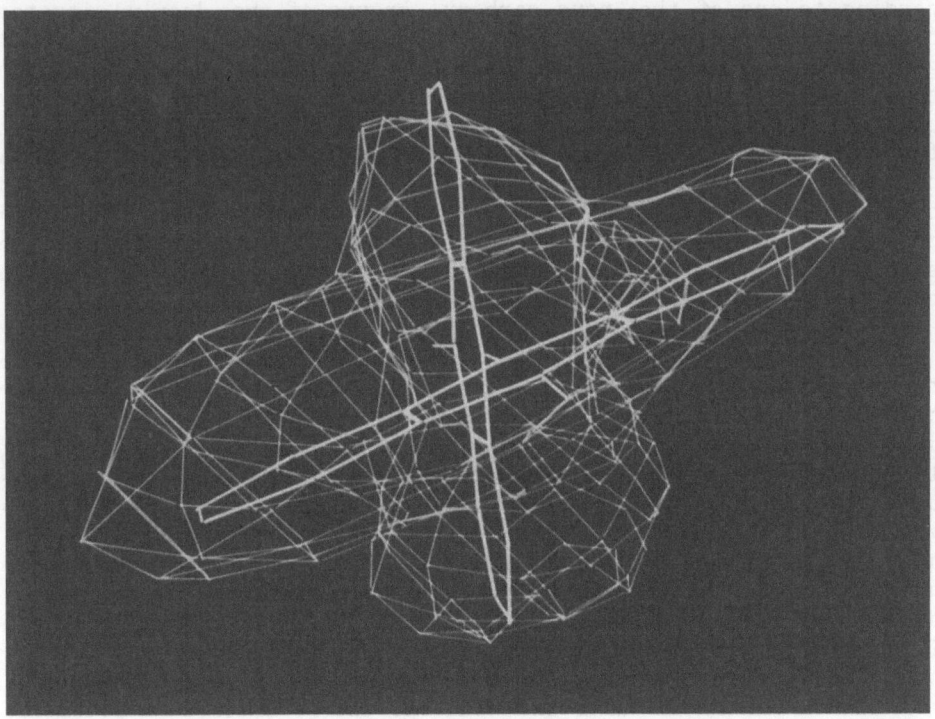

**Fig. 7.**  'Cage' electron density representing lucigenin, as bound
in the main cavity of the Mcg Bence-Jones dimer.  Lucigenin was
diffused into the crystal, and the bound ligand's position, shape
and size were determined by difference Fourier analysis at 2.7Å
resolution.  Skeletal models of the two *N*-methylacridine rings were
fitted to the cage density by interactive computer graphics.  The
angle between these two rings was probably influenced by the interactions with the protein during the binding process.

5- and 6-isomers, which were separated by TLC. Spectroscopic measurements, besides the 2.7 Å difference maps, indicated that both isomers and the mixture behaved as dimers under the conditions used for ligand binding in the crystals. The dimer of 5-isomers, in which the two carboxyl groups are in *meta*-positions on the phenyl ring, occupied a shallow site in the main cavity. It could be quantitatively removed from the cavity by perfusion of the crystal with the ammonium sulphate crystallizing medium. In contrast, the dimer of 6-isomers, with the carboxyl groups in *para*-positions, filled the cavity and could not be washed out of the crystal to any measurable extent in 63 days.

*Isomeric discrimination by the Mcg binding cavity.* - The difference Fourier maps provided a clear explanation for the discrimination between the 5- and 6-isomers of CTMR. Electron density corresponding to one of the *para*-carboxyl groups of the deeply situated 6-isomer was continuous with the position assigned to the hydroxyl group of Ser 91, monomer 2. This indicated a strong, linear hydrogen bond which partially compensated for the presence of a negatively charged group in a hydrophobic cavity. It is not sterically possible for a carboxyl group in the *meta*-position to participate in such a hydrogen bond. Moreover, the side-chain of Phe 99, monomer 2, would collide with a *meta*- but not a *para*-carboxyl group. Thus the power of a combining site to discriminate among similar ligands may involve such subtle factors as the potential directionality of a hydrogen bond or steric interference by a single side-chain.

*Conformational changes during the binding process.* - In response to the presence of fluorescein, lucigenin and particularly the dimer of 6-CTMR, the binding site constituents clearly moved in ways conducive to complementarity. These movements included rotations of large aromatic side-chains and expansion of the cavity by translations of the polypeptide backbone in the second and third hypervariable loops [2]. For such 'primitive' combining sites and perhaps for sites showing cross-reactivity for ligands other than those used to elicit an immune response, a limited neo-instructive theory of binding seems to be compatible with the experimental results. These results are not in conflict with most of the general tenets of the clonal selection theory. However, we feel that conformational adjustments in the binding site enlarge the repertoire of potential ligands, especially in low-affinity Ab's. In high-affinity systems possible adjustments may be viewed in terms of conformational 'fine-tuning' of the combining sites and/or their specific antigens.

*Binding of chemotactic peptides by the Mcg dimer.* - Peptides are good examples of flexible ligands which may have conformational changes imposed upon them during the binding process. Recently, we found that *N*-formylated (N-f) peptides bind to the trigonal form with surprising specificity (A.B. Edmundson & K.R. Ely, unpublished); they are chemoattractants for which human and rabbit neutrophils have high-affinity receptors [33, 34].

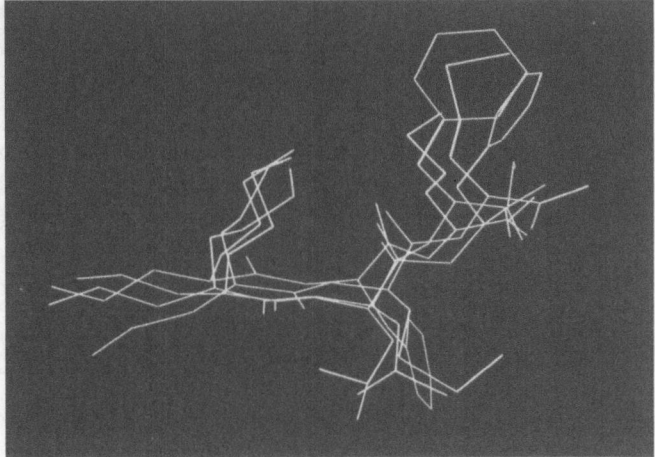

**Fig. 8.** Skeletal model of 4 chemotactic N-f tripeptides displayed in the conformations which they assumed in binding to the Mcg dimer. Each peptide was independently fitted to the electron density in a difference Fourier map at 2.7 Å resolution. The sequences were N-f-Met-Leu-Phe, N-f-Met-Phe-Met, N-f-Met-Met-Met and N-f-Nle-Leu-Phe (Nle = norleucine). Note the marked conformational similarities.

The binding modes were determined for 4 N-f tripeptides (Fig. 8 legend) and N-f-Met-Trp. Under the same conditions N-f-Met peptides were bound with higher occupancies than N-f-Nle derivatives. With t-butyloxycarbonyl (t-BOC) in place of N-f the binding fell to near-background; likewise if the α-amino group were underivatized. The 4 N-f tripeptides gave maps indicating marked similarities in binding pattern, as shown in Fig. 8 which simultaneously displays skeletal models of the ligands in orientations they assumed in the binding cavity. Another view of the prototype ligand, N-f-Met-Leu-Phe, is shown in Fig. 9. The bound peptide has adopted a wedge shape, with its side chains swept back toward the entrance of the cavity.

The wedge shape is characteristic of all the tripeptides in this series. In Fig. 10, a van der Waals surface dot representation depicting N-f-Met-Phe-Met binding in the cavity, note that the peptide conforms closely to the cavity's general topography and effectively fills the site. When not under such conformational restraints, as in solution or in single crystals, the chemotactic peptides adopt a rigid, extended conformation typical of β-pleated sheets [35, 36].

In the binding cavity there were key interactions common to all N-f tripeptides. The *N*-formyl group was located near the floor of the cavity and was equidistant to the two Tyr residues (No. 38 in each monomer; see Fig. 2). Continuous electron density between the

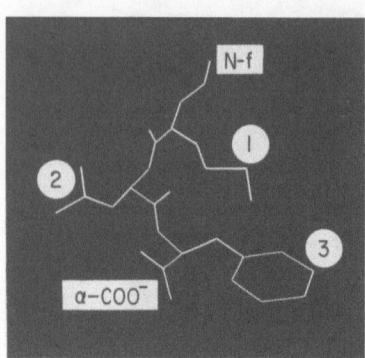

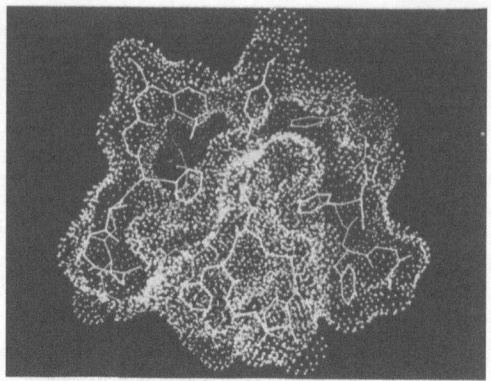

**Fig. 9.** Skeletal model of a prototype N-f tripeptide in a view accentuating the characteristic wedge shape assumed during binding to the Mcg dimer. Side-chains: 1, Met; 2, Leu; 3, Phe.

**Fig. 10.** Cutaway view *(at top)* of a chemotactic tripeptide, N-f-Met-Phe-Met, as bound in the main cavity of the Mcg dimer (van der Waals surface dot representations). Like Fig. 9 this picture emphasizes the ligand's wedge shape matching the cavity's topography.

carbonyl oxygen of the $N$-formyl group and the phenolic group of Tyr 38, monomer 1, indicated the presence of a specific, nearly linear, hydrogen bond. The intricate requirements for its formation explain why the $N$-formyl group was necessary for binding (t-BOC derivatives or free α-amino groups would not suffice). Two Phe rings (No. 99 in each monomer) impinged on opposite sides of the $N$-formyl group and provided additional stabilization.

The Met or Nle side-chain in ligand position 1 (Fig. 9) extended along the wall of the cavity and made close contacts with Tyr 93, monomer 2. This residue had moved to a position of greater complementarity with the ligand side-chain. Ligand residue 2 occupied a hydrophobic pocket which imposed fewer steric requirements. Various bulky side-chains (e.g. those of Leu, Phe and Met) could be accommodated, and the relative orientations were not as firmly fixed as in residue 1. Conformational adjustments occurred even in this case, as evidenced by the forced expansion of the sub-site by outward movements of the backbone and side-chains of Tyr 51 and Glu 52 (second hypervariable loop of monomer 2).

The ligand's third side-chain was situated on the rim of the cavity in a pocket between two molecules in the crystal lattice. The Phe side-chains of N-f-Met-Leu-Phe and N-f-Nle-Leu-Phe were particularly well defined in the difference maps, an indication that the rings were held firmly in one orientation. The α-carboxyl group was inside the rim of the cavity, but was facing the solvent.

Ligands longer than N-f tripeptides would not fit inside the
cavity.  Moreover, the crystal packing was very tight; as confirmation of our prediction that a fourth residue could not be inserted
into the pocket between molecules, N-f-Met-Leu-Phe-Lys did not bind.

*Binding of N-f-Met-Trp by the Mcg dimer.-* The 2.7 Å difference
map for this ligand clearly indicated the presence of two mutually
exclusive sub-sites of approximately equal occupancy.   Sub-site 1
was deep, with the *N*-formyl group and Met side-chains in nearly the
same locations as those in the N-f-Met  tripeptides.  The Trp side-
chain was too large for the hydrophobic region occupied by residue 2
in the tripeptides, but it was readily accommodated in a more spaci-
ous part of the cavity.  In the outer sub-site (2) the Trp ring was
found in a position similar to that of Phe 3 in a tripeptide (i.e.
in the pocket between molecules).  The *N*-formyl group, the Met side-
chain and the α-carboxyl group were situated inside the cavity.

**Cis-trans** *isomerization of peptide bonds in the ligands.-* Skeletal models fitted into the electron-density maps normally have standard bond lengths and angles, and the peptide bonds are displayed in
the energetically more favourable *trans*-configurations.  Fitting of
the electron density for the carbonyl group of Met-1 in the tripeptides was not completely satisfactory with peptide bonds in either
the *cis*- or the *trans*-configuration. Possibly this peptide bond had
partial *cis*  character.  For N-f-Met-Trp in the deep sub-site (1),
the corresponding peptide bond could be fitted to the density only
in the  *cis*-configuration (Fig. 11).  In the outer sub-site the peptide bond was clearly *trans*.  It seems likely that the protein-ligand
interactions in the inner site  resulted in torsional strain in the
peptide bond.  In the dipeptide this strain was relieved by isomerization to the *cis*-configuration. In the tripeptides, however, distal
interactions involving residues 2 and 3 were sufficiently strong to
prevent conversion of this bond to the *cis*-form.

It is now apparent that the binding of flexible ligands can
involve distortion of the geometry of a peptide bond.  Full isomerization from  *trans*  to *cis* is possible in favourable cases such as a
dipeptide or a peptide containing Pro (where the energy barrier is
lower for the conversion).

*The Mcg binding cavity as a model for a chemotactic receptor.-*
The binding characteristics described above are strikingly similar
to those described for specific neutrophil receptors [33-35]. We have
long regarded the Mcg binding cavity as a model for a primitive Ab
site with wide specificity for hydrophobic ligands like the chemotactic peptides.  Bacterial proteins, with *N*-formyl Met at the amino
terminus, were probably among the earliest antigens challenging
primitive immune systems. At the outset of these studies we reasoned
that an ancient recognition function might be retained in a primitive, albeit modern, binding cavity.  We are now convinced that Ab's

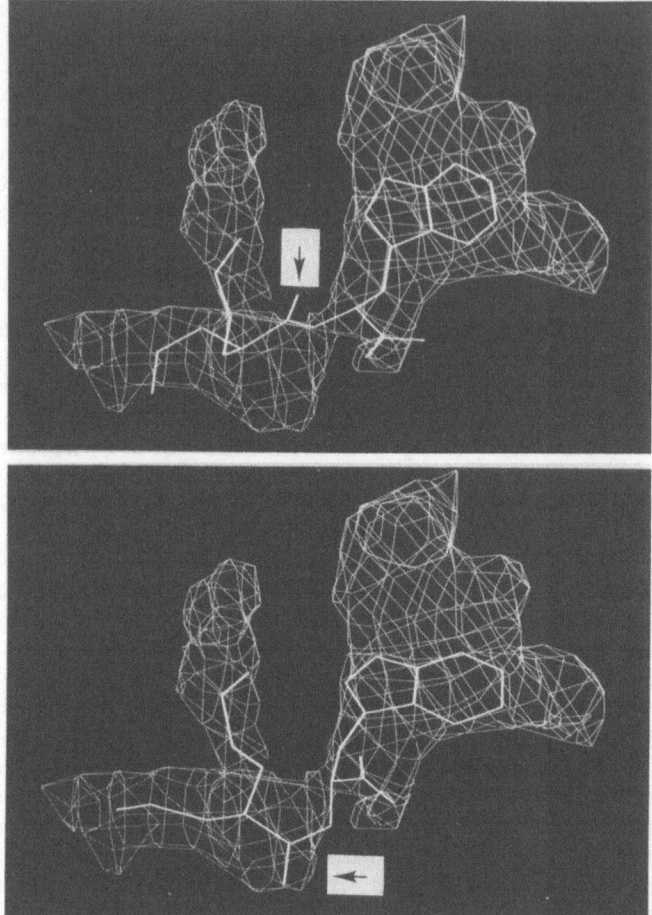

**Fig. 11.** *Cis vs. trans* geometry: binding of N-f-Met-Trp in the
deeper of two sub-sites in the Mcg Bence-Jones dimer's cavity.
*Upper panel.* - Cage electron density co-displayed with the skeletal
model of the fitted dipeptide. The peptide bond between Met and
Trp was built in the energetically favourable *trans*-configuration,
which was evidently unlikely because the carbonyl *(arrow)* was outside the electron-density envelope. *Lower panel.* - *Cis* construction
whereby the carbonyl *(arrow)* could be confidently placed inside.

raised against chemotactic peptides will have properties very like
those of neutrophil receptors. But only a few Ab's can be expected
to show the high affinity associated with receptors. It will be
interesting to compare highly specific sites with a primitive cavity
which inherently has so many features favourable for the binding of
chemotactic peptides.

*Acknowledgements*

We gratefully acknowledge the gifts of anti-fluorescyl and anti-DNA Ab's from David Kranz, Dean Ballard and Edward Voss, Jr., of the University of Illinois at Urbana; the graphics work of Barbara Staker and the photography of Brad Nelson; and the skill and suggestions of Maurine Vaughan in the preparation of the manuscript for publication.

The investigation was supported by Grant CA19616, awarded by the National Cancer Institute, Department of Health and Human Services, and by post-doctoral fellowships to J.N.H.: PDF-8166016 from the National Science Foundation and GMO 8843 National Research Training Award from the National Institutes of Health.

*References*

1.  Edmundson, A.B., Ely, K.R., Girling, R.L., Abola, E.E., Schiffer, M., Westholm, F.A., Fausch, M.D. & Deutsch, H.F. (1974) *Biochemistry 13*, 3816-3827.
2.  Edmundson, A.B., Ely, K.R. & Herron, J.N. (1984) *Mol. Immunol. 21*, 561-576.
3.  Kranz, D.M., Herron, J.N. & Voss, E.W., Jr. (1982) *J. Biol. Chem. 257*, 6987-6995.
4.  Gibson, A.L., Herron, J.N. & Edmundson, A.B. (1984) *Fed. Proc. 43*, 1546.
5.  Navia, M.A., Segal, D.M., Padlan, E.A., Davies, D.R., Rao, N., Rudikoff, S. & Potter, M. (1979) *Proc. Nat. Acad. Sci. 76*, 4071-4074.
6.  Colman, P.M., Gough, K.H., Lilley, G.G., Blagrove, R.J., Webster, R.G., & Laver, W.G. (1981) *J. Mol. Biol. 152*, 609-614.
7.  Rose, D.R., Seaton, B.A., Petsko, G.A., Novotný, J., Margolies, M.N., Locke, E. & Haber, E.(1983) *J. Mol. Biol. 165*, 203-206.
8.  Ballard, D.W., Lynn, S.P., Gardner, J.F. & Voss, E.W., Jr. (1984) *J. Biol. Chem. 259*, 3492-3498.
9.  Fett, J.W., Deutsch, H.F. & Smithies, O. (1973) *Immunochemistry 10*, 115-118.
10. Sarma, V.R., Silverton, E.W., Davies, D.R. & Terry, W.D. (1971) *J. Biol. Chem. 246*, 3753-3759.
11. Silverton, E.W., Navia, M.A. & Davies, D.R. (1977) *Proc. Nat. Acad. Sci. 74*, 5140-5144.
12. Marquart, M., Deisenhofer, J., Huber, R. & Palm, W. (1980) *J. Mol. Biol. 141*, 369-391.
13. Ely, K.R., Colman, P.M., Abola, E.E., Hess, A.C., Peabody, D.S., Parr, D.M., Connell, G.E., Laschinger, C.A. & Edmundson, A.B. (1978) *Biochemistry 17*, 820-823.
14. Edmundson, A.B., Wood, M.K., Schiffer, M., Hardman, K.D., Ainsworth, C.F., Ely, K.R. & Deutsch, H.F. (1970) *J. Biol. Chem. 245*, 2763-2764.

15. Rajan, S.S., Ely, K.R., Abola, E.E., Wood, M.K., Colman, P.M., Athay, R.J. & Edmundson, A.B. (1983) *Mol. Immunol. 20*, 787-799.

16. Herzberg, O. & Sussman, J.L. (1983) *J. Appl. Cryst. 16*, 144-150.

17. Firca, J.R., Ely, K.R., Kremser, P., Westholm, F.A., Dorrington, K.J. & Edmundson, A.B. (1978) *Biochemistry 17*, 148-158.

18. Peabody, D.S., Ely, K.R. & Edmundson, A.B. (1980) *Biochemistry 19*, 2827-2834.

19. Kabat, E.A., Wu, T.T. & Bilofsky, H. (1977) *J. Biol. Chem. 252*, 6609-6616.

20. Ponstingl, H. & Hilschmann, N. (1971) *Z. Physiol. Chem. 352*, 859-877.

21. Fett, J.W. & Deutsch, H.F. (1974) *Biochemistry 13*, 4102-4114.

22. Putman, F.W., Whitley, E.J., Jr., Paul, C. & Davidson, J.N. (1973) *Biochemistry 12*, 3763-3780.

23. Bertram, J., Gualtieri, R.J. & Osserman, E.F. (1980) in *Amyloid and Amyloidosis* (Glenner, G.G., Costa, P.P. & de Freitas, A.F., eds.), Excerpta Medica, Amsterdam, pp. 351-360.

24. Jabusch, J.R. & Deutsch, H.F. (1982) *Mol. Immunol. 19*, 901-906.

25. Ely, K.R., Peabody, D.S., Holm, T.R., Cheson, B.D. & Edmundson, A.B. (1985) *Mol. Immunol.*, in press.

26. Ely, K.R., Wood, M.K., Rajan, S.S., Hodsdon, J.M., Abola, E.E., Deutsch, H.F. & Edmundson, A.B. (1985) *Mol. Immunol.*, in press.

27. Abola, E.E., Ely, K.R. & Edmundson, A.B. (1980) *Biochemistry 19*, 432-439.

28. Ely, K.R., Herron, J.N. & Edmundson, A.B. (1983) *Prog. Immunol. 5*, 61-66.

29. Schiffer, M., Girling, R.L., Ely, K.R. & Edmundson, A.B. (1973) *Biochemistry 12*, 4620-4631.

30. Edmundson, A.B., Ely, K.R., Abola, E.E., Schiffer, M. & Panagiotopoulos, N. (1975) *Biochemistry 14*, 3953-3961.

31. Poljak, R.J., Amzel, L.M., Chen, B.L., Phizackerley, R.P. & Saul, F. (1974) *Proc. Nat. Acad. Sci. 71*, 3440-3444.

32. Connolly, M.L. (1983) *Science 221*, 709-713.

33. Schiffmann, E., Showell, H.J., Corcoran, B.A., Ward, P.A., Smith, E. & Becker, E.L. (1975) *J. Immunol. 114*, 1831-1837.

34. Freer, R.J., Day, A.R., Muthukymaraswamy, N., Pinon, D., Wu, A., Showell, H.J. & Becker, E.L. (1982) *Biochemistry 21*, 257-263.

35. Becker, E.L., Bleich, H.E., Day, A.R., Freer, R.J., Glasel, J.A., Latina, M. & Visintainer, J. (1979) *Biochemistry 18*, 4656-4668.

36. Eggleston, D.S., Jeffs, P.W., Chodosh, D.F. & Heald, S.L. (1984) *Abs., Am. Crystallogr. Assoc., Annual Mtg., Lexington, KY, PB4*, 36.

#A-4

# COMPARATIVE STRUCTURES OF MOUSE ANTIBODY COMBINING SITES

**David R. Davies, Talapady N. Bhat, Gerson H. Cohen
and Eduardo A. Padlan**

Laboratory of Molecular Biology
National Institute of Arthritis, Diabetes, and Digestive
and Kidney Diseases
Building 2, Room 316
National Institutes of Health
Bethesda, MD 20205, U.S.A.

*The crystal structures of two mouse myeloma antibody-binding
fragments (Fab's) have been determined: McPC603, an anti-phosphocholine immunoglobulin (anti-PC Ig), and J539 which is specific for
galactan. A comparison has been made between the sequences of other
mouse Ig's that have similar binding properties. In both heavy and
light chains of the PC-binding proteins, the residues in contact
with the PC remain invariant or highly conserved. Examination of the
pattern of charge conservation at the combining site reveals a possible coupling of somatic mutations in order to maintain specificity.*

*In J539, the actual site of galactan binding is undefined since
galactan cannot be bound in the crystal because of interference by
a neighbouring molecule. Yet a tentative assignment can be made
based on the identification of a central cavity that is connected to
a pair of clefts. A comparison of the sequences of galactan-binding
mouse Ab's shows a situation similar to that for McPC603. The light
chain sequences reveal the use of a single subgroup. The heavy
chains, as in the PC-binding proteins, show scattered amino acid
substitutions with considerable variation in the D region, while
maintaining the total number of residues in this loop constant. We
have attempted to correlate these observations with the observed
3-dimensional structure.*

X-ray diffraction is the only technique capable of providing a
complete 3-dimensional model of a protein molecule at atomic resolution. The quality of such a model is largely dependent on the nature
of the particular protein crystals being examined: crystals that
exhibit high degrees of order and diffract to beyond a resolution of

2 Å can provide models that are correspondingly of high precision and accuracy. At lower resolutions, the structures are less accurate but even at resolutions approaching 3 Å, modern constrained and restrained least-squares refinement procedures, coupled with accurate amino acid sequence information, can still yield good stereochemically acceptable structures in which there are unlikely to be any significant errors.

The crystal structures of over a dozen Ab molecules or fragments have been determined. These include three intact cryoglobulins; Kol, Dob and Mcg, and a variety of fragments: 4 Fab's, Fc and Fc bound to fragment B of protein A from *S. aureus*, together with a variety of light-chain and $V_L$ dimers. Although so far these results have been obtained on myeloma proteins, the next generation of structural results will undoubtedly involve Ab's prepared by hybridoma methods [1] to specific macromolecular antigens [2, 3]. However, at present our principal structural data concerning Ab combining sites and the nature of Ab-antigen interactions comes from the study of the Fab's with known binding specificity.

Here we briefly describe the refined X-ray structures of two mouse myeloma Fab's with known binding specificities: McPC603, specific for PC, and J539, specific for β-(1,6)-D-galactan. We shall analyze the nature of the interaction of PC with McPC603 and compare this with the presumed mode of binding of J539 to galactan. Each of these proteins belongs to a large family of myeloma and hybridoma Ab's with very similar binding specificities, and a comparative study is being made in the light of the known combining site structures.

## STUDIES WITH McPC603

The structure of the Fab of McPC603 has been refined by the application of Hendrickson-Konnert restrained least-squares procedures together with extensive molecular modelling on a computer graphics system to a set of 2.7 Å resolution intensity data collected photographically by oscillation methods (Y. Satow, G.H. Cohen, E.A. Padlan & D.R. Davies, in preparation). The refined structure (whose atomic coordinates have been deposited in the Protein Data Bank at Brookhaven National Laboratory [4]) has been used as a basis for independently refining the crystal structure of the complex with PC at 3.1 Å resolution (E.A. Padlan, G.H. Cohen & D.R. Davies, in preparation).

Fig. 1a shows an α-carbon skeletal model of McPC603 $F_v$ viewed in a direction perpendicular to the combining site. The loops in the foreground consist of complementarity-determining regions (CDR). The pair of variable domains are related to each other by an approximate two-fold axis. $V_H$ can be rotated into superposition with $V_L$ by a rotation of 173 degrees with an r.m.s. mismatch of 1.08 Å between 323 matched pairs of main-chain atoms [r.m.s. = root-mean-square].

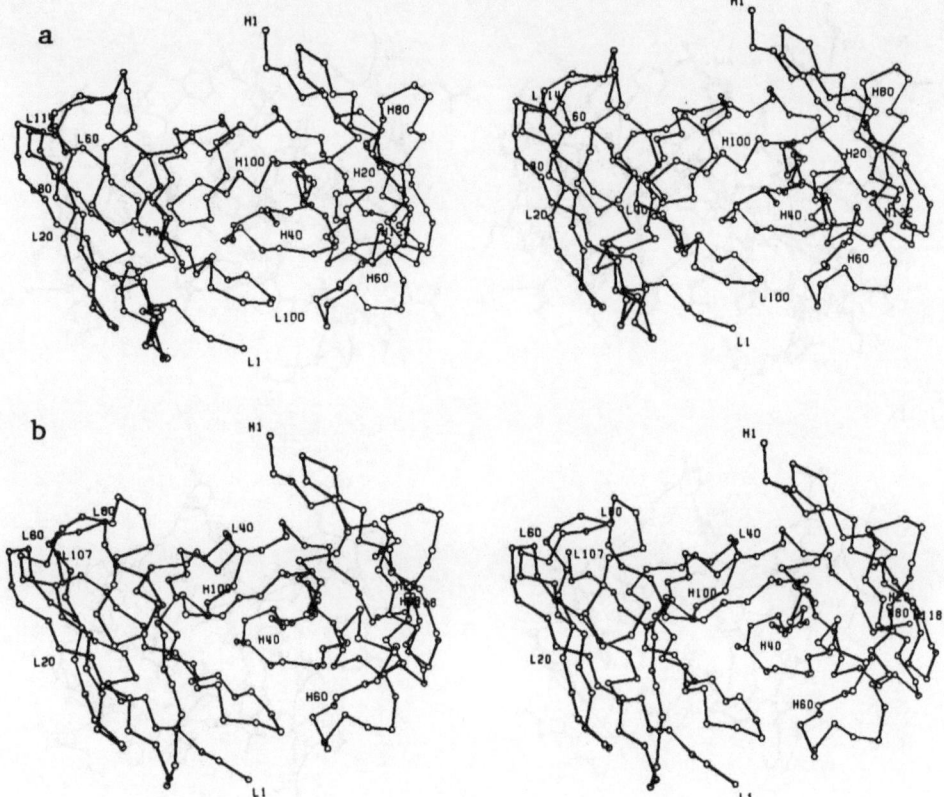

**Fig. 1.** α–Carbon backbone trace for the $F_V$ of (**a**) McPC603, and (**b**) J539. *Left:* light chains.   *Right:* heavy chains.

Fig. 2a shows the hypervariable amino acid residues.   These define a pocket in which the PC binds [5–8].   The amino acids most closely in contact with PC are  Tyr 33, Arg 52, Asn 101[*] and Trp 107 from the heavy chain, together with Asp 97, Tyr 100, Leu 102 and main-chain atoms of residue 98 of the light chain.   The phosphate is anchored on the surface by hydrogen bonds from Tyr 33 and Arg 52.

---

[*] The numbering used here is sequential and differs from that in [9] as follows:

McPC603   Heavy chain: Asn 101 = 95; Trp 107 = 100a.
          Light chain: Asp 97 = 91; His 92 = 98; Tyr 100 = 94;
                       Leu 102 = 96.

J539      Heavy chain: Leu 99 = 95; Tyr 101 = 97; Tyr 104 = 100;
                       Residues 54–57 = 53–56.
          Light chain: Ser 30 = 31; Ser 31 = 32; Glu 49 = 50;
                       Try 90 = 91; Tyr 92 = 93; Ile 95 = 96.

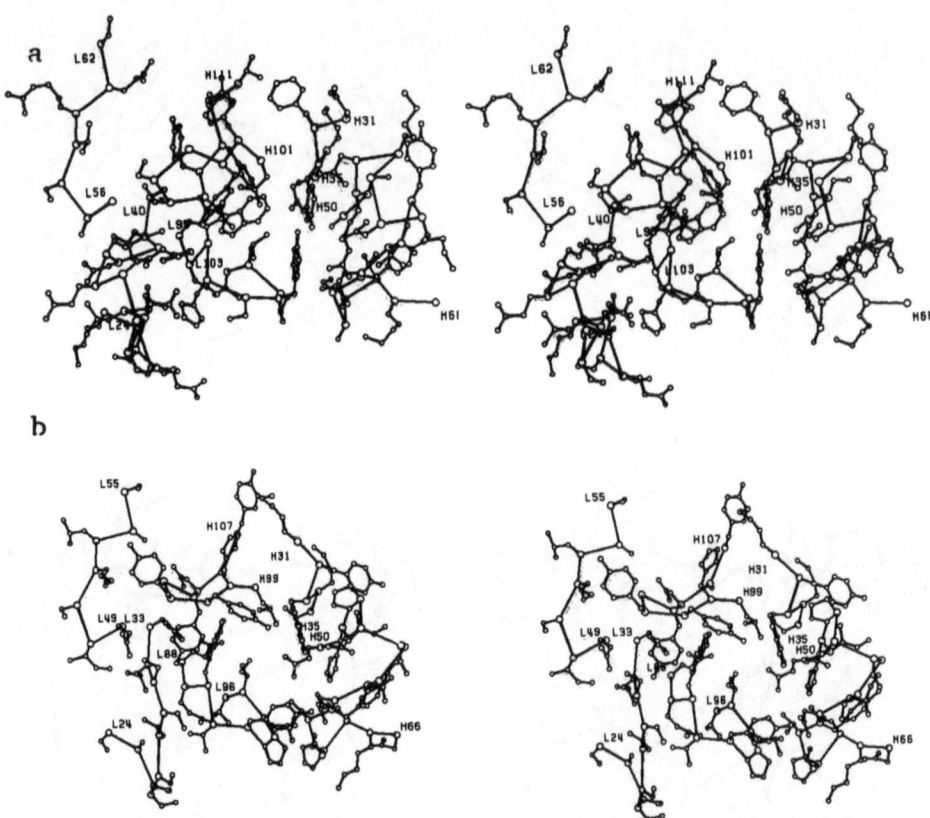

**Fig. 2.** Amino acid residues of the 6 CDR's for (a) McPC603, and (b) J539. The orientation is the same as in Fig. 1. The numbering is sequential as in the text and is related to the numbering in [9] as described in the footnote on the preceding page. The CDR's are defined as in [9].

Charge neutralization is partially achieved by having the positively charged guanidinium group of Arg 52 in the vicinity of the phosphate, while the positive charge of the choline is near to the side-chain of the Asp 97 of the light chain (Fig. 3).

Tables 1 and 2 contain the hypervariable sequence data for a number of mouse PC-binding Ab's [9]. The close similarities in heavy chain sequences support and extend the finding of Gearhart et al. [10] that these proteins all derive from the same $V_H$ germ-line gene. With only a few exceptions the hapten-contacting residues found in McPC603 are invariant in Table 1, extending the suggestion [7] that these proteins share the same molecular mode of PC binding.

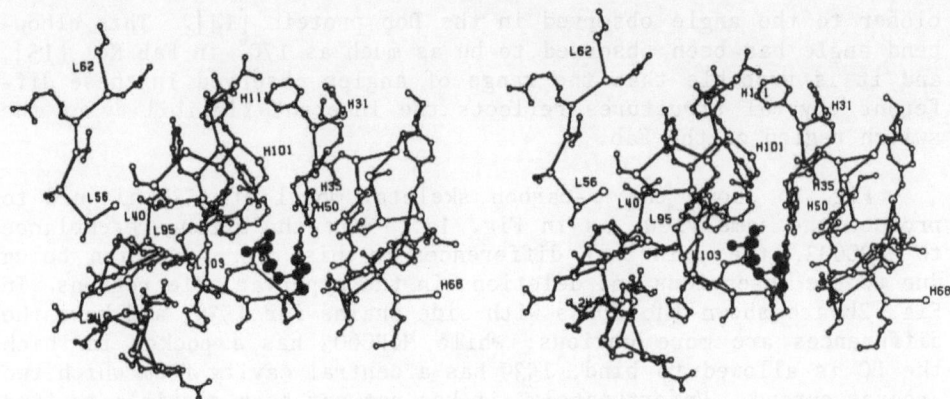

**Fig. 3.** The CDR's of McPC603 with the PC ( ● ) superimposed. The phosphate is on the surface, with the choline residue buried in the pocket. *Left:* light chains. *Right:* heavy chains.

The light chain sequences in Table 2, although at first sight quite different, with three different classes (McPC603, S107 and MOPC167 [11], nevertheless show that the contacting residues Tyr 100 and Leu 102 are invariant, thus maintaining the general shape of the specificity pocket.

Asp 97 in McPC603, noted above as a charge–neutralizing residue for the choline, is replaced in other light chain sequences by the neutral residues Phe or Leu. However, in these proteins Asn 101 of the heavy chain in McPC603 is replaced by an aspartic acid, thus restoring the charge-neutralization effect. This suggests the possibility of a remarkable coupling of somatic mutations, one in the light chain and the other in the heavy chain, to maintain specificity.

## STUDIES WITH J539

The crystal structure of the Fab of J539, previously described at 4.5 Å resolution [12], has been extended to 2.8 Å resolution, and the structure is being refined by Hendrickson-Konnert least-squares procedures. In general it resembles that of McPC603 Fab. Very little difference exists between the two pairs of constant domains which have the same sequence. Minor differences partly reflect the uncertainties of refinement at this resolution and are also partly due to the different packing arrangements in the crystals that result in different contacts between neighbouring molecules.

One major difference between J539 and McPC603 occurs in the angle between the two local dyad axes relating $V_H$ to $V_L$ and $C_H1$ to $C_L$, the 'elbow bend' of the Fab. This angle is 135° in McPC603, similar to the angle observed in Fab New [13]. In J539 it is 145°, much closer to the angle observed in the Dob protein [14]. This elbow-bend angle has been observed to be as much as 170° in Fab Kol [15], and it is probable that the range of angles observed in these different crystal structures reflects the inherent flexibility of the switch region of the Fab.

Fig. 1b shows the α-carbon skeletal model of J539 aligned to produce the same view as in Fig. 1a. Note the strong resemblance to McPC603, the principal differences in this representation being due to the insertions and deletions in the hypervariable regions. In Fig. 2b are shown the CDR's with side chains for J539, and here the differences are more obvious; while McPC603 has a pocket in which the PC is allowed to bind, J539 has a central cavity from which two grooves extend. Unfortunately, it has not yet been possible to find a derivative of galactan that binds to J539 in the crystal, probably because the combining site is at least partially blocked by the constant domain of another molecule. Thus we have no direct experimental evidence to show the precise mode of binding of galactan. It has been shown by binding experiments [16] that J539 binds to galactose residues in the middle of the polymer but not to the terminal residues. Modelling studies with galactan suggest that this is quite feasible with one residue in the central cavity and the residues on either side binding in the grooves. However, the great flexibility of the polymeric carbohydrate molecule makes it impossible to rule out structures on stereochemical grounds alone, and we are now trying different galactan oligomers to get some facts on the binding site.

Tables 3 & 4 list the heavy and light chain sequences for J539 and for other mouse antigalactan Ab's. It is noteworthy that in McPC603, despite the invariance of the amino acid side chains lining the PC pocket, there is marked variability in the lengths of the hypervariable loops, implying that although PC is the major antigenic determinant, considerable variation can be expected in the carrier specificity. In J539 this is not the case, and the data in Tables 3 & 4 indicate an invariance in the length of the 6 CDR's which reflects strict structural requirements extending over a greater part of the combining site to maintain specificity for this larger antigenic determinant.

Several other aspects of the comparison of anti-PC and anti-galactan sequences have been discussed by Rudikoff [11]. He notes that whereas three different light chain subgroups are commonly used in the response to PC, only a single subgroup is found in the anti-galactan response. Also, while all three light chains in the anti-PC response use the same J region, the antigalactan light chains use all 4 J segments.

Four of the Ig's in Tables 3 & 4 have been sequenced in both the heavy and the light chains, and large differences are found between these sequences and J539. TEPC601, XRPC24 and XRPC44 differ from

J539 in the CDR's of the combined light and heavy chain sequences by 10, 10 and 8 amino acid residues respectively. A closer examination of these differences with reference to their potential to interact directly with antigen reveals a more restricted picture.

The central cavity of J539 is lined principally by aromatic residues: Trp33H, Glu50H, Leu99H, Tyr101H and Tyr104H; Ser30L (partly), Trp90L, Tyr92L, Ile95L, together with some main-chain atoms from CDR3H and CDR3L. Those that line the clefts are more variable, consisting of His52H, Asp54H, Ser55H, Thr57H, Asn59H, i.e. all from CDR2H lining one cleft. The other is lined by Ser30L, Ser31L and Glu49L. When only these residues are considered, then TEPC601, XRPC24 and XRP44 differ from J539 by only 2, 3 and 3 residues respectively.

## CONCLUSION

The above discussion reveals a surprising absence of variation in the combining-site residues for both anti-PC and antigalactan Ab's. Whether this restriction can be generalized to other antigens and other species is debatable. Examination of the sequence data for mouse anti-$\alpha$(1,3)dextran and anti-$\beta$(2,1)fructosan [9] shows much the same pattern of restriction. However, in the rabbit, for example, the Ab's against type III pneumococcal polysaccharide show considerable variation in both length and sequence of the CDR's.

**Table 1.** Heavy-chain hypervariable region sequences of PC-binding antibodies. For #1, #2, etc., see foot of page.

```
      CDR 1              CDR 2                          CDR 3

      31    35 50      55      60      65      101   105     107   110
 #1  D F Y M E   A S R N K G N K Y T T E Y S A S V K G   N Y Y G S T - - - W Y F D V
 #2  D F Y M E   A S R N K A N D Y T T E Y S A S V K G   D Y Y G S S - - Y W Y F D V
 #3  D F Y M E   A S R B K A N D Y T T Z Y S A S V K G   D Y Y G N S - - Y W Y F D V
 #4  D F Y M E   A S R N K A N D Y T T E Y S A S V K G   D Y Y G S S - - Y W Y F D V
 #5  D F Y M E   A S R N K A N D Y T T E Y S A S V K G   D Y Y G S S - - Y W Y F A V
 #6  D F Y M E   A S R N K A N D Y T T E Y S A S V K G   D Y Y G S S - - Y W Y F D V
 #7  D F Y M E   A S R N K A N D Y T T E Y S A S V K G   D Y Y G S S - - Y W Y F D V
 #8  D F Y M E   A S R N K A N D Y T T E Y S A S V K G   D Y Y G S S - - Y W Y F D V
 #9  D F Y M E   A S R N K A N D Y T T E Y S A S V K G   D Y Y D Y P - - H W Y F D V
#10  D F Y M E   A S R N K A F D Y T T E Y S A S V K G   D Y Y G S R - - Y W Y F D V
#11  D F Y M E   A S R N K A N D Y T T E Y S A S V K G   D Y Y G S S - - Y W Y F D V
#12  D F Y M E   A S R N K A N D Y T T E Y S A S V K G   D Y Y G S S - - Y W Y F D V
#13  D F Y M E   A S R B K A N D Y T T Z Y S A S V K G   N Y Y K Y D - - L W Y V D V
#14  A F Y M E   A S R N K A N D Y T T E Y S A S V K G   D V Y Y G Y - - D W Y F D V
#15  B F Y M E   A S R N K A N D Y T T E Y S A S V K G   D G D Y G S S - Y W Y F D V
#16  D F Y M E   A S R N K V Y D Y T T E Y S A S V K G   D A Y Y G S - - Y W Y F D V
#17  D F Y M E   A S R S K A H D Y R T E Y S A S V K G   D A D Y G D S Y F G Y F B V
#18  D F Y M E   A S R N K A N D Y T T E Y S A S V K G   D Y Y G S S - - Y W Y F D V
#19  D F Y M E   A S R N K A N D Y T T E Y S A S V K G   D Y Y G S S - - Y W Y F D V
#20  D F Y M E   A S R N K A N D Y T T E Y S A S V K G   N Y Y G S A D Y Y W Y F D V
#21  D F Y M E   A S R N K A N D Y T T E Y S A S V K G   D A Y Y G N - Y G G Y F D V
#22  D F Y M E   A S R N K A N D Y T T E Y S A S V K G   N Y Y G S S - - Y W Y F D V
#23  D F Y M E   A S R N K A N D Y T T E Y S A S V K G   G G Y - - - - - Y Y T M D Y
#24  D F Y M E   A S R N K A N D Y T T E Y S A S V K G   N Y Y D G S - - Y W Y F D V
#25  D F Y M E   A S R N K A N D Y T T E Y S A S V K G   D Y Y D G S - - H W Y F D V
```

**Table 2.** Light-chain hypervariable region sequences of PC-binding antibodies.

```
                  CDR 1                    CDR 2           CDR 3

        25      30      35      40 56      60      95      100
 #1  K S S Q S L L N S G N Q K N F L A   G A S T R E S   Q N D H S Y P L T
 #4  T A S E S L Y S S K H K V H Y L A   G A S N R Y I   A Q F Y S Y P L T
#17  R S S K S L L Y K - D G K T Y L N   L M S T R A S   Q Q L V E Y P L T
#21  R S S K S L L Y K - D G K T Y L N   L M S T R A S   Q Q L V E Y P L T
#15  R S S K S L L Y K - D G K T Y L N   L M S T R A S   Q Q L V E Y P L T
```

**Key for Tables 1 & 2:**

| #1 | MCPC603 | #14 | HPCG14 |
| #2 | TEPC15 | #15 | MOPC511 |
| #3 | HOPC8 | #16 | HPCG13 |
| #4 | S107 | #17 | MOPC167 |
| #5 | S107.U1 | #18 | CBBPC-3 |
| #6 | HPCM1 | #19 | C57BL 293 |
| #7 | HPCM2 | #20 | C57BL 1613 |
| #8 | HPCM3 | #21 | C57BL 2857 |
| #9 | HPCM6 | #22 | C57BL 23169 |
| #10 | HPCG8 | #23 | CBA/N 1B8E5 |
| #11 | S63 | #24 | CBA/J 6F9 |
| #12 | Y5236 | #25 | CBA/J 7C6 |
| #13 | W3207 | | |

**Table 3.**  Heavy-chain hypervariable region sequences of
β(1,6)-D-galactan-binding antibodies.  For #1, #2, etc., see foot of
page.

|  | CDR 1 | | CDR 2 | | | | CDR 3 | |
|---|---|---|---|---|---|---|---|---|
|  | 31 | 35 | 50 | 55 | 60 | 65 | 100 | 105 |
| #1 | K Y W M S | | E I H P D S G T I N Y T P S L K D | | | | L H Y Y G Y N A Y | |
| #2 | R Y W M S | | E I N P D S S T I N Y T F S L K D | | | | L G Y Y G Y F D V | |
| #3 | R Y W M S | | E I N P G S S T I N Y T F S L K D | | | | L G Y Y G Y F D Y | |
| #4 | R Y W M S | | E I N P D S S T I N Y T F S L K D | | | | L H Y Y G Y A A Y | |

**Table 4.**  Light-chain hypervariable region sequences of
β-(1,6)-D-galactan-binding antibodies.

|  | CDR 1 | | CDR 2 | | CDR 3 | |
|---|---|---|---|---|---|---|
|  | 25 | 30 | 50 | 55 | 90 | 95 |
| #1 | S A S S S V S S L H | | E I S K L A S | | Q Q W T Y P L I T | |
| #5 | S A S S S V S Y M H | | E I S K L A S | | Q Q W N Y P L I T | |
| #6 | S A S S S V S Y M H | | E I S K L A S | | Q Q W N Y P L I T | |
| #7 | S A S S S V S Y M H | | E I S K L A S | | Q Q W N Y P L I T | |
| #8 | S A S S S V S Y M H | | E I S K L A S | | Q Q W N Y P L I T | |
| #9 | S A S S S V S Y M H | | E I S K L A S | | Q Q W N Y P L I T | |
| #10 | S T S S S V S Y M H | | E I S K L A S | | Q Q W N Y P L I T | |
| #11 | S T`S S S V S Y M H | | E I S K L A S | | Q Q W N Y P L I T | |
| #12 | S A S S S V S Y M H | | E I S K L A S | | Q Q W N Y P L I T | |
| #13 | S A S S S V S Y M H | | E I S K L A S | | Q Q W N Y P L I T | |
| #14 | S A S S S V S Y M H | | E I S K L A S | | Q Q W N Y P L I T | |
| #15 | S A S S S V S Y M H | | E I S K L A S | | Q Q W N Y P L I T | |
| #16 | S A S S S V S Y M H | | E I S K L A S | | Q Q W N Y P L I T | |
| #2 | S A S S S V S Y M H | | E I S K L A S | | Q Q W N Y P L I T | |
| #3 | S A S S S V S Y M H | | E I S K L A S | | Q Q W N Y P L I T | |
| #4 | S A S S S V S Y M H | | E I S K L A S | | Q Q W N Y P L W T | |

**Key for Tables 3 & 4:**

| #1 | J539 | #9 | HY GAL6 |
|---|---|---|---|
| #2 | TEPC 601 | #10 | HY GAL7 |
| #3 | XRPC 24 | #11 | HY GAL9 |
| #4 | XRPC 44 | #12 | HY GAL10 |
| #5 | HY GAL1 | #13 | HY GAL11 |
| #6 | HY GAL2 | #14 | HY GAL12 |
| #7 | HY GAL3 | #15 | SAPC 10 |
| #8 | HY GAL4 | #16 | TEPC 191 |

*References*

1.  Köhler, G. & Milstein, C. (1975) *Nature 256*, 495–497.
2.  Silverton, E.W., Davies, D.R., Smith-Gill, S. & Potter, M. (1983) *Abs. Am. Crystallographic Assn. 11 (2)*, 28.
3.  Mariuzza, R.A., Jankovic, D.L., Boulot, G., Amit, A.G., Saludjian, P., LeGuern, A., Mazie, J.C. & Poljak, R.J. (1983) *J. Mol. Biol. 170*, 1055–1058.
4.  Bernstein, F.C., Koetzke, T.F., Williams, G.J.B., Meyer, E.F., Brice, M.D., Rogers, J.R., Kennard, O., Shimanouchi, T. & Tasumi, M. (1977) *J. Mol. Biol. 112*, 535–542.
5.  Padlan, E.A., Segal, D.M., Rudikoff, S., Potter, M., Spande, T.F. & Davies, D.R. (1973) *Nature New Biol. 245*, 165–167.
6.  Segal, D.M., Padlan, E.A., Cohen, G.H., Rudikoff, S., Potter, M. & Davies, D.R. (1974) *Proc. Nat. Acad. Sci. 71*, 4298–4302.
7.  Padlan, E.A., Davies, D.R., Rudikoff, S. & Potter, M. (1976) *Immunochemistry 13*, 945–949.
8.  Davies, D.R. & Metzger, H. (1983) *Ann. Rev. Immunol. 1*, 87–117.
9.  Kabat, E.A., Wu, T.T., Bilofsky, H., Reid-Miller, M. & Perry, H. (1983) *Sequences of Proteins of Immunological Interest*, National Institutes of Health, Bethesda, MD. [Cf. E.A. Kabat, this vol.–*Ed.*]
10. Gearhart, P., Johnson, N.D., Douglas, R. & Hood, L. (1981) *Nature 291*, 29–34.
11. Rudikoff, S. (1983) in *Contemporary Topics in Molecular Immunology*, Vol. 9 (Inman, F.P. & Kindt, T.J., eds.), Plenum, New York, pp. 169–209.
12. Navia, M.A., Segal, D.M., Padlan, E.A., Davies, D.R., Rao, D.N., Rudikoff, S. & Potter, M. (1979) *Proc. Nat. Acad. Sci. 76*, 4071–4074.
13. Saul, F., Amzel, L.M. & Poljak, R.J. (1978) *J. Biol. Chem. 253*, 585–597.
14. Silverton, E.W., Navia, M.A. & Davies, D.R. (1977) *Proc. Nat. Acad. Sci. 74*, 5140–5144.
15. Marquart, M., Deisenhofer, J., Huber, R. & Palm, W. (1980) *J. Mol. Biol. 141*, 369–391.
16. Manjula, B.N. & Glaudemans, C.P.J. (1976) *Immunochemistry 13*, 469–471.

#NC(A)

NOTE and COMMENTS related to the foregoing topics


Comments related to particular contributions:

#A-1 & #A-2, p. 69
#A-3 & #A-4, p. 70

#NC(A)-1

*A Note on*

# AN APPROACH TO THE STUDY OF ANTI-PROTEIN ANTIBODY COMBINING SITES

M.J. Darsley, P. de la Paz, D.C. Phillips,
A.R. Rees and *B.J. Sutton

Laboratory of Molecular Biophysics
Department of Zoology, University of Oxford
South Parks Road, Oxford OX1 3QU, U.K.

The main aims of this study are to investigate the nature of the immune response to a single defined antigenic region of a protein and to explore in atomic detail the interaction between antibody (Ab) and a protein antigen. Virtually no structural data are yet available for anti-protein Ab combining sites, since X-ray crystallographic studies have hitherto been restricted to myeloma proteins that bind small-molecule haptens, and only recently have structural studies of monoclonal antibodies (MAb's) to protein antigens been undertaken [1, 2]. Here we describe briefly a procedure for modelling Ab combining sites based on sequence data alone. This approach has been applied to a series of MAb's and their complexes with a single antigenic region of the lysozyme molecule.

## ANTI-LYSOZYME MONOCLONAL ANTIBODIES

The antigenic properties of hen egg-white lysozyme (HEL) have been studied extensively (for review see [3]). One region known as 'loop', residues 57-84, elicits Ab's either as the isolated peptide or as part of the native protein [4]. A series of 5 MAb's has been prepared in response to the loop peptide coupled to bovine serum albumin (BSA) [5], and all 5 Ab's, termed Gloop 1-5, recognize native HEL. Their isotypes are given in Table 1.

## THE MAPPING OF EPITOPE BOUNDARIES

The boundaries of the epitopes (antigenic sites) recognized by each of the Ab's were defined by measuring the relative affinity of

---

* Author to whom any correspondence should be addressed

**Table 1.** Mapping of epitope boundaries for anti-lysozyme MAb's. The maximum extent of residues in the epitope is 57-83.

|  | Group A | | Group B | | Group C |
|---|---|---|---|---|---|
| MAb's | Gloop 1 γ2a,κ | | Gloop 3 γ2a,κ | | Gloop 5 γ1,κ |
|  | Gloop 2 γ2b,κ | | Gloop 4 γ2b,κ | | |
| Excluded residues* | 57 to 59 | | 57 to 59 | | 57 to 59 |
|  | 62 to 63 | | 62 to 63 | | 62 to 63 |
|  | | | 68 to 72 | | |
| Excluded residues † | 73 | | 73 | | 73 |
|  | | | 79 | | 79 |

\* deduced from measurements of HEL activity in the presence of Gloop Ab

† deduced from inhibition studies with tryptic fragment of loop peptide and with lysozymes of related sequence

**Table 2.** CDR regions in crystal structures of V region fragments and models of anti-lysozyme Ab's. The entries for L1-H3 are CDR's as defined by Wu & Kabat [6]; model structures used for CDR's of Gloop Ab's shown thus, (Rei). PC, phosphorylcholine; DNP, dinitrophenyl.

|  | New | McPC 603 | Kol | Mcg | Rei | Rhe | Gloop 2 | Gloop 5 |
|---|---|---|---|---|---|---|---|---|
| Source | Human | Mouse | Human | Mouse | Human | Human | Mouse | Mouse |
| Composition | Fab | Fab | Fab | L chain dimer | $V_L$ dimer | $V_L$ dimer | Fv | Fv |
| Class | λ, γ1 | κ, α | λ, γ1 | λ | κ | λ | γ2b, κ | γ1, κ |
| Specificity | Vit. K1-OH & others | PC | | DNP & others | | | Loop of lysozyme | |
| Resolution | 2.0 Å | 2.7 Å | 1.9 Å | 2.3 Å | 2.0 Å | 1.6 Å | – | – |
| Ref. | [7] | [8] | [9] | [10] | [11] | [12] | | |
| L1 | 14 | 17 | 13 | 14 | 11 | 13 | 11 (Rei) | 17 (McPC) |
| L2 | 0 | 7 | 7 | 7 | 7 | 7 | 7 (McPC) | 7 (McPC) |
| L3 | 9 | 9 | 11 | 10 | 9 | 11 | 9 (Rei) | 9 (Rei) |
| H1 | 5 | 5 | 5 | – | – | – | 5 (New) | 5 (New) |
| H2 | 15 | 18 | 16 | – | – | – | 16 (Kol) | 16 (Kol) |
| H3 | 9 | 11 | 17 | – | – | – | 5 (New)Ø | 5 (New)Ø |

Ø significant modification of model structure now required

each Ab for a series of related lysozymes, and for a tryptic sub-
fragment of loop [5]. Table 1 summarizes the results. The 5 Ab's
segregate into 3 groups (A, B, C), each of which recognises one of 3
distinct but overlapping epitopes within what was previously considered to be a single antigenic site.

## MODELLING OF THE ANTI-'LOOP' ANTIBODY COMBINING SITES

The known high-resolution crystal structures for Ab combining
sites (Table 2) have shown that the combining site consists of three
hypervariable or 'complementarity-determining' regions (CDR's) from
each chain (L1-L3 and H1-H3), brought into proximity by the formation
of a remarkably constant 9-stranded 'barrel' of β-strands which comprise the less variable 'framework' regions of the $V_L$ and $V_H$ domains
[13]. In general, the CDR's form loops between the framework β-strands
so that, by the appropriate choice of model for each CDR according
to length and sequence, the structure of the combining site of any
Ab of known  sequence may be predicted (reviews: [14, 15]). These
predicted structures can be refined and tested as described below.

The cDNA sequence of the Fv part of each of the Gloop Ab's has
been determined, and preliminary modelling of the combining sites of
Ab's Gloop 2 and 5 completed. The amino acid sequences were aligned
with those of known crystal structures, and the models chosen for
each CDR on the basis of length and maximum sequence homology are
shown in Table 2. The substitutions, deletions and insertions were
performed on an interactive molecular graphics system; non-bonded
contacts were checked and corrected, and stereochemistry was regu-
larized at the break points.

The resulting models for Gloop 2 & 5 at this stage are shown in
Fig. 1, a & b. One of the striking similarities is the equally short
length of H3 in these two Ab's, compared with the three of known
structure. This opens up the binding site considerably.  The remaining CDR's are relatively conserved between Gloop 2 and Gloop 5, apart
from L1, which displays significant differences in both length and
sequence. (Ref. [16] gives details of recent studies; cf. [5].)

## FUTURE WORK

The next step in the modelling procedure is to refine these
model structures by energy minimization to ensure their consistency
with stereochemical constraints.  This is now in progress using the
method of Levitt [17]. A similar technique  has recently been applied to the predicted structure of the galactan-binding Ab J539 [18].
This step is particularly important for the model CDR's which have
involved building insertions and deletions.

It will then be possible to model the Ab-antigen complexes by
matching the shape and nature of the two surfaces, guided by the

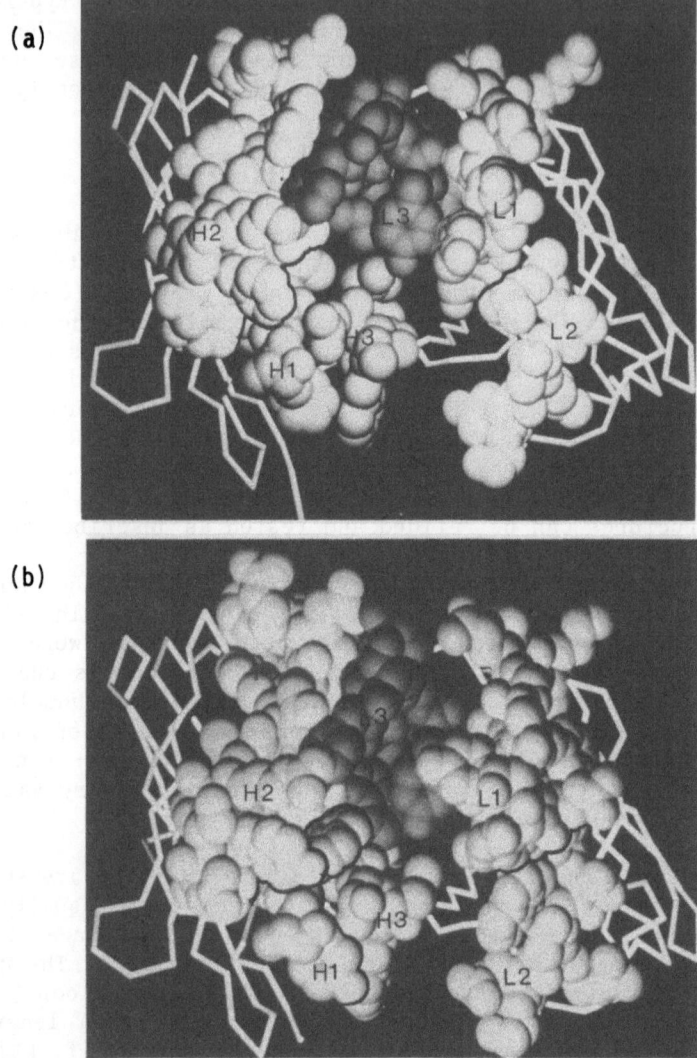

Fig. 1.  Space-filling representations of the modelled hyper-variable loops for the Ab's (**a**) Gloop 2 and (**b**) Gloop 5, viewed down the axis of the β-barrel framework.  Loops denoted L1 - L3, H1 - H3.

detailed knowledge of the epitope boundaries and the binding data. From the resulting models of the complexes, it will also be possible to predict the effect of amino acid substitutions in the CDR's on structure, specificity and affinity. These changes can be introduced by the technique of site-directed mutagenesis, and the predictions tested by binding and structural studies.

It is clearly important, however, to test at least one of the predicted combining-site models, and crystallographic studies are already under way for one Gloop Ab, together with further analysis of the epitope residues for each complex in solution by nuclear magnetic resonance.    Other crystallographic studies are in progress on myeloma proteins for which the combining sites have been predicted [18-20].

*Acknowledgements*

X-ray coordinates  for the modelling of hypervariable loops (Table 2) were obtained from the Brookhaven Protein Data Bank.  We thank Dr. D.R. Davies (National Institutes of  Health, U.S.A.) for the refined coordinates of McPC603, Drs. J. Burridge and A. Morffew (IBM Scientific Centre, U.K.) for the computer graphics drawings and Mr. L. Mangold for the photography.   We thank the Medical Research Council for supporting this work; B.J.S. holds a University Research Fellowship of the Royal Society.

*References*

1.  Colman, P.M., Gough, K.H., Lilley, G.G., Blagrove, R.J., Webster, R.G. & Laver, W.G. (1981) *J. Mol. Biol. 152*, 609-614.
2.  Mariuzza, R.A., Jankovic, D.L., Boulot, G., Amit, A.G., Saludjian, P., Le Guern, A., Mazie, J.C. & Poljak, R.J. (1983) *J. Mol. Biol. 170*, 1055-1058.
3.  Benjamin, D.C., Berzofsky, J.A., East, I.J., Gurd, F.R.N., Hannum, C., Leach, S.J., Margoliash, E., Michael, J.G., Miller, A., Prager, E.M., Reichlin, M., Sercarz, E.E., Smith-Gill, S.J., Todd, P.E. & Wilson, A.C. (1984) *Ann. Rev. Immunol. 2*, 67-101.
4.  Arnon, R. & Sela, M. (1969) *Proc. Nat. Acad. Sci. 62*, 163-170.
5.  Darsley, M.J. & Rees, A.R. (1985) *EMBO J. 4*, 383-392.
6.  Wu, T.T. & Kabat, E.A. (1970) *J. Exp. Med. 132*, 211-250.
7.  Saul, F., Amzel, L. & Poljak, R. (1978) *J. Biol. Chem. 253*, 585-597.
8.  Segal, D., Padlan, E., Cohen, G., Rudikoff, S., Potter, M. & Davies, D.R. (1974) *Proc. Nat. Acad. Sci. 71*, 4298-4302.
9.  Marquart, M., Deisenhofer, J., Huber, R. & Palm, W. (1980) *J. Mol. Biol. 141*, 369-391.
10. Edmundson, A.B.,  Ely, K.R., Girling, R.L., Abola, E.E., Schiffer, M., Westholm, F.A., Fausch, M.D. & Deutsch, H.F. (1974) *Biochemistry 13*, 3816-3827.
11. Epp, O., Latham, E., Schiffer, M., Huber, R. & Palm, W. (1975) *Biochemistry 14*,  4943-4952.
12. Furey, W., Wang, B., Yoo, S. & Sax, M. (1983) *J. Mol. Biol. 167*, 661-692.
13. Novotny, J., Bruccoleri, R., Newell, J., Murphy, D., Haber, E. & Karplus, M. (1983) *J. Biol. Chem. 258*, 14433-14437.
14. Padlan, E., Davies, D.R., Pecht, I., Givol, D. & Wright, C. (1976) *Cold Spring Harbor Symp. Quant. Biol. 41*, 627-637.
15. Davies, D.R. & Metzger, H. (1983) *Ann. Rev. Immunol. 1*, 87-117.

16. Darsley, M.J. & Rees, A.R. (1985) *EMBO J.* *4*, 393-398.
17. Levitt, M. (1974) *J. Mol. Biol.* *82*, 393-420.
18. Mainhart, C.R., Potter, M. & Feldmann, R.J. (1984) *Mol. Immunol.* *121*, 469-478.
19. Navia, M.A., Segal, D.M., Padlan, E.A., Davies, D.R., Rao, N., Rudikoff, S. & Potter, M. (1969) *Proc. Nat. Acad. Sci.* *76*, 4071-4074. [See also #A-4 (preceding art., this vol.).]
20. Aschaffenburg, R., Phillips, D.C., Rose, D.R., Sutton, B.J., Dower, S.K. & Dwek, R.A. (1979) *Biochem. J.* *181*, 497-499.

## Comments on material in #A

*Comments on* #A-1, E.A. Kabat - ANTIBODY BINDING SITES

**Question by P. Marrack.-** Taking account of Tonegawa's sequence and some of the β-chain J-region sequences where Ala-X-Gly replaces Gly-X-Gly, what might be the consequences of the Ala-for-Gly substitution? **Reply.-** This substitution will cause severe restrictions on the properties of the protein. **Question to E.A. Kabat. -** What do you feel about the multi-specificity of Ab's? The frequency of 1:10⁵ antigen-binding cells, besides numerous examples of the binding of two antigens to one Ab combining site, indicate that the multi-specificity is a real phenomenon. **Reply.-** I don't think there is any case in which the evidence for multi-specificity is convincing. Most of the cases reported could have some structural similarity. I have discussed the question (1978) in *Adv. Prot. Chem. 32*, 43-44.

**Remark by B.F. Erlanger.-** In view of recent evidence (from the laboratories of Alt and of Hood) for transcription and translation of a DJC segment, DNA sequence conservation may be necessary to assure production of a specific regulatory protein - not only for the integrity of a combining site. **E.A. Kabat's response.-** I agree that the synthesis of a 'mini-protein' made up of DJC opens new vistas in regulation and control of Ig's and may prove of importance in Ab diversity and to the mini-gene hypothesis.

**M.M. Davis asked:** why couldn't the observed conservation of Framework 2 be the result of evolutionary selection? **Reply.-** We are not dealing with the preservation of one framework. There are many alternatives in both the mouse and the rabbit which occur only infrequently, and this hierarchy of usage has existed for 80 million years. If all of the various FR 2's were linearly distributed in V regions throughout the genome, one would not have expected the occurrence of these FR 2 segments in such a proportion. **Davis, continuing.-** One out of the 12 possibly functional T-cell receptor beta chain J-region elements has the sequence Ala-X-Gly *vs.* the more canonical Gly-X-Gly. We have not seen this expressed in any cDNA clones thus far.

*Comment on* #A-2, J.M. Thornton - THE Ig FOLD

**Reply to P. Marrack.-** Of the hypervariable regions, only that around position 96 is in a hairpin turn.

**B.F. Erlanger asked**: is there any possibility of *cis* peptide bonds in the loop regions? **J.M. Thornton's reply.**- In proteins, *cis* peptide bonds occur infrequently, usually for proline residues. They are non-compatible with helix and sheet structures (since the appropriate hydrogen bonds cannot be made) and therefore they will be found predominantly in the loop regions between secondary structures. **Question by I.M. Roitt.**- You originally suggested that the hairpin sequence may affect conformation, but in the analysis of bends with two amino acids you said that the demands of 2 H-bonded chains in the β-sheet were such that only the extremely flexible Gly could be tolerated as the second amino acid, indicating a subservient role of the hairpin.   Can you clarify? **Reply.**- In the very tight hairpin loops with 2 residues, the conformations are such that only Gly is allowed at specific residue positions. If the Gly were replaced by a different amino acid, the conformation of the loop region would have to change – which may in turn modify the β-strand arrangement. We are currently comparing homologous β-hairpins to assess the effects of such replacements.

*Comments on* #**A**-3, A.B. Edmundson et al. – BINDING SITES IN Ig CRYSTALS
#**A**-4, D.R. Davies et al. – MOUSE Ab COMBINING SITES

**I.M. Roitt, to A.B. Edmundson.**- I gather that you are making the interesting suggestion that certain Ab's are inherently low-affinity because they may have 'wobbly' or easily deformable hyper-variable loops while high-affinity Ab's presumably have more rigid complementarity-determining sequences.   Evidence for the former is based on the change in conformation induced by crystallization from ammonium sulphate; but this might be an exceptionally strong deforming force with certain structures.   I am not sure if there is any evidence for the latter.   **Question by A. Feinstein.**- What is the maximum displacement of the combining site you have observed when the ligands are bound? **Reply.**- We have seen movement of ~2Å. With A, side-chains can swing around without entailing any other movement.

**D.P. Lane, to both contributors.** – Could you both comment, in the light of your recent work on the flexibility of the Ab combining site, on the results obtained following immunization with synthetic peptides corresponding to regions of mature proteins?   Lerner et al. have reported that up to 50% of anti-peptide hybridomas will react with the mature protein antigen.   **Response.**- Although the results obtained with anti-peptide Ab's were initially a surprise to us, they became easier to understand in the context of the studies on the binding site of the Mcg Bence-Jones protein.   In particular, the binding of the *N*-formylated chemotactic peptides illustrates the flexibility of conformation open to peptides in binding Ab.   This mobility must therefore be reflected in regions of the protein structure corresponding to the peptide sequence. **Remark by C.M. Lewis to D.R. Davis,** concerning Ab's to peptides recognizing the native protein.- We have raised a number of MAb's to peptides, and only a proportion of these react with the native protein. It is my belief that the method of immunization, usually by conjugation of peptide to protein in a random fashion, leads to the peptide assuming a number of different conformations, and that some of these may be related to the structure of that particular sequence in the parent protein.

**Questions by K.L. Knight.**- (1) Is the difference in Mcg binding site according to whether water or ammonium sulphate is used for the crystallization still found when this is done in the presence of DNP?  **A.B. Edmundson's reply.**- We cannot consider the question because the DNP bound in solution would interfere with key crystal-packing interactions between proteins in ammonium sulphate. In crystals, however, not all of the space available for DNP binding is obstructed by protein-protein interactions.  (2) How can we interpret the structure of the binding site from crystallized proteins if they differ depending on how they are crystallized? **Reply.**- This very interesting question will be clarified when the water form's high-resolution structure is completed. At present we know that the over-all geometries and micro-environments differ in the two solvent systems, but most of the contact residues for ligands remain the same. *Re* (1), **D.R. Davies added:** one must observe the complex with antigen.

**Question by I.M. Roitt.**- With a small ligand one would expect cavity-type binding to generate enough attractive force, but with a globular protein antigen, for example, could not ample binding energy be provided by binding across several of the 'peaks' presented by the hypervariable loops?  **Edmundson's reply.**- We agree that even in a molecule with a cavity, considerable binding energy can be contributed by interactions around the cavity rim and neighbouring loops. **Added by Davies.**- Agreed, provided there is antigen-Ab complementarity.  **A. Feinstein, to Davies.**- Is there any computation of the segmental mobility (i.e. temperature factors) of the hypervariable regions for the combining sites of Ab's examined at high resolution by X-ray crystallography?  This would give a feeling for the extent of possible adaptive fit.  **Reply.**- I believe not.  In our own case the data are not at the resolution which would permit the calculations.

**E.F. Hounsell, to Davies.**- In connection with flexible protein sequences you  touched on the flexible nature of oligosaccharides. Experience in the carbohydrate antigen field using information from NMR analysis to help in molecular model-building has shown that one can often get a good idea of the spatial arrangement of atoms in the antigenic determinant.  Did you actually try to make a molecular model of the galactan,  to see if it would help to predict something of the combining site topography?  **Reply.**- We do plan to further investigate the binding of galactan to JS39. If we fail to observe this experimentally we shall then try modelling.

Section #B

# PRODUCTION AND CHARACTERIZATION OF ANTIBODIES, IN CONTEXTS SUCH AS RECEPTOR INVESTIGATION

#B-1

# INVESTIGATING THE SPECIFICITY OF MONOCLONAL ANTIBODIES TO PROTEIN ANTIGENS USING β-GALACTOSIDASE FUSION PROTEINS

D.P. Lane, J. Gannon, S.P.J. Brennan and S.E. Mole

Department of Biochemistry
Cancer Research Campaign
Eukaryotic Molecular Genetics Research Group
Imperial College
London SW7 2AZ, U.K.

*Numerous monoclonal antibodies (MAb's) to the oncogene, large T, of the small double-stranded DNA (dsDNA) tumour virus SV40 have been produced [1-4, and J. Yewdell & D.P.Lane, unpublished work). Detailed analysis of these Ab's[\*]has provided insights into both the structure and the function of large T, besides protein-antigen immunochemistry. The Ab's can be subdivided by the following criteria.- (a) Ability of the Ab to recognize the protein after denaturation. (b) Recognition of only a sub-set of the T antigen molecules. (c) Inhibition of biological activity of large T. (d) Location of the binding site of the Ab on the linear structure and the 3-dimensional structure of the protein. Each subdivision can be enhanced by subcloning fragments of the large-T coding information into bacterial and eukaryotic expression vectors that cause the large-T fragment to be incorporated at the carboxy terminus of β-galactosidase [5, 6].*

*Using such large T-β-galactosidase fusion proteins we have very closely mapped Ab binding sites on large T, and identified immunologically silent regions. Immunization with the fusion proteins permitted the isolation of MAb's and polyclonal antibodies (PAb's) to these silent regions [6]. Our results with large T and similarly with other protein antigens seem widely applicable, as do the methods used.*

The advent of MAb's and then of Ab's[\*] to short synthetic peptides has enabled the nature of the antigenic determinants present on protein antigens to be analyzed in much more detail than hitherto. Some general rules and procedures are now emerging, and here we discuss, using a few defined systems, the genesis and application of these general rules.

* Ab connotes MAb unless the context indicates otherwise *(Ed.'s term).*

## CLASSIFICATION OF MAb's ON BASIS OF DENATURATION

MAb's raised to 'native' protein antigens can be subdivided universally into two distinct classes on the basis of their ability to recognize the antigen after exposure to denaturing conditions. The Ab's will all react with the antigen in native form, but only a subset will react with the antigen after it has been subjected to boiling in 2% SDS and reducing agent and subjected to polyacryl- amide gel electrophoresis (PAGE). The ability to bind 'denatured' antigen can be assessed using the Western blotting technique [7] or by direct binding in solution [8].

The relative proportion of Ab's in a library that will react with SDS-resistant as opposed to SDS-sensitive determinants depends on the protein antigen under study. In the case of SV40 large T, a DNA-binding protein that has 708 amino acids and is encoded by the dsDNA tumour virus (for recent review see [9]), ~10% of the MAb's recognize SDS-resistant epitopes whereas out of 8 Ab's recognizing the host p53 protein that binds large T, 6 recognize SDS-denatured p53 and 2 do not [2, 3, 10, 11; J. Yewdell, J. Gannon & D.P. Lane, unpublished work]. In the case of another dsDNA tumour virus onco- gene, the Polyoma virus middle T, all four existing Ab's to the pro- tein [12] recognize denaturation-resistant epitopes. In making the distinction between sensitive and resistant epitopes it should not be assumed that the latter represent completely linear sequences of amino acids. Kinetic studies of the binding of Ab's to resistant epitopes, both in solution and in Western blot analysis, indicate that Ab-driven localized renaturation of the antigen may be impor- tant for detection of binding. Thus, it was found that prolonged incubations of the nitrocellulose-bound antigen with the MAb greatly increased the sensitivity of the Western blot procedure for p53 and T regardless of the Ab concentration used (Yewdell, Gannon & Lane, unpublished results).

## EPITOPE MAPPING

Mapping of the epitopes recognized by MAb's is important for localizing functional domains on protein antigens (e.g. active sites of enzymes, sites for virus neutralization). Epitope mapping by steric hindrance of sets of MAb's has been carried out for a reason- able number of protein antigens, as exemplified by our recent map of the murine p53 protein defining 5 sterically discrete epitopes (Fig. 1). The approach here is to immobilize the antigen on a solid phase and then to assess the degree of inhibition of the binding of labelled Ab A with unlabelled Ab's B, C, D, etc. The results are usually quite clear-cut and allow the demarcation of a series of sterically discrete sites. Several anomalies do arise in this type of study, some of which are fairly readily explained whereas others are more complex. The first phenomenon is manifestation of serial inhibition, viz. Ab A competes for B but not C, while B competes for both A and

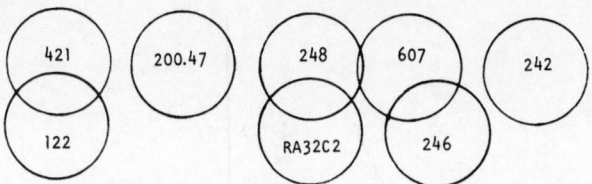

Fig. 1. Schematic steric inhibition map of anti-mouse p53 Ab's.
The binding of each was measured by RIA. A complex of SV40 large T
and mouse p53 extracted from the SV40-transformed cell line SVA31 E7
was immobilized on plastic microtitre dishes through an anti-T MAb,
PAb 419. The binding of each iodinated anti-p53 Ab was measured in
the presence of excess amount of every other p53 Ab. The diagram
illustrates the result. Ab's that inhibit each other's binding are
depicted by overlapping circles. The slight overlap between 607
and 248 indicates only a weak mutual interference.


C, and finally as expected C competes only for B. This situation is
illustrated in Fig. 2, depicting the binding of the MAb's PAb122,
PAb421 and PAb1005 [13] to human p53.

    The second phenomenon, 'enhancement' is observed stimulation of
the binding of Ab A in the presence of Ab B rather than inhibition.
This has been described for instance with MAb's to the Human Class
One gene products [14, 15].

    The explanation for this phenomenon may be either a conforma-
tional shift in the antigen induced by B such that A now has a
higher affinity for the antigen, or alternatively, aggregation of
two antigen molecules by Ab B, such that A is now more likely to
engage in divalent rather than monovalent binding to the antigen.
These two possibilities can be distinguished by using monovalent Ab
fragments in the assays. In at least two cases bivalency has been
shown to be essential for enhancement, supporting the aggregation
rather than the conformational model [15].

    The third phenomenon seen in these solid-phase assays is non-
reciprocal binding, as illustrated for murine p53 in Fig. 3. Thus
when Ab A is the solid phase, labelled B is unable to detect the


Fig. 2. Human p53 epitopes. The
binding of 3 different MAb's to Human
p53 was assessed as in Fig. 1 except that
Human p53 T complex was used as the solid-
phase antigen. Evidently PAb 421 inhibits
the binding of both PAb 1005 and PAb 122 but
PAb 122 and PAb 1005 do not inhibit each
other's binding reactions.

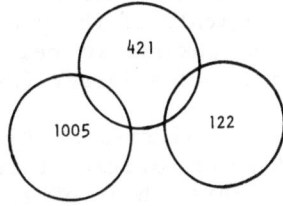

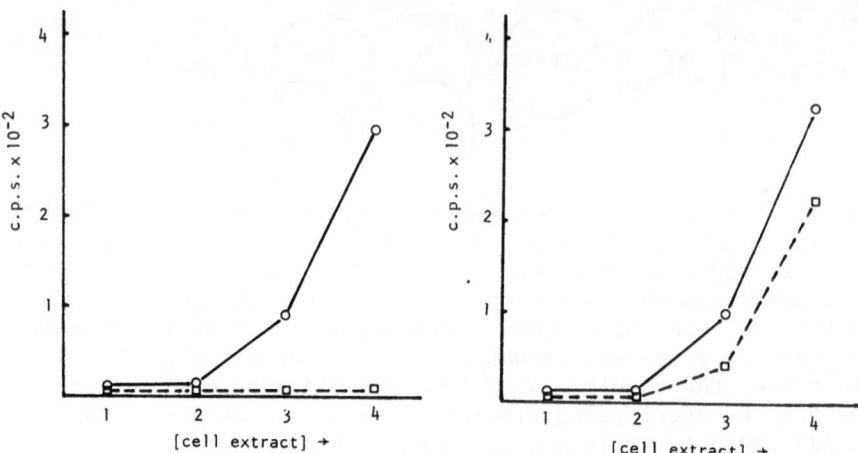

Fig. 3.  An example of non-reciprocal binding.  The concentration
of p53 in two cell extracts was measured using a solid-phase RIA.
One cell extract from spontaneously transformed 3T3 cells contains
p53 in the free state (--□--), whereas the other extract, from
SV40-transformed 3T3 cells, contains p53 complexed to SV40 large T
(—o—).  When PAb 248 is the solid phase and PAb 246 is labelled,
p53 is detected only if it is complexed to T.  Merely reversing the
Ab's so that PAb 246 is the solid phase and PAb 248 the label
permits the detection of p53 in both cell extracts.

antigen.  However, when B is the solid phase, labelled A readily
detects it.  Careful analysis showed that in this case, the epitope
recognized by B was very unstable and deteriorated during the course
of the assay, whilst that recognized by A was completely stable;
binding of the B epitope by Ab B protected it against deterioration.
Thus the assay was non-reciprocal since only when epitope B was protected throughout the assay by solid-phase Ab could the antigen be
detected.

## USE OF RECOMBINANT DNA TECHNOLOGY

     Mapping of the steric relationships between Ab binding sites,
whilst useful for the design of immunoassays, is somewhat restricted
unless the sites can be localized within the linear amino acid
sequence of the protein antigen.  This mapping process can be greatly
facilitated by use of recombinant DNA technology.  In this approach
subfragments of the protein-encoding gene are expressed in prokaryotic or eukaryotic cells either as isolated fragments  or else as
elements of a larger fusion protein.

     To illustrate this approach it is instructive to consider a
particular example, that of the anti SV40 large T antigen MAb PAb204
[1].  This Ab is of special interest because it inhibits the ATPase

activity of large T [16] and because it cross-reacts very specifically with an ubiquitous growth-regulated host protein of mol. wt.
68,000 [1, 17].

Initial mapping of the Ab's binding site utilized naturally
occurring Adeno-SV40 hybrid viruses [1] and, later, synthetically
produced SV40 deletion mutant viruses which removed the carboxy-
terminal amino acids from the protein [18]. Thus PAb204 was shown
to recognize the Ad2 ND2 proteins of mol. wt. 56,000 and 42,000 and
the mol. wt. 64,000 truncated T antigen of the SV40 large T deletion
mutant dl 1058. This localized the binding site to the overlap of
the amino terminus of the 42,000 protein of the Adeno-SV40 hybrid
virus and the deletion mutant T's carboxy terminus, a region of ~100
amino acids. Mapping of the PAb204 binding site was further refined
by constructing fusion proteins between sections of large T and β-
galactosidase, using plasmid vectors. Since this procedure is versa-
tile and has wide-ranging practical benefits, it will be discussed
here in more detail.

## Construction and use of fusion proteins

The fusion protein vectors chosen, pUR290, 291 and 292 [5],
consist of the bacterial *lac* Z gene driven by the *lac* promoter. The
*lac* operator region is intact, so that *lac* Z expression is still
inducible. At the naturally occurring 3' *Eco*RI site in the *lac*Z
gene, a polylinker fragment has been inserted containing several
plasmid-unique restriction sites. The polylinker is phased one nucleotide apart in 290, 291 and 292; thus any inserted restriction
fragment derived from the coding region of a gene will be in frame
with *lac* Z in one of the plasmids. The pUR plasmids reach very high
copy numbers, such that on induction the fusion proteins produced
come to represent as much as 10% of the total cell protein. Because
they are such major species and are of such high mol. wt., they are
very easily identified and purified by PAGE.

We have successfully produced, using these vectors, fusion
proteins between β-galactosidase and SV40 large T antigen [6], Polyoma large, middle and small T (T. Dale, S.E. Mole & D.P. Lane, in
preparation), Murine p53 (R. Bartsch, S.E. Mole & D.P. Lane, in preparation), Yeast CDC28 and Mouse H-2 class 1 gene fragments (unpublished observation). The design of the large T construction is
shown in Fig. 4. In all cases the fusion proteins were (a) produced
in abundant amount, and (b) shown to react with MAb's known to
recognize SDS-resistant epitopes within the insert fragment. The
stability of protein fragments inserted at the carboxy terminal end
of β-galactosidase seems to be the result of the fusion proteins
precipitating within the *E. coli* cells, thus protecting them from
proteolysis [19].

**Fig. 4.** Large T fusion protein constructions. The *Hind*III A and D fragments of
SV40 were cloned into the unique *Hind*III
site of pUR 290, 291 and 292. The A
fragment in pUR 290 and the D fragment
in pUR 292 resulted in the synthesis of
β-galactosidase Large T fusion proteins.
The A construct encodes the carboxy terminus of Large T (amino acids 446–708),
and the D construct encodes a central
section of Large T (amino acids 271–
446).

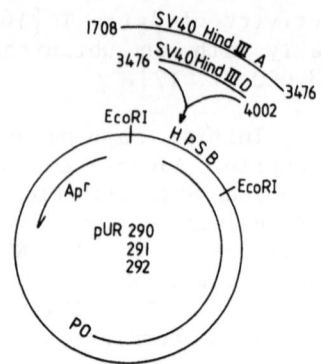

PAb204 was found to react with the *Hind*III A fragment of SV40
and not with the *Hind*III D fragment, localizing its binding site
within 63 amino acids. Further constructions using smaller restriction fragments will permit us to localize the site down to 20 amino
acids.

At this point precise definition of the site can be economically
confirmed using either peptide or oligonucleotide synthesis. A similar approach to mapping Ab binding sites using phage rather than
plasmid vectors has also been successful [20], implying the general
utility of the method.

We assessed the binding of the available libraries of anti-T
MAb's to the *Hin*III A and *Hin*III D fusion proteins which encode
amino acids 271 – 446 and 446 – 708 respectively. Surprisingly, while
many Ab's known to react with sterically distinct sites bound to the
A fusion protein, none reacted with the D species. The position is
as follows concerning the binding of Ab's to the β-galactosidase T
antigen fusion proteins.-
- Ab's binding the *Hind*III A fragment, amino acids 446 – 708 of
  SV40 Large T:  PAb 204 [1], 423 [3], 404 [3] & 409 [3].
- No Ab's bind the *Hin*dIII  fragment, amino acids 271 – 446 of
  SV40 Large T.
- Ab's that bind the Adeno SV40 Hybrid proteins from Ad2 ND2 but do
  not bind either fusion protein:  PAb 203 [16] & 205 [16].

The binding sites of the PAb203 and PAb205 MAb's should be
present in either the A or the D fragment, because of their pattern
of reaction with the Adeno–SV40 hybrid proteins [16]. Since these non-
scoring MAb's recognize SDS-sensitive sites, it may merely be that the
fusion protein's insert fails to fold up or be modified correctly in
the prokaryotic cell. To confirm this, the β-galactosidase fusion
protein constructions are now being transferred into vectors that
will promote their expression in eukaryotic cells [21].

Since no Ab's recognized the *Hind*III D fragment, the reason for this immunological silence was investigated. The β-galactosidase – *Hind*III D fusion protein was isolated by preparative PAGE and used to immunize mice and rabbits. Analysis of the mouse sera and of MAb's derived from the spleen of one of the immunized mice established, through immunoprecipitation, RIA and immunocytochemistry procedures, that the fusion protein had induced a good anti-T response. The Ab's were able to recognize both native and SDS-denatured T. Thus while the normal anti-T response does not elicit Ab's to these epitopes, they can be detected using the fusion protein as immunogen.

This result mirrors the kind of observations made using synthetic peptides derived from the influenza virus HA molecule [22] and reiterates the message that the normal Ab response to native protein antigens is incomplete and selective. This has important practical implications both for detailed immunochemical analysis of protein antigens and for vaccine design.

## CONCLUDING COMMENT

The use of recombinant DNA methodology and, in particular, bacterial fusion proteins provides a cheap and rational complement to the use of synthetic peptides in the study of the immunochemistry of protein antigens.

*Acknowledgements*

This research was supported by a grant from the Cancer Research Campaign. Ms. S.E. Mole holds a postgraduate training award from the Medical Research Council.

*References*

1.  Lane, D.P. & Hoeffler, W.K. (1980) *Nature 288*, 167–170.
2.  Gurney, E.G., Harrison, R.O. & Fenno, J. (1980) *J. Virol. 34*, 752–763.
3.  Harlow, E., Crawford, L.V., Pim, D.C. & Williamson, N.M. (1981) *J. Virol. 39*, 861–869.
4.  Ball, R.K., Siegl, R., Quellhorst, S., Branpner, G. & Braun, D.G. (1984) *EMBO J. 3*, 1485–1491.
5.  Rüther, U. & Müller-Hill, B. (1983) *EMBO J. 2*, 1791–1794.
6.  Mole, S.E. & Lane, D.P. (1985) *J. Virol. 52*, in press.
7.  Towbin, H., Staehlin, T. & Gordon, J. (1979) *Proc. Nat. Acad. Sci. 76*, 4350–4354.
8.  Lane, D.P. & Robbins, A.K. (1978) *Virology 87*, 182–193.
9.  Rigby, P.W.J. & Lane, D.P. (1983) in *Advances in Viral Oncology*, Vol. 3 (Klein, G., ed.), Raven Press, New York, pp. 31–57.
10. Benchimol, S., Pim, D. & Crawford, L.V. (1982) *EMBO J. 1*, 1052–1062.

11. Dippold, W.G., Jay, G., De Leo, A.B., Khoury, G. & Old, L.J. (1981) *Proc. Nat. Acad. Sci. 78*, 1695–1699.
12. Dilworth, S.M. & Griffen, B.E. (1982) *Proc. Nat. Acad. Sci. 79*, 1059–1062.
13. Thomas, R., Kaplan, L., Reich, N., Lane, D.P. & Levine, A.J. (1983) *Virology 131*, 502–517.
14. Holmes, N.J. & Parham, P. (1983) *J. Biol. Chem. 256*, 1580–1586.
15. Thomson, R.J. & Jackson, A.P. (1984) *Trends in Biochem. Sci. 9*, 1–3.
16. Clark, R., Lane, D.P. & Tjian, R. (1981) *J. Biol. Chem. 256*, 11854–11858.
17. Crawford, L., Leppard, K., Lane, D. & Harlow, E. (1983) in *Tumor Viruses and Differentiation* (Scolnick, E.M. & Levine, A.J., eds.), Alan Liss, New York, pp. 421–425.
18. Clark, R., Peden, K., Pipas, J.M., Nathans, D. & Tjian, R. (1983) *Mol. Cell. Biol. 3*, 220–228.
19. Stanley, K.K. (1983) *Nucleic Acids Res. 11*, 4077–4092.
20. Nunberg, J.H., Rodgers, G., Gilbert, J.H. & Snead, R.M. (1984) *Proc. Nat. Acad. Sci. 81*, 3675–3679.
21. Hall, C.V., Jacob, P.E., Ringold, G.M. & Lee, F. (1983) *J. Mol. Appl. Genet. 2*, 101–109.
22. Wilson, I.A., Niman, H.L., Houghten, R.A., Cherenson, A.R., Connnolly, M.L. & Lerner, R.A. (1984) *Cell 37*, 767–778.

#B-2

# ANTIBODY-SCREENING cDNA LIBRARIES

[1]Frank A. Simmen, [2*]Denis R. Headon, [1]Tanya Z. Schulz,
[1]Mohan Cope, [3]David A. Wright, [4]Graham Carpenter and
[1]Bert W. O'Malley

[1]Department of Cell Biology
Baylor College of Medicine
Houston, TX 77030, U.S.A.

[2]Department of Biochemistry
University College
Galway, Ireland

[3]Department of Genetics
M.D. Anderson Hospital and
Tumor Institute
Houston, TX 77030, U.S.A.

[4]Department of Biochemistry
Vanderbilt University
School of Medicine
Nashville, TN 37232, U.S.A.

*Molecular cloning and ancillary techniques offer an attractive alternative in studying many biological phenomena. Successful molecular cloning of genes encoding proteins for which amino acid sequence data are not available can be achieved by antibody (Ab) screening. Requisite precautions in adopting immunological screening methods include interaction of Ab preparations with host bacterial proteins (i.e. host bacteria for plasmid uptake). This should be assessed by Western blotting of host bacterial proteins or by performing immunoprecipitations following* in vivo *labelling of the host bacterial proteins.*

*Another need, especially for a glycoprotein antigen, is to ensure that the Ab or Ab's to be used react with the peptide portion of the molecule, as glycosylation should not be expected following translation in host bacterial cells. Additionally, especially with monoclonal rather than polyclonal Ab's for Ab-screening cDNA libraries, one must consider the region of the peptide that is recognized by the Ab, viz. amino- or carboxy-terminal regions, so that one might realistically expect expression of those regions of the gene that code for the immunogenic sites. An Ab cocktail is desirable if monoclonals are used in primary screening. If screening entails use of a secondary Ab and/or protein A, specificity controls are needed where the primary Ab is omitted.*

* Author to whom any enquiries should be addressed.

*With such precautions, Ab screening of cDNA libraries provides a highly sensitive and specific method for the detection of cloned sequences in expression vectors.*

Screening cDNA libraries can be carried out by a variety of methods. These include hybrid-selected translation [1] which is difficult with low-abundance mRNA or when the efficiency of *in vitro* translation is low. An alternative to such screening is to prepare oligonucleotide probes, which have a nucleotide sequence deduced from amino acid sequence data for the gene product, and screening by colony hybridization [2]. The latter requires purified protein for amino acid sequence analysis and the preparation of a number of oligonucleotide probes to cover the degeneracy of the genetic code. Purification of low-abundance proteins for amino acid sequencing can be difficult. The ability to raise polyclonal Ab's to a purified antigen preparation or to raise monoclonal Ab's to an antigen-enriched preparation provides a means of overcoming the problems referred to above.

The cloning of cDNA in a bacterial expression vector and the screening of the resulting recombinant clones for antigenic determinants expressed *in situ* requires only an Ab. A number of suitable vectors have been developed for this purpose [3-9]. Indeed the molecular cloning of the progesterone receptor B subunit of chick oviduct has been successfully accomplished using pBR322 as vector and Ab screening procedures [10], and the same vector has been used in the molecular cloning of proinsulin [11].

The use of expression vectors which give rise to insoluble hybrid protein products as in [8] or with bacterial hosts which are defective in protein degradation [7] increases the possibilities for detecting expressed cloned sequences by Ab screening methods.

## MATERIALS AND METHODS

*Cell culture.* Human A431 cells were grown in Dulbecco's Modified Eagle Medium (GIBCO) containing 10% (v/v) foetal calf serum. The growth of the cells and their labelling with [$^{35}$S]methionine have been previously described [12].

*Immunoprecipitation.* The $^{35}$S-labelled proteins solubilized in RIPA [10 mM Tris-HCl, pH 8.5, 0.15 M NaCl, 5 mM EDTA, 1% (v/v) Triton X-100, 1% (w/v) Na deoxycholate, and 0.1% SDS] were incubated with polyclonal Ab's against the EGF-R of A$_{431}$ cells according to a double-immunoprecipitation protocol [12]. The Ab's used in this study were IgG fractions prepared as previously described [13, 14] from rabbit antisera to the purified EGF receptor. Antisera 450 and 986 were raised to the affinity-purified receptor as previously described [13]. Other antisera were produced in response to

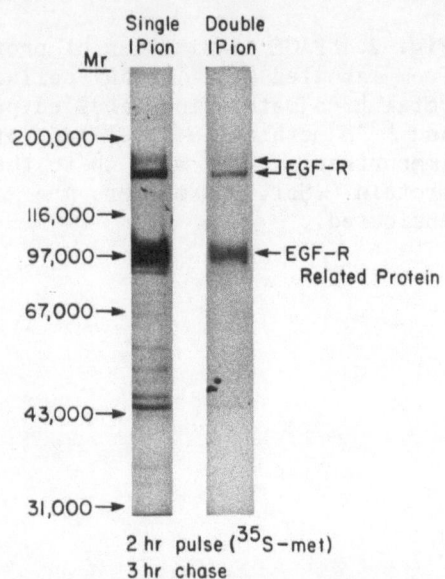

**Fig. 1.** PAGE examination of immunoprecipitates from labelled A431 cells, which were pulsed with [$^{35}$S]methionine for 2 h and chased for 3 h prior to solubilization in RIPA. Details are given in ref. [12]. Mol. wt. markers are indicated.

affinity-purified and SDS gel-purified EGF receptor from A431 cells and their properties are summarized in [14]. RNA isolation and translation were carried out as described in [12].

*Molecular cloning procedures.* Poly(A)$^+$-RNA isolated from A431 cells was size-fractionated and cDNA generated by procedures previously reported [10]. The cDNA was inserted into either pBR 322 [10] or λgt 11 [7].

## RESULTS AND DISCUSSION

The specificity of the Ab preparations for the EGF receptor is shown in the immunoprecipitation data of Fig. 1. Double immunoprecipitation results in the precipitation of two protein bands of mol. wt. 160K and 90K. These correspond to the EGF receptor's precursor and to an EGF-receptor protein respectively. A faint band of higher mol. wt., ~170K, is also evident; it corresponds to the EGF receptor. As Fig. 1 demonstrates, the double immunoprecipitation protocol yields clearer immunoprecipitation patterns. The presence of antigenic material in $^{35}$S-labelled *E. coli* RR 1 is evident in Fig. 2. *E. coli* RR1 cells grown in L-broth in the presence of [$^{35}$S]methionine were solubilized in RIPA and subjected to immunoprecipitation. The immunoprecipitates were then subjected to SDS-PAGE.

Fig. 2 shows that a protein band of mol. wt. 45K is immunoprecipitated. Such data reinforce the importance of determining Ab interactions with host bacterial proteins. At this point IgG was purified by affinity chromatography on a column of protein A-Sepha-

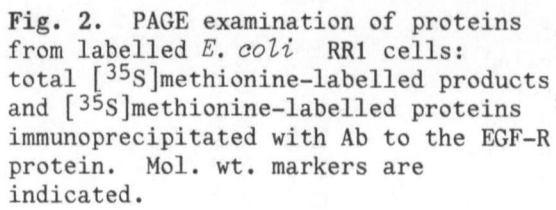

Fig. 2.  PAGE examination of proteins
from labelled *E. coli*  RR1 cells:
total [$^{35}$S]methionine-labelled products
and [$^{35}$S]methionine-labelled proteins
immunoprecipitated with Ab to the EGF-R
protein.  Mol. wt. markers are
indicated.

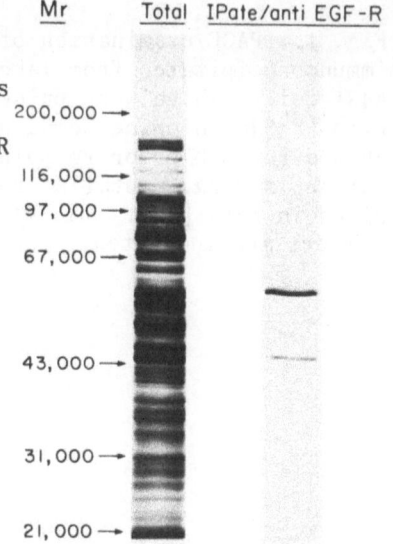

rose 4B and then pre-absorbed against a bacterial extract coupled to
Sepharose (CNBr).  If secondary Ab and/or protein A are to be used,
then their interaction with host bacterial proteins should be evaluated.

Data obtained using four Ab preparations to the EGF-receptor in
the immunological screening of a cDNA library in pBR 322  prepared
from poly(A)$^+$-RNA of ·A431 cells are shown in Fig. 3.  Evidently the
four different Ab preparations clearly distinguish the same colony
reacting above background,  and the results would be interpreted as
being a positive recombinant colony.

This colony contained an insert in excess of 2 kb.  However,
exhaustive analysis of this insert by hybrid selection followed by
translation in Xenopus oocytes failed to identify it as an EGF
receptor clone.   Indeed, with subsequent analysis and sequencing
this colony proved to be negative for the EGF receptor.  These data
further highlight the necessity for caution in the identification of
Ab-positive recombinant colonies.

In determining the sensitivity of immunological assay procedures
it is important to apply the antigen preparation to the agar plates
when are then overlaid with nitrocellulose paper.  Such an application allows one to determine the Ab sensitivity under conditions of
antigen pick-up from lysed  bacterial colonies or plaques, and is a
better indication of the sensitivity than direct application of the
antigen to the nitrocellulose filter itself.  It is also important
to optimize Ab-antigen interaction and Ab detection at this point.

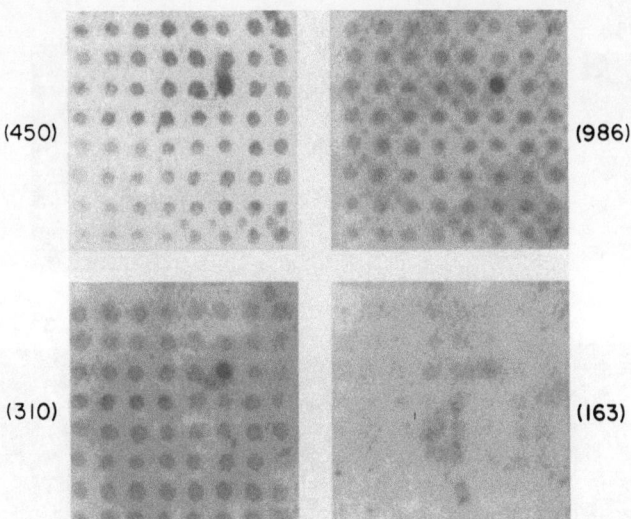

(450)                                            (986)

(310)                                            (163)

Fig. 3. Antibody-screening a cDNA library in λgt11 prepared from
A431 cells. Colonies were picked in a grid pattern of 64 colonies
per petri dish. The Ab screening was carried out with 4 Ab preparations raised in 4 different animals by injection of 4 different
receptor preparations. These are labelled 450, 986, 310 and 163.

Suitable dilutions of primary Ab and of secondary AB, if used, should
be determined to obtain maximum sensitivity with acceptable back-
grounds. Ideally these experiments should be carried out with antigen applied to agar plates containing colonies or plaques. Dilutions
depend upon the Ab preparation and can range from 1:1000 to 1:10000
for IgG preparations.

Fig. 4 shows the results obtained with Ab-screening an A431 cDNA
library in the high-expression vector λgt 11. Ab 986 readily detects
EGF receptor down to 0.25 ng, and the specificity is evident by the
fact that a polyclonal Ab to avidin (αAu) does not react with the
antigen preparation from A431 cells. Primary, secondary and tertiary
screenings of the λgt 112 library are also shown in Fig. 4. The
plaque indicated by the arrow on the primary screen was subjected to
secondary and tertiary screening and proved to be a clone containing
sequences of the EGF receptor gene. A point of importance in screening cDNA libraries with Ab's is the necessity to carefully examine
the autoradiographic or stained patterns so as to compare any
possible colony with the background in its immediate vicinity. The
importance of this is evident from Fig. 4, where a variable background is evident along a nitrocellulose filter. Such a variable
background is very frequently detected. Thus the comparison of a
colony with its immediate neighbours is important in selecting

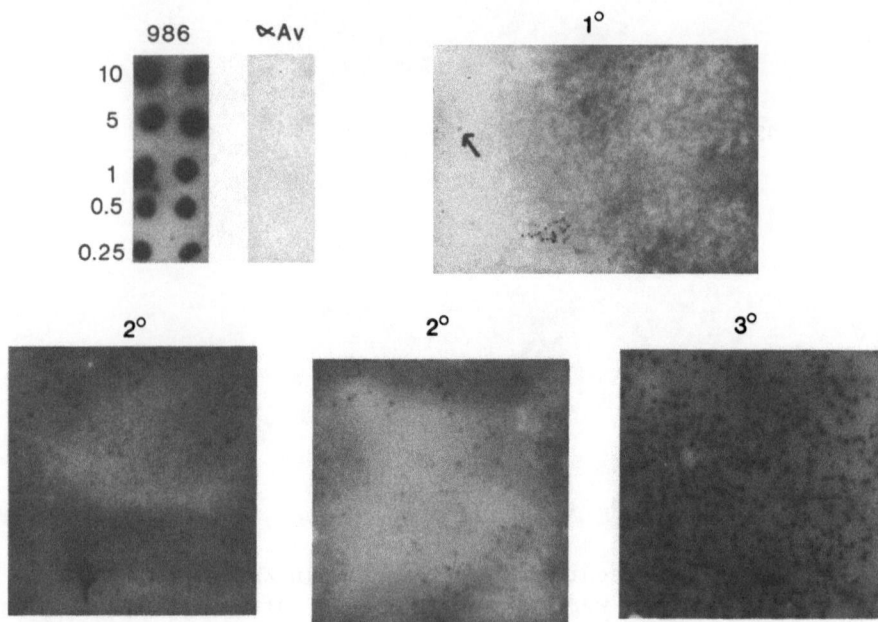

**Fig. 4.** Antibody screening a cDNA library in gt 11 prepared from A431 cells (screening data as indicated in the text). *Arrow* denotes a colony that was subjected to secondary and tertiary screening and was identified as a positive clone.

possible positive recombinants. The cDNA insert was subcloned from λgt 11 to the Pst 1 site of pBR 322.

The authenticity of the cDNA insert identified in the λgt 11 cloning was verified by hybrid selection of mRNA from a poly(A)$^+$-RNA preparation from A431 cells followed by translation in Xenopus oocytes [12]. Fig. 5 shows the data obtained. Track 1 refers to oocytes injected with buffer that did not contain RNA. Track 2 was from oocytes injected with poly(A)$^+$-RNA grown in [$^{35}$S]methionine and subjected to immunoprecipitation. Evidently in this track the EGF-R and the EGF-R-related protein have been immunoprecipitated. Track 3 was prepared from oocytes injected with hybrid-selected poly(A)$^+$-RNA using plasmid EGF-R1 grown in [$^{35}$S]methionine and subjected to immunoprecipitation. The major band manifests the EGF-R protein with smaller amounts of the EGF-R-related protein. Track 4 represents the use of pBR 322 in hybrid selection of poly(A)$^+$-RNA as a negative control. Sequencing of the insert in ph EGF-R1 yielded sequence data identical with part of the published sequence [15] for EGF-R.

The protocol for Ab screening of cDNA libraries is set out below. Using such procedures with high-expression vectors enables one to screen for hybrid proteins containing antigenic sites to Ab's

**Fig. 5.** Translation in Xenopus oocytes of poly(A)+-RNA, hybrid-selected mRNA and controls, and immunoprecipitation with anti-EGF-R. Clone ph EGF-R1 clearly selects EGF-R mRNA. For additional information, see [12].

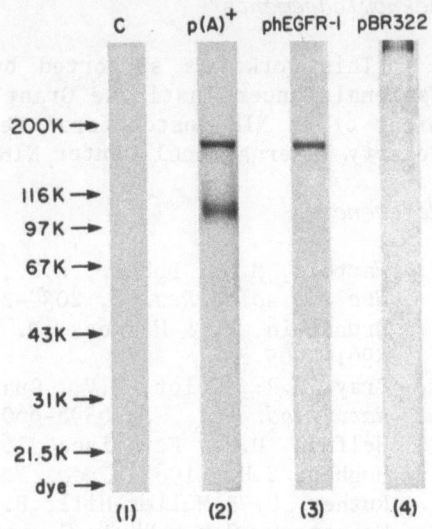

raised to gene products of interest. Using such systems it is now possible to screen >50,000 recombinants per nitrocellulose filter. Screening such large numbers of recombinants allows one to rapidly search in cDNA libraries for cDNA sequences from low-abundance mRNA's.

## PROTOCOL FOR IMMUNOLOGICAL SCREENING OF cDNA LIBRARIES

1. Lysed bacterial colonies or phage plaques are transferred to nitrocellulose filters by overlaying agar plates with the filters for 1 h at room temperature.

2. The filters are removed and washed free of debris in 'TBS', viz. 50 mM Tris-HCl, pH 7.5, containing 150 mM NaCl.

3. The washed filters are incubated for 2 h in a shaking incubator in TBS containing 3% BSA to block unreacted sites on the nitrocellulose. All incubations in TBS or TBS-BSA were at 20-25°.

4. The blocked filters are transferred to TBS-BSA containing an appropriate dilution of Ab (IgG or DEAE-purified ascites fluid) and incubated for several hours or overnight. The appropriate dilution of Ab should be determined before screening the recombinant bacteria.

5. Following incubation with Ab the filters are washed in TBS, 5 x 10 min washes, and incubated for 2 h in TBS-BSA. Filters were incubated in [125]I-labelled protein A (2 ml/filter; 2 x 10 cpm/ml); for Ab's that do not bind protein A an appropriate second Ab is added followed by [125]I-protein A. The required dilution of the second Ab should be determined prior to screening recombinant bacteria. A 4 h incubation in protein A is sufficient.

6. The filters are washed extensively in TBS and, following drying, are mounted against X-ray film for autoradiography at -70°.
*Preparation of the labelled protein A* (usually 20-30 µCi/µg): by a published procedure [16] using iodo beads (Pierce) and [125]I (Amersham).

*Acknowledgements*

This work was supported by USPHS Grant RP-05425 (TZS) and a National Cancer Institute Grant CA 24071 (G.C.). F.A.S. is a recipient of an NIH postdoctoral fellowship and D.R.H. a recipient of a Fogarty International Center NIH Fellowship 1 F05 TW 03360-01.

*References*

1. Harpold, M.M., Dobner, P.R., Evans, R.M. & Bancroft, F.C. (1978) *Nucleic Acids Res. 5*, 2039-2053.
2. Grunstein, M. & Hogness, D. (1975) *Proc. Nat. Acad. Sci. 72*, 3961-3965.
3. Gray, M.R., Colot, H.V., Guarente, L. & Rosash, M. (1982) *Proc. Nat. Acad. Sci. 79*, 6598-6602.
4. Helfman, D.M., Feramisco, J.R., Fiddes, J.C., Thomas, G.P. & Hughes, S.H. (1983) *Proc. Nat. Acad. Sci. 80*, 31-35.
5. Ruther, U. & Müller-Hill, B. (1983) *EMBO J. 2*, 1791-1794.
6. Weinstock, G.M., Rhys, C. ap., Berman, M.L., Hampar, B., Jackson, D., Silhavy, T.J., Weisemann, J. & Sweig, M. (1983) *Proc. Nat. Acad. Sci. 80*, 4432-4436.
7. Young, R.A. & Davis, R.W. (1983) *Proc. Nat. Acad. Sci. 30*, 1194-1198.
8. Stanley, K.K. & Luzio, J.P. (1984) *EMBO J. 3*, 1429-1434.
9. Broome, S. & Gilbert, W. (1978) *Proc. Nat. Acad. Sci. 75*, 2746-2749.
10. Zarucki-Schulz, T., Kulomaa, M.S., Headon, D.R., Weigel, N.L., Baez, M., Edwards, D.P., McGuire, W.L., Schrader, W.B. & O'Malley, B.W. (1984) *Proc. Nat. Acad. Sci. 81*, 6358-6362.
11. Villa-Komaroff, L., Efstratiadis, A., Broome, S., Lomedico, P., Tizard, R., Naber, S.P., Chick, W.L. & Gilbert, W. (1978) *Proc. Nat. Acad. Sci. 75*, 3727-3731.
12. Simmen, F.A., Schulz, T.Z., Headon, D.R., Wright, D.A., Carpenter, G. & O'Malley, B.W. (1984) *DNA 3*, 393-400.
13. Stoscheck, C.B. & Carpenter, G. (1983) *Arch. Biochem. Biophys. 227*, 457-468.
14. Stoscheck, C.B. & Carpenter, G. (1983) *Cell Biol. Internat. Repts. 7*, 529-530.
15. Downward, J., Yarden, Y., Mayes, E., Scrace, G., Totty, N., Stockwell, P., Ullrich, A., Schlessinger, J. & Waterfield, M.D. (1984) *Nature 307*, 521-527.
16. Markwell, M.A. (1982) *Anal. Biochem. 125*, 427-432.

#B-3

# ANTI-RECEPTOR ANTIBODIES BY THE AUTO–ANTI–IDIOTYPIC ROUTE

B.F. Erlanger, W.L. Cleveland, N.H. Wassermann, B.L. Hill,
*A.S. Penn, H.H. Ku and R. Sarangarajan

Departments of Microbiology and (*) Neurobiology and the
Cancer Center/Institute of Cancer Research
Columbia University
New York, NY 10032, U.S.A.

*A procedure was developed for preparing antibodies (Ab's) to the acetylcholine receptor (AChR) on the supposition that, regardless of functional differences, macromolecules of the same specificity will show binding-site homologies. Ab's prepared in rabbits to a structurally constrained agonist of AChR (BisQ) mimicked the binding specificity of AChR in its activated state; agonist binding affinities parallelled biological activity. Rabbits immunized with a specifically purified preparation of anti-BisQ produced anti-idiotypic (anti-Id) Ab's that cross-reacted with determinants on AChR preparations from* Torpedo californica, Electrophorus electricus *and* rat muscle. *Moreover, some rabbits showed signs of transient experimental myasthenia gravis, manifesting circulating AChR Ab's.*

*In a more direct route to monoclonal anti-receptor Ab's, mice were immunized with a bovine serum albumin conjugate of BisQ (BisQ-BSA). Fusion of the spleen cells with an appropriate myeloma line yielded monoclonal anti-AChR Ab's besides anti-BisQ Ab's. One monoclonal Ab (MAb), F8-D5, bound anti-BisQ and AChR, and binding to one was inhibited by the other. Moreover, binding to both was inhibited by BisQ, decamethonium and α-bungarotoxin. Immunofluorescence studies on* Torpedo *tissue gave patterns identical to that produced by anti-AChR serum raised by immunization with AChR. In preliminary experiments it inhibited* $^{134}Cs$ *influx in a reconstituted vesicle system and, when immobilized, selectively bound AChR. The same approach, based on Jerne's idiotypic network theory, furnished MAb's specific for the adenosine receptor.*

Specificity in biological systems is almost invariably mediated by binding to proteins. Thus, there are specific enzymes, carrier proteins, receptors, Ab's, allosteric subunits, etc. Specificity of binding occurs through a constellation of amino acid functional groups

arranged in a suitable geometric array within one or more binding
sites. Because short-range forces are involved in binding, e.g.
electrostatic forces, H bonds, van der Waal's forces and hydrophobic
interactions, high affinity requires that interactions with ligands
take place over short distances and in a directed manner; thus, in H
bonds, the donor and receptor atoms ideally should be colinear. In
other words, 'close fit' or 'complementarity' is required.

Since the same variety of forces and functional groups confer
specificity in all proteins, it should not be surprising to find that
there are binding homologies among proteins that differ in function
but that have prime specificity for the same ligand. (Since nature
need not solve the problem of specificity the same way for all proteins specific for the same ligand, *sequence* homology need not occur;
e.g. chymotrypsin, carboxypeptidase A and subtilisin all bind aromatic amino acids but binding site sequences differ [1].) Binding
homologies have been reported in the past. Tanenbaum, Beiser and
their colleagues [cf. 2] presented evidence for homology in binding
properties of enzymes and Ab's specific for the same ligand. Correla-
tions in specificity were found between (a) protocatechuoyl BSA and the
induction of protocatechuic acid oxidase of Neurospora, (b) Ab to
β-D-galactosido-phenylazo-BSA and the enzyme β-D-galactosidase of
*E. coli* and *B. megaterium* as well as the permease system of *E. coli*,
and (c) Ab to β-D-glucosido-phenylazo-BSA and the inducible β-D-
glucosidase of *Rhodotorula minute* as well as the constitutive β-D-
glucosidase of *Stachybotrys atra* [2].

Support for this type of homology emerged [3] during the development of a radioimmunoassay (RIA) for thaumatin, a plant protein
~200,000 times sweeter than sucrose on a molar basis. Cross-reactions
were found with several non-protein sweet compounds and a correlation
of 'sweetness' with binding affinity to Ab was noted [3]. It was concluded [3] that Ab binding-site structure and 'sweet-taste' receptor
structure may be fundamentally related; the system was suggested to
be useful for *in vitro* recognition of novel sweet substances.

These and more recent observations (see below) led to our own
work on receptors, which involves the compound 'BisQ', *trans*-3,3'-
bis[α-(trimethylammonio)methyl]azobenzene bromide. We were led to
synthesize it through a series of serendipitous events, starting in
1967 with the development in our laboratory of photochromic molecules
that could be used to photoregulate enzyme activities [4], viz.
molecules that can assume two distinct configurations having different absorption spectra, whose relative concentrations can be controlled by light of selected wavelengths [5]. All the compounds we
used were azobenzene derivatives; they exist as *cis* or *trans* isomers, whose relative concentrations can be regulated by exposure to
long-UV or to visible (blue) light. Both the *cis* and *trans* isomers
of the compounds we synthesized could function as enzyme substrates
[6], inactivators [4, 7, 8] or inhibitors [7, 8] but the *cis* and

*trans* stereoisomers differed markedly in activity. As is evident from models, the *trans* isomer is planar; in the *cis* isomer the benzene rings are not co-planar but are at 90° to each other. Molecules so very different in structure would be unlikely to bind with the same affinity to enzymes and affect activity equally. Thus, photoregulation of their relative concentrations would, in effect, photoregulate the activities of enzymes for which they are specific, as demonstrated with trypsin, chymotrypsin [4] and acetylcholinesterase (AChE) [7-9]. This work has been reviewed [10, 11].

Whilst these studies were in progress, conversations with Dr. David Nachmansohn, a pioneer AChR investigator [12], got us interested in synthesizing photochromic agonists of the nicotinic AChR. BisQ was synthesized and its *trans* isomer was found to be a notably potent agonist of AChR in the *Electrophorus electricus* electroplax system. Surprisingly, the *cis* isomer lacked measurable activity.

Molecular models of *trans*-BisQ reveal that it is a highly constrained, planar molecule [13, 14] which has few degrees of freedom. Hence, if its biological activity depends upon interaction with the binding site of a macromolecular receptor, it can be either highly active or near-inactive, depending on whether its conformation (or small number of possible conformations) allows complementary binding to the target receptor. The extremely high activity of *trans*-BisQ as an agonist suggested that its structure was complementary to the combining site of AChR when the latter was in the activated state.

## STRUCTURE OF AChR [15]

AChR is made up of 5 subunits, $2\alpha$, $\beta$, $\gamma$, and $\delta$. The ACh binding sites are on the two $\alpha$ subunits; in concert with these, the other subunits make up the total structure of the receptor, including an ionophore responsible for the ion flux that occurs when ACh binds to receptor. Formalistically presented, the receptor exists in two states which are in equilibrium with each other [16]. (Cf. S.M. Sine & P. Taylor, #D-1 in Vol. 13, this series.- *Ed.*)

Resting state    $\xrightarrow[\text{antagonists}]{\text{agonists}}$    Activated state
(ionophore closed)              (ionophore open)

Agonists stabilize the activated state, in which the ionophore is open, and antagonists the resting state in which it is closed. As BisQ is a potent agonist, its structure can be envisaged as being complementary to the AChR binding site when AChR is activated.

## ANTIBODIES TO BisQ MIMIC AChR

As noted earlier, macromolecules that bind the same ligand but have different biological functions can show binding-site homology [2, 3]. It might follow, therefore, that the combining sites of Ab's

**Table 1.**  Inhibition of binding of $trans$-3,3'-[$^3$H]BisQ to antibody.

| Inhibitor | $IC_{50}$, µM | Relative inhibitory concn. |
|---|---|---|
| $trans$-3,3'-BisQ* | 0.15 | (1) |
| 3,3'-BisQ amide | 0.25 | 1.7 |
| Methylene stigmine† | 0.55 | 3.7 |
| $trans$-4,4'-BisQ | 0.9 | 6.0 |
| Decamethonium ion | 2 | 13.3 |
| Acetylcholine | 20 | 133 |
| Succinoylcholine | 25 | 167 |
| Butyltrimethylammonium ion | 100 | 667 |
| Carbamoylcholine | 200 | 1,333 |
| Hexamethonium ion | 1,000 | 6,665 |
| d-Tubocurarine | 1,000 | 6,665 |

* The $cis$ isomer has not been purified.  Preliminary studies using a 70:30 $cis/trans$ mixture indicate little if any binding of the $cis$ isomer.

† $m$-(dimethylcarbamoyloxy)benzyltrimethylammonium bromide, a highly active agonist and homologue of neostigmine (unpublished)

raised to  $trans$-BisQ would show homology  with the binding site of AChR when the latter is in the activated state.  If so, these Ab's should be able to bind agonists of AChR in accordance with their activities in the AChR system, and should bind antagonists poorly. Put simplistically,  Anti-BisQ ≡ AChR.

A BisQ derivative [17] coupled to BSA via the mixed anhydride reaction [18] — the conjugate containing 8 molecules of BisQ per BSA — was used to immunize New Zealand White rabbits.  Binding specificities of the antisera were determined by a competitive RIA using [$^3$H]$trans$-BisQ.   The order of agonist  binding activities was similar to that seen with AChR (Table 1).  Exact comparisons are precluded because many of the AChR experiments were done with electroplax or membrane preparations as well as with soluble AChR [19,20].   There were other striking findings.—
(1) The $IC_{50}$ value for  $trans$-BisQ (0.15 µM) was close to the 50% point on the dose-response curve in $Electrophorus$ $electricus$ electroplax experiments (0.06 µM) [21].
(2) Decamethonium, a potent agonist, was bound with an apparent affinity 500 times that of hexamethonium, a potent antagonist.   It would not be expected $a$ $priori$  that an Ab would distinguish between two molecules of such similar structure.
(3) d-Tubocurarine, a potent antagonist, was bound poorly.

Evidently we had succeeded in showing binding homologies in AChR and anti-BisQ (i.e. that  anti-BisQ ≡ AChR).

## ANTIBODIES TO  Anti-BisQ CROSS-REACT WITH AChR

The above findings suggested that anti-Id Ab's [22] directed at determinants of the combining sites of anti-BisQ might share characteristics of anti-AChR. In other words,

since anti-BisQ $\equiv$ AChR, then anti-anti-BisQ $\equiv$ anti-AChR.

Rabbits immunized with purified AChR make specific anti-AChR Ab's and simultaneously show signs of experimental myasthenia gravis [23]. Although the disease in humans was long suspected of being an an autoimmune disease [24, 25], it was the experiments in rabbits that encouraged the search and led to discovery of anti-AChR Ab's in myasthenic patients. Subsequently the passive transfer of anti-AChR serum was shown to transfer the syndrome to experimental animals [26]. If we were to succeed in raising anti-Id Ab's directed against the combining sites of anti-BisQ (i.e. anti-anti-BisQ) and if, indeed, these Ab's cross-reacted with AChR, the rabbits in which these Ab's were raised should show signs of myasthenia gravis.

Anti-BisQ Ab's purified from antisera by affinity chromatography on BisQ-aminohexyl-Sepharose columns [17] were used to immunize three New Zealand White rabbits. Primary and booster multiple intradermal injections were given with complete Freund's adjuvant (days 1, 21, 40, 72 & 130; animals bled at various times up to 145 days and observed regularly for signs of muscular weakness as they moved in their cages or on a smooth floor). Limb strength was tested by repeatedly extending the limb and assessing the force and rapidity with which it was retracted to the body. One rabbit (#524) never manifested muscular weakness. One (#522) showed slight weakness in the left hind-leg at 7 days after the first booster injection, and increasingly severe weakness from 9 days, maximal at 15-20 days, in both hind-legs (no retraction after a single extension of the limbs). With a second booster injection 21 days after the first, the rabbit improved progressively and appeared normal 8-10 days later. Rabbit #523 responded as for #522 but less severely: 8 days after the first booster injection the left hind-leg showed some weakness, worse 2-3 days later but then tapering off although persisting for 5 days.

Anti-AChR activity of the sera was assayed by complement fixation and by an enzyme immunoassay. Fixation results, with a purified rat-receptor preparation [27], are shown in Fig. 1a for rabbit #522 serum taken 15 days after the first booster injection when muscle weakness was acute, and for serum from a myasthenia gravis patient. The rabbit curves shown are typical of those obtained at excess Ab concentration. Unfortunately the serum became anti-complementary before higher serum dilutions could be tested. Evidently, however, the anti-AChR activity was at least as high as found in the patient. Anticomplementary activity could be removed by DEAE-cellulose chromagraphic isolation of serum IgG [28]. With IgG from the rabbit serum (Fig. 1b) conventional complement fixation curves indicated a titre

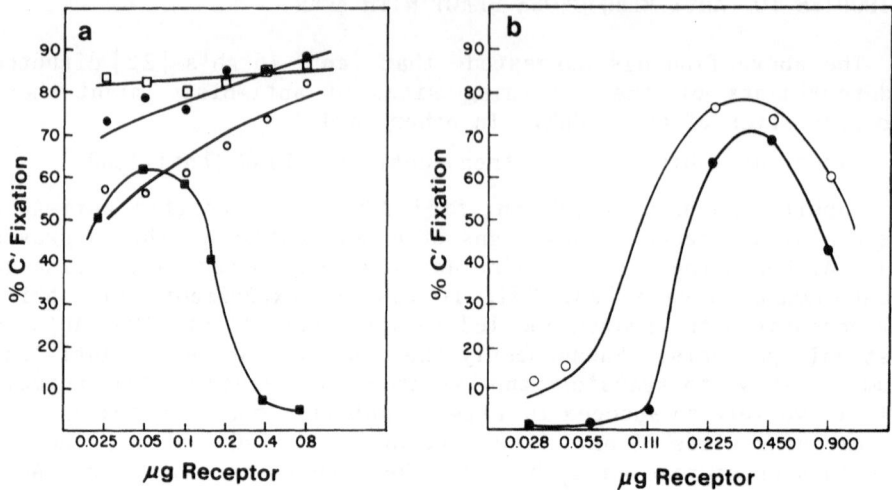

Fig. 1.  Complement-fixation assays with rat receptor.
(a) Serum 6/23 from rabbit 522 diluted 1:25 (□), 1:50 (●) and
1:100 (o) is compared with patient's serum (K.C.) diluted 1:100 (■).
(b) IgG fraction of serum 6/23 from rabbit 522 diluted 1:8 (o) and
1:10 (●). Original protein concentration of IgG was 0.6 mg/ml;  the
vol. of assay solution used in (b) was 10 times that used in (a).
*From ref. [17], by permission; similarly for Figs. 2 & 3.*

~70% that of whole serum, presumably because of the absence of IgM.
Similar dilution of an IgG fraction prepared from serum taken just
prior to the first booster injection of rabbit 522 (serum 6/8) did
not fix complement (data not shown).

    A more direct and convenient assay of anti-AChR activity was by
enzyme immunoassay.  Fig. 2 shows the binding of a rat receptor preparation [27] by serum 6/29 from rabbit 522, drawn just prior to the
second booster injection (i.e. when signs of muscle weakness were
still severe).  Three different sera from non-immunized rabbits were
also assayed, as was inhibition by soluble receptor and by BisQ.
Serum 522-6/29 showed considerable binding, well above that of non-
immunized animals.  Binding was inhibited by incubation of the serum
with excess soluble rat receptor or by 10 µM BisQ.  Binding of purified receptor preparations from *Torpedo* (not shown) and *Electrophorus*
(Fig. 3) was similarly demonstrable.  On the other hand, serum drawn
prior to the first booster injection had a low titre, near that for
non-immunized rabbits (latter illustrated in Fig. 2).   For  rabbits
523 and 524 there were finite titres, 30-40% of that found in rabbit
522, after the first booster injection; although  muscle weakness
was evident in 523 but not 524, the titres hardly differed.  Finally
some evidence of a correlation betweeen muscle weakness and titre
was obtained in rabbit 522 [26]: thus the titre was highest when,

**Fig. 2.** Enzyme immunoassay with peroxidase label and rat receptor. #522 just before 2nd boost: o, ● (2 separate runs); □ (with 4 μg/ml soluble receptor present). Non-immunized rabbits (3): △, ▲, & ▽.

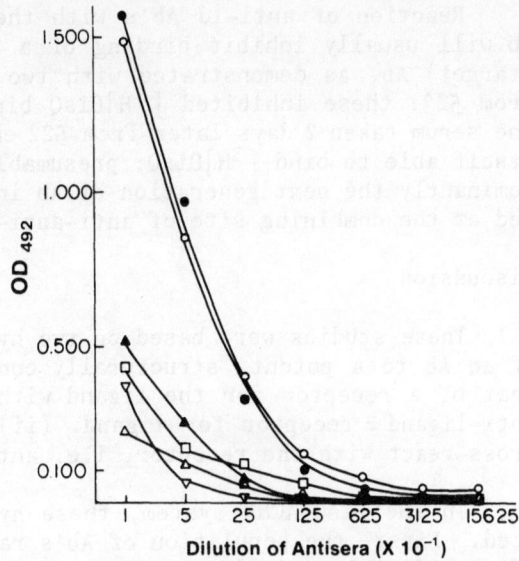

**Fig. 3.** Enzyme immunoassay with alkaline phosphatase label. #522, just before 1st boost, eel receptor: □; 6 days before 2nd boost, eel receptor: o, and rat receptor, ●. Normal (non-immunized) rabbit serum, eel receptor: ■.

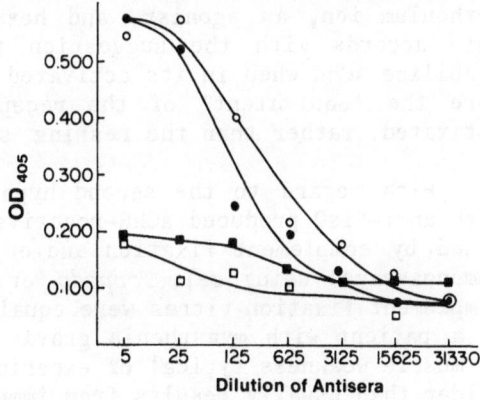

just before the second booster injection, severe weakness was seen, and lower thereafter, falling to normal or below normal. Signs of muscle weakness decreased in parallel with the decrease in titre.*

The above booster injections were given in complete Freund's adjuvant. Rabbit 523 was given anti-BisQ in saline intradermally as a 5th booster injection 1 month after the 4th, and 5-9 days thereafter it showed deteriorating muscle function, e.g. curtailed grooming. When, 2 weeks thereafter, neostigmine (0.4 mg) and atropine (0.1 mg) i.m., within 30 min there was a dramatic improvement which, however, lasted only 1 h. Amongst other immunized rabbits one which showed slowed responses and poor grooming after the first booster injection showed post-tetanic hindlimb muscle exhaustion after 50 Hz (but not 3 or 5 Hz) stimulation of the sciatic nerve.

_____

* This and ensuing 'clinical' text has been abridged.- *Ed.*

Reaction of anti-Id Ab's with the combining site of the target Ab will usually inhibit binding of a ligand by the ligand-specific (target) Ab, as demonstrated with two sera from rabbit 522 and one from 523: these inhibited [$^3$H]BisQ binding by anti-BisQ. However, one serum taken 2 days later from 522 enhanced this binding, and was itself able to bind [$^3$H]BisQ; presumably this serum contained predominantly the next generation of Ab in the anti-Id network, directed at the combining site of anti-anti-BisQ, as amplified below.

## Discussion

These studies were based on two hypotheses. (i) The binding site of an Ab to a potent, structurally constrained ligand will resemble that of a receptor for the ligand with respect to specificity, i.e. anti-ligand ≡ receptor for ligand. (ii) Ab's to the anti-ligand will cross-react with the receptor, i.e. anti-anti-ligand ≡ anti-receptor.

In the BisQ-AChR system, these hypotheses have been substantiated. First, the population of Ab's raised to BisQ showed a pattern of specificities similar to that of AChR. The most striking aspect of the Ab specificity was the ability to distinguish between decamethonium ion, an agonist, and hexamethonium ion, an antagonist. This accords with the suggestion [16] that agonists bind to and stabilize AChR when in its activated state. The antisera are therefore the 'equivalent' of the receptor when the latter is in the activated, rather than the resting, state: anti-ligand ≡ AChR (activ$^d$.).

With regard to the second hypothesis, immunization of rabbits with anti-BisQ produced AChR-reactive Ab's in all of them, as determined by complement fixation and/or direct binding studies (enzyme immunoassays) using eel, *Torpedo* or rat receptor preparations. The complement fixation titres were equal to or greater than those found in a patient with myasthenia gravis. Several rabbits showed signs of muscle weakness typical of experimental myasthenia gravis, albeit milder than usually results from immunization with purified receptor preparations [23]. The severity of the signs in the rabbits corelated well with their anti-AChR titres. Yet rabbit 524 had a significant titre but showed no signs of muscle weakness.

Seemingly, then, the immunization of rabbits with rabbit anti-BisQ provoked a set of anti-idiotypic responses directed at determinants near or at the binding site of the first Ab, and the ability of anti-anti-BisQ to bind the AChR is due to determinants on the binding site of the AChR that cross-react with determinants on anti-BisQ. The inhibition by BisQ of binding to the AChR accords with the earlier suggestion that the Ab resembles the receptor when the latter is bound to an agonist, i.e. is in the activated state.

Interestingly, the response to anti-BisQ was transient, with respect both to anti-AChR titre and to signs of experimental myas-

thenia gravis. The titre and muscle weakness were maximal after the
first booster injection and remained high until the second, after
which the titre fell to 25-35% of the earlier value and signs of
weakness disappeared. Subsequent booster injections actually caused
refractoriness: the anti-AChR titre decreased to the range found in
non-immunized animals. Such behaviour could well be caused by an
idiotypic response to the anti-anti-BisQ Ab, hence suppression of
the response [29]. The subsequent transient appearance of a BisQ-
binding Ab in rabbit 523 supports this suggestion, as do the results
obtained by Couraud et al. [30] with anti-Id Ab's reactive with the
β-adrenergic receptor.

Our findings accord with an earlier report [31] that Ab's raised
to anti-insulin Ab's react with insulin receptors. While this work
was in progress, similar studies were reported for β-adrenergic receptors [32,33], the neutrophil's chemotactic receptors [34], the TSH receptor [35], reovirus receptor [36] and the dopamine receptor [37].

## THE MONOCLONAL APPROACH TO Anti-AChR

In order to devise a more direct route to anti-receptor MAb's,
we developed a strategy based on the anti-idiotypic [21, 38] network
theory of Jerne [39]. According to this theory, injection of an
antigen elicits, besides Ab's to the antigen, other populations that
include anti-Id Ab's directed at the combining sites of the antigen-
specific Ab's. Thus, when a mouse is immunized with BisQ-BSA conjugate, it should be possible to expand populations of spleen cells
that secrete Ab's which bind anti-BisQ besides populations that produce anti-BisQ. Fusion of the spleen cells with an appropriate myeloma line should yield monoclonal anti-anti-BisQ Ab's, a subset of
which should bind AChR. Here we report the success of this approach
and its implications.

BALB/cCr mice were immunized i.p. with 0.1 ml of BisQ-BSA conjugate (1 mg/ml) in complete Freund's adjuvant. After a boosting
repeat 23 days later, spleen cells were collected 5 days later and
fused with a non-secreting myeloma line (P3x63-Ag 8.653) [40] essentially by the protocol of Köhler & Milstein [41] as modified by Sharon
et al. [42]. Supernatants from hybridoma clones were obtained by a
replica transfer technique and screened for activity against BisQ-
RSA, rabbit anti-BisQ and AChR (*Torpedo*) by EIA using peroxidase-
labelled goat anti-mouse Ig as the second Ab. Among 480 wells tested
for specificity towards BisQ-RSA, 14.0% were positive. Among 741
tested with anti-BisQ (the specifically purified rabbit Ab), 7.4% were
positive. Among 933 tested with AChR, 2.4% were positive - and were
also positive for anti-BisQ, i.e. they were a subset of the wells
positive for anti-BisQ activity. (Purified *Torpedo* and rat-muscle
receptor preparations gave the same results.) Thus, not all anti-
BisQ-reactive supernatants also bound receptor. Although a significant number of the wells might have contained more than a single

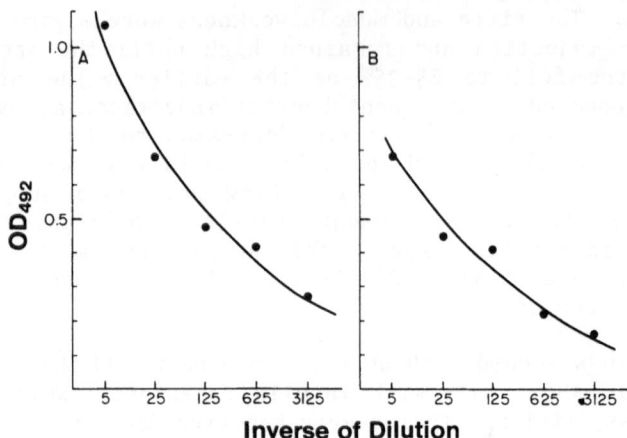

**Fig. 4.** Binding of *Torpedo* AChR (**a**) and rabbit anti-BisQ (**b**) by F8-D5. *From ref. [43], by permission of Macmillan Journals Ltd.*

clone and, certainly, the same clones could have been distributed in more than one well, we can conclude that immunization with BisQ-BSA induced a significant anti-anti-BisQ response.

Positive wells were then subcloned by limiting dilution, and several of the clones that produced Ab reactive with anti-BisQ *and Torpedo* receptor were examined. Fig. 4 shows titrations by enzyme immunoassay of an ammonium sulphate precipitate of a supernatant from one clone, F8-D5. The Ab produced by this clone was able to bind AChR of *Torpedo* as well as specifically purified rabbit anti-BisQ. This property was retained by clones obtained after two sub-sequent subclonings by limiting dilution. No attempt can be made to relate the respective binding affinities because of the characteristics of enzyme immunoassay procedures in general. Thus, there is no way to ensure that the same number of determinant sites of receptor and of anti-BisQ adhere to the plastic wells.

To confirm that the same MAb bound receptor and anti-BisQ, reciprocal inhibition experiments were performed [43]. Thus, with reactants as in Fig. 4, inhibition of binding to anti-BisQ was 67% with the highest concentration of *Torpedo* receptor tested (25 µg) and 50% with ~8 µg/well. With a 5-fold decrease in Ab amount to 2 µg per well, 50% inhibition was achieved with 0.9 µg of receptor. In analogous experiments, 50% binding of F8-D5 to the receptor was accomplished with 3 µg of purified rabbit anti-BisQ.

The binding of F8-D5 to *Torpedo* receptor and to rabbit anti-BisQ could be inhibited by BisQ and other agonists, as well as by the inhibitors hexamethonium and α-bungarotoxin (Table 2). These results reinforce the conclusion that the specificity of F8-D5 is for deter-

**Table 2.**  Inhibition of binding of F8–D5, as $IC_{50}$ values (mM), viz.
the concentration that caused 50% inhibition.

| Inhibitor | to *Torpedo* receptor | to anti–BisQ |
|---|---|---|
| BisQ | 0.04 | 0.04 |
| Decamethonium Br | 0.06 | 0.07 |
| Carbamoylcholine Cl | 0.05 | 0.1 |
| Hexamethonium Br | 0.37 | 0.29 |
| α–Bungarotoxin | $5.5 \times 10^{-3}$ | $0.7 \times 10^{-3}$ |

minants intimately associated with the combining sites of both the
*Torpedo* receptor and rabbit anti–BisQ, consistent in the latter case
with a specificity for an idiotypic determinant.  It is noteworthy
that BisQ is a small molecule (mol. wt. 486) and not likely to block
reaction with non–idiotypic determinants near the combining site as
a result of steric interference.

We have not yet studied the binding of F8–D5 with idiotypic
determinants of the various monoclonal mouse anti–BisQ Ab's detected
in the above–mentioned screening assay.  Moreover, as F8–D5 arose in
mice by immunizing with BisQ–BSA, rather than with rabbit anti–BisQ,
it can be concluded that an auto–anti–Id immune response occurred.
Thus it seems that idiotypic determinants are shared by rabbit and
mouse anti–BisQ.  Similar quantitative studies with 3 other clones
that bind both anti–BisQ  and receptor are now being done.

## WORK IN PROGRESS

The strategy described here provides a powerful route to anti–
receptor Ab's which are likely to be directed at determinants associ–
ated with the combining sites of the receptor.  As we found with
rabbits, purified receptor is not required for immunization; in fact
we have preliminary evidence that these anti–Id Ab's can be used to
isolate receptor. A crude preparation of *Torpedo* receptor was incub–
ated with F8–D5 immobilized on diaminodipropylamine agarose (Pierce
Chemical Co.)  using glutaraldehyde as linking agent [44] and then
washed. Almost all bound to the column, and could be eluted by hexa–
methonium bromide (assays by binding to $[^{125}I]$–α–bungarotoxin).  We
have not yet attained purity, as judged by SDS–PAGE examination.

The utility of Ab's thus produced is also  illustrated  by  evi–
dence from immunofluorescence that F8–D5 binds to  *Torpedo*  tissue,
giving a pattern similar to that with rabbit antisera raised by
immunization with purified AChR (Fig. 5).

In collaboration with J. Lindstrom of the Salk Institute, F8–D5
was found to block function in a vesicle system containing reconsti–
tuted *Torpedo*  AChR [45].  At an Ab:receptor ratio of 2:1, with 1 mM

**Fig. 5.** Immunofluorescence of sectioned *Torpedo* tissue after reaction with F8-D5 and fluorescein-labelled rabbit anti-mouse Ig. Mag. 700X.

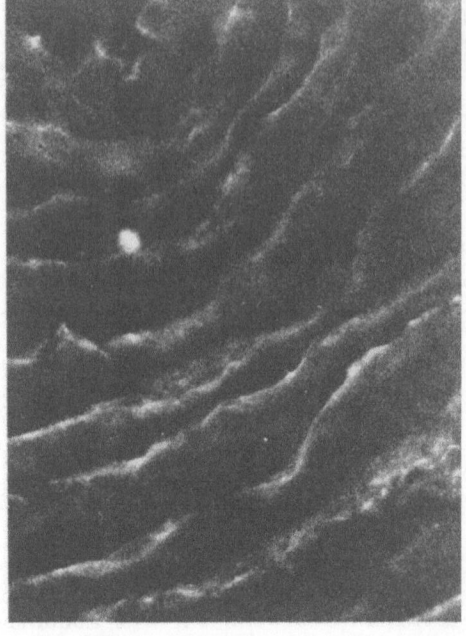

carbamylcholine present, $^{134}$Cs influx was inhibited by ~80%, supporting our view that F8-D5 reacts with AChR's binding site(s).

In an extension of the monoclonal auto-anti-Id procedure to the adenosine receptor, Ab's were prepared in rabbits to $N$-6-adenosine-caproyl-BSA (AdC-BSA), purified on an aminohexyl-Sepharose-AdC column and then used to screen for mouse MAb's prepared by immunizing Balb/c mice with AdC (KLH conjugate of the acid). The products of each of two clones inhibited the binding of [$^3$H]Ad to the purified rabbit anti-Ad, as shown in Fig. 6. Both Ab's bound to a partially purified preparation of Ad receptor [46] (Fig. 7). For the binding of each to the receptor, preliminary results indicated inhibition by $N^6$-cyclohexyladenosine, 2-chloroadenosine and theophylline.

**GENERAL DISCUSSION**

Jerne's postulate [39] of a functioning idiotypic network is supported by our results, besides our earlier rabbit findings [17] - notably the transient anti-BisQ response when immunizing with anti-BisQ - and the observation of Schechter et al. [47] that insulin-receptor-specific Ab's appeared spontaneously after immunization with insulin. Similarly, auto-anti-Id Ab appeared during a normal human immune response to tetanus toxoid [48]. We now find, too, that in the initial screening of the hybridomas the no. of wells scoring +ve for anti-Id Ab's was as much as $\frac{1}{2}$ of the no. +ve for anti-antigen Ab's. Thus, in a normal immunization process the idiotypic network is functioning not merely normally [49] but very actively, as early

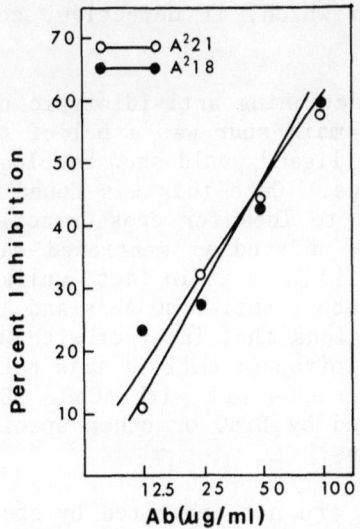

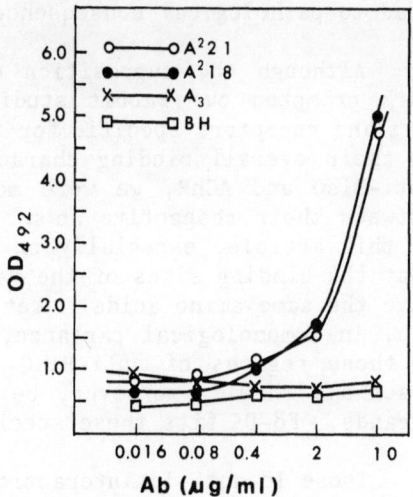

**Fig. 6.** RIA results: inhibition of binding of [$^3$H]-Ad to anti-Ad by anti-Id Ab's: o, monoclonal A$^2$21; ●, monoclonal A$^2$18.

**Fig. 7.** Anti-Id Ab binding to bovine brain membrane preparation. Mouse MAb's as controls: A$_3$, anti-Ad; BH, an Ab specific for TSH.

as 5 days after a first booster injection. Our ready detection of auto-anti-Id Ab's was notably dependent on the use of hybridoma technology. By 'immortalizing' the clones active at the time of cell harvest, the cellular events are 'frozen' in time, making it possible to study clones that may have only a transient existence *in vivo*. Moreover, these clones being separated from one another, the approach avoids the technical problems which arise from the formation of immune complexes and are a handicap in serological studies.

   Our findings also bear on the origins of autoimmune diseases in which anti-receptor Ab's are implicated. Some such diseases [17, 47] might originate from an anti-idiotypic response to Ab's formed against biologically active ligands normally present *in vivo*, e.g. insulin or TSH. An anti-idiotypic aetiology is especially likely if the patient's Ab's are directed at a receptor's combining site, as in myasthenia gravis patients particularly if severely ill [50], and support comes from our induction of the disease in rabbits via the anti-idiotypic route. The nature of the primary immunogen is, of course, not known. In Grave's disease, however, circulating anti-thyroid receptor Ab's are *almost always* specific for the TSH receptor's combining site [51] and so may well be directed at idiotypes of anti-TSH Ab [e.g. 35]. Previously [17, 47] only low titres of circulating anti-receptor Ab's were detected and, in our case, the anti-idiotypic response appeared to be transient. Such a response now seems a crucial aspect of the response to an antigen; normally

there must be some form of modulation which, if defective, could
lead to pathological consequences.

Although the supposition of a functioning anti-idiotypic network prompted our rabbit studies, the main spur was a belief that
Ab's and receptors specific for the same ligand would show homologies
in their overall binding characteristics.  Once this was found for
anti-BisQ and AChR, we were motivated to look for cross-reactions
between their respective Ab's.  In view of studies mentioned early
in this article, especially on enzymes [1], it is in fact  unlikely
that the binding sites of the various rabbit anti-BisQ Ab's and AChR
have the same amino acids.  Yet the regions that interact with BisQ
can, in immunological parlance, share epitopes: anti-Id Ab's raised
to those regions of anti-BisQ should cross-react with AChR.  Both
reactions should, moreover, be inhibited by BisQ or other specific
ligands.  F8-D5 fits these specifications.

Those Id-anti-Id interactions that are not inhibited by specific ligands have been explained  as "non-combining site-specific"
[29], i.e. involving amino acids not directly participating in the
binding interaction.  This conclusion reflects a simplistic view of
an Ab combining site, usually diagrammatically represented as a cleft
or groove in a solid form.  In reality, the same amino acids that
directly participate in the interactions can also provide idiotopes
that yield reactions with anti-Id Ab's which are *not* inhibited by
specific ligands. In Fig. 8, representing the Fab of a mouse myeloma
protein specific for phosphocholine  [52], most of the complementarity-determining amino acid residues, which, obversely, function in
binding are, on their reverse side, exposed to the 'outside' solution
and available as antigenic determinants.   Moreover, the reverse is
as idiotypic and reflective of specificity as the obverse.  Anti-Id
Ab's directed at the reverse determinants *might or might not* be inhibited by specific ligands, depending upon the effect of the binding
interaction on the integrity of the combining site.

Some anti-Id Ab's have been designated as "internal images" or
homobodies [53].  In the BisQ system they would be reckoned to mimic
BisQ's topography, being theoretically identifiable by their capacity
to compete with BisQ for the binding sites  of 'all' Ab's specific
for BisQ. In practice, a necessary condition would be their ability
to completely inhibit the reaction of a polyclonal Ab with BisQ,  as
awaits investigation - though $A^2 21$, anti-anti-Ad, does completely inhibit the binding of rabbit anti-Ad to [$^3$H]Ad.  *Functional* 'internal
images' have been reported, e.g. Ab's that functionally mimic a hormone
(insulin: [31]). Whether these are also *structural* images is unknown.

The implied requirement that an Ab insert itself into the 'cleft'
of another Ab or receptor has been held against the internal image
concept [54].  Evidently, however, an 'internal image' Ab could just
as well perturb the combining site by  reacting with reverse-side

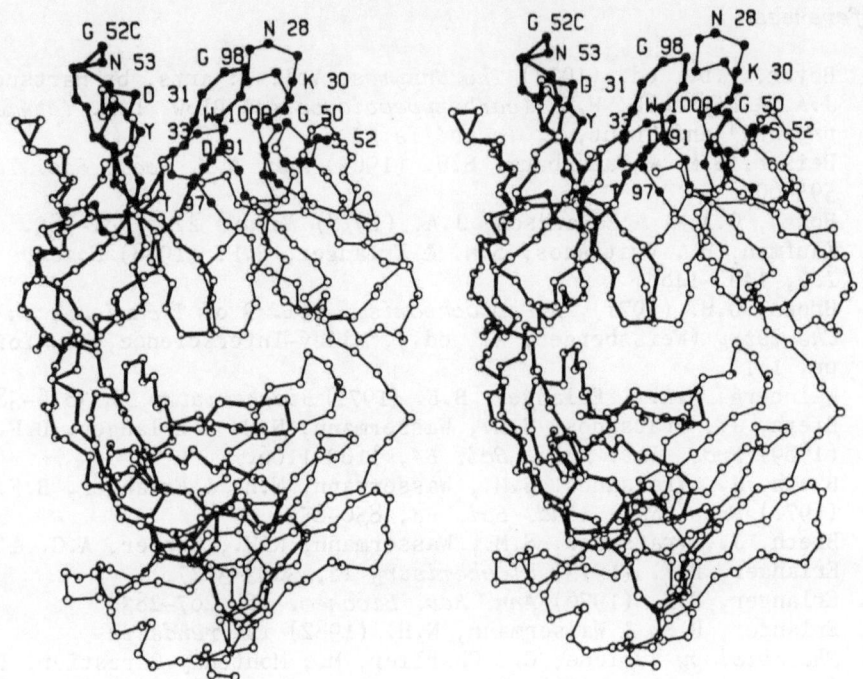

**Fig. 8.** Stereoscopic presentation of α-carbon backbone of McPC603 Fab. Complementarity-determining residues shown ●. *From ref. [52], by permission of Annual Reviews Inc.*

determinants, dependent on BisQ-specificity as for those facing the combining site. The ability of Ab's, as proteins, to mimic small pharmacologically active compounds such as BisQ has also been held against the internal image concept. In fact the endorphins, which are peptides, bind to the same receptor as the alkaloid morphine [55], i.e. one mimics the other.

In summary, the auto-anti-idiotypic route potentially offers a new way to make Ab's to receptors of all kinds and with many functions and, through the possible elicitation of 'internal images', to prepare globulins that mimic pharmacologically active agents. To what extent the potential can be achieved will be settled quite soon.

*Acknowledgements*

The research described here was supported by the NIH (NS-15581, NS-17904 & AI-17949) and the Muscular Dystrophy Association. We had skilled help from Mrs. F. Schneider in preparing the manuscript.

*References*

1. Boyer, P.D., ed. (1977) *The Enzymes*, Vol. 3: arts. by Hartsuck, J.A. & Lipscomb, W.N. *(carboxypeptidase A)*, Blow, D.M. *(chymotrypsin)* and Kraut, J. *(subtilisin)*.
2. Beiser, S.M. & Tanenbaum, S.W. (1963) *Ann. N.Y. Acad. Sci. 103*, 595-609.
3. Hough, C.A.M. & Edwardson, J.A. (1978) *Nature 271*, 381-383.
4. Kaufman, H., Vratsanos, S.M. & Erlanger, B.F. (1968) *Science 162*, 1487-1488.
5. Brown, G.H. (1971) in *Photochromism*, Vol. 3 of *Techniques in Chemistry* (Weissberger, A., ed.), Wiley-Interscience, New York, pp. 1-11.
6. Wainberg, M.A. & Erlanger, B.F. (1971) *Biochemistry 10*, 3816-3819.
7. Bieth, J., Vratsanos, S.M., Wassermann, N.H. & Erlanger, B.F. (1969) *Proc. Nat. Acad. Sci. 64*, 1103-1106.
8. Bieth, J., Vratsanos, S.M., Wassermann, N.H. & Erlanger, B.F. (1970) *Proc. Nat. Acad. Sci. 66*, 850-854.
9. Bieth, J., Vratsanos, S.M., Wassermann, N.H., Cooper, A.G. & Erlanger, B.F. (1973) *Biochemistry 12*, 3023-3027.
10. Erlanger, B.F. (1976) *Ann. Rev. Biochem. 45*, 267-283.
11. Erlanger, B.F. & Wassermann, N.H. (1982) in *Trends in Photobiology* (Helene, C., Charlier, M., Montenay-Garestier, Th. & Laustriat, G., eds.), Plenum, New York, pp. 81-92.
12. Nachmansohn, D. & Neumann, E. (1975) *Chemical and Molecular Basis of Nerve Activity*, Academic Press, New York, 403 pp.
13. Wassermann, N.H., Bartels, E. & Erlanger, B.F. (1979) *Proc. Nat. Acad. Sci. 76*, 256-259.
14. Wassermann, N.H. & Erlanger, B.F. (1981) *Chem.-Biol. Interactions 36*, 251-258.
15. Lindstrom, J., Tzartos, S. & Gullick, W. (1981) *Ann. N.Y. Acad. Sci. 377*, 1-19.
16. Heidmann, T. & Changeux, J.-P. (1978) *Ann. Rev. Biochem. 47*, 317-357.
17. Wassermann, N.H., Penn, A.S., Freimuth, P.I., Treptow, N., Wentzel, S., Cleveland, W.L. & Erlanger, B.F. (1982) *Proc. Nat. Acad. Sci. 79*, 4810-4814.
18. Erlanger, B.F. (1973) *Pharmacol. Rev. 25*, 271-280.
19. Karlin, A. (1974) *Life Sci. 14*, 1385-1415.
20. Meunier, J.C. & Changeux, J.-P. (1973) *FEBS Lett. 32*, 143-148.
21. Bartels, E., Wassermann, N.H. & Erlanger, B.F. (1971) *Proc. Nat. Acad. Sci. 68*, 1820-1823.
22. Oudin, Y. & Michael, M. (1963) *C.R. Acad. Sci., Paris 257*, 805-808.
23. Patrick, J. & Lindstrom, J.M. (1973) *Science 180*, 871-872.
24. Simpson, J.A. (1960) *Scott. Med. J. 5*, 419-436.
25. Nastuk, W.L., Plescia, D. & Ossermann, K.E. (1960) *Proc. Soc. Exp. Biol. Med. 105*, 177-184.
26. Tokya, K.V., Drachman, D., Griffin, D., Pestronk, A., Winkelstein, J.A., Fishbeck, Jr., K.H. & Kao, I. (1977) *N. Engl. J. Med. 296*, 125-131.

27. Brockes, J.P. & Hall, Z.W. (1975) *Proc. Nat. Acad. Sci. 79*, 188-192.
28. Fahey, J.L. & Terry, E.W. (1978) in *Handbook of Experimental Immunology* (Weir, D.M., ed.), Blackwell Scientific Publications, Oxford, pp. 8.1-8.16.
29. Eichmann, K. (1978) *Adv. Immunol. 26*, 195-254.
30. Couraud, P.-O., Lu, B.-Z. & Strosberg, A.D. (1983) *J. Exp. Med. 157*, 1369-1378.
31. Sege, K. & Peterson, P.A. (1978) *Proc. Nat. Acad. Sci. 75*, 2443-2447.
32. Schreiber, A.B., Couraud, P.O., Andre, C., Vray, B. & Strosberg, A.D. (1980) *Proc. Nat. Acad. Sci. 77*, 7385-7389.
33. Homcy, C.J., Rockson, S.G. & Haber, E. (1982) *J. Clin. Invest. 69*, 1147-1153.
34. Marasco, W.A. & Becker, E.L. (1982) *J. Immunol. 128*, 963-968.
35. Farid, N.R., Pepper, B., Urbina-Briones, R. & Islam, N.R. (1982) *J. Cell. Biochem. 19*, 305-313.
36. Nepom, J.T., Weiner, H.L., Dichter, M.A., Tardieu, N., Spriggs, D.R., Gramm, C.F., et al. (1982) *J. Exp. Med. 155*, 155-163.
37. Schreiber, M., Fogelfield, L.F., Souroujon, M.C., Kohen, F. & Fuchs, S. (1983) *Life Sci. 33*, 1519-1526.
38. Kunkel, H.G., Mannick, M. & Williams, R.C. (1963) *Science 140*, 1218-1219.
39. Jerne, N.K. (1974) *Ann. Inst. Pasteur, Paris 125c*, 373-389.
40. Kearney, J.F., Radbruch, B.L. & Rajewsky, K. (1979) *J. Immunol. 123*, 1548-1550.
41. Köhler, G. & Milstein, C. (1975) *Nature 256*, 495-497.
42. Sharon, J., Morrison, S.L. & Kabat, E.A. (1979) *Proc. Nat. Acad. Sci. 76*, 1420-1424.
43. Cleveland, W.L., Wassermann, N.H., Sarangarajan, R., Penn, A.S. & Erlanger, B.F. (1983) *Nature 305*, 56-57.
44. Reichlin, M. (1980) *Meth. Enzymol. 70A*, 159-165.
45. Suarez-Isla, B.A., Wan, Kee, Lindstrom, J. & Montal, M. (1983) *Biochemistry 22*, 2319-2323.
46. Daly, J.W., Butts-Lamb, P. & Padgett, W. (1983) *Cell Molec. Neurobiol. 3*, 69-80.
47. Schechter, Y., Maron, R., Elias, D. & Cohen, I.R. (1982) *Science 216*, 542-544.
48. Geha, R.S. (1982) *J. Immunol. 129*, 139-144.
49. Binion, S. & Rodkey, L.S. (1982) *J. Exp. Med. 156*, 860-872.
50. Fulpius, B.W., Lefvert, A.K., Cuenoud, S. & Mourey, A. (1981) *Ann. N.Y. Acad. Sci. 377*, 307-315.
51. Rees-Smith, B. & Hall, R. (1974) *Lancet ii*, 427. [Cf. #C-3 by Buckland, Creagh & Rees Smith in Vol. 13, this series.- *Ed.*]
52. Davies, D.R. & Metzger, H. (1983) *Ann. Rev. Immunol. 1*, 87-117.
53. Lindenmann, J. (1979) *Ann. Immunol. Paris 130C*, 311-319.
54. Rodkey, L.S. (1980) *Microbiol Rev. 44*, 631-659.
55. Guillemin, R. (1977) *New Engl. J. Med. 296*, 226-228.

#B-4

# PRODUCTION AND CHARACTERIZATION OF ANTI-MORPHINE ANTI-IDIOTYPIC ANTIBODIES

**Gary E. Isom**

Department of Pharmacology and Toxicology
School of Pharmacy and Pharmacal Science
Purdue University
W. Lafayette, IN 47907, U.S.A.

*Anti-morphine antibodies (AM-Ab's) have now been generated in rabbits by immunization with 3-0-carboxymethylmorphine-bovine serum albumin (BSA). The polyclonal antibodies (PAb's) mimicked the opioid binding characteristics of opiate receptors since they bound morphine stereospecifically and with high affinity. Anti-idiotypic antibodies (anti-Id Ab's) which cross-react with the opiate receptor were produced by immunizing guinea pigs and mice with purified AM-Ab. Guinea pig antisera produced opiate-like responses in the isolated ileal longitudinal muscle and vas deferens preparations. With antisera present the binding of [$^3$H]naloxone to mouse-brain homogenate was reduced. Scatchard analysis indicated that its binding was to two different sites, each affected by antisera. Anti-Id antisera administered to mice attenuated biological responses to opiates.*

Anti-receptor antibodies have proven to be powerful reagents for detailed structural and functional studies of membrane-bound receptors. Through recent advances in MAb technology, Ab's highly specific for various receptor components are available in large quantities [1]. Anti-receptor Ab's have been used in receptor isolation, immunofluorescent examination of intact membranes, and receptor structural-phylogenetic studies [2]. The elegant work of Linstrom and colleagues [3] typifies the utility of anti-receptor Ab's. A library of over 180 MAb's was developed and used to map antigenic components of the 4 subunits of the acetylcholine (ACh) nicotinic receptor. Fraser & Venter [4] have used MAb's to β-adrenergic receptors for studies on purification and structural aspects. Certain MAb's recognized both $\beta_1$ and $\beta_2$ receptors, suggesting that these two subunits have some structural homology. These Ab's were specific for the β-receptor; thus, the cholinergic, α-adrenergic and dopamine receptors were not recognized.

Another immunological approach to receptor study is the use of ligand-specific anti-Id Ab's [5, 6]. The anti-Id network theory of Jerne [7] predicts that anti-Id Ab's can be generated against the idiotype of an antigen-specific Ab. Anti-Id Ab's would be complementary to the antigen binding site and essentially be an immunological image of the antigen. With Ab's against specific receptor ligands, the anti-Id Ab would be a mirror image of the ligand and may bind its receptor.

This has been achieved with various receptors. Thus, Strosberg et al. [2] demonstrated that anti-alprenolol anti-Id Ab's can bind to the surface of turkey erythrocytes and, by use of immunofluorescent Ab techniques, these receptors were visualized histologically. The anti-Id Ab's inhibited the binding of alprenolol to the β-receptor in a dose-dependent, non-competitive manner. Anti-Id Ab's can also exhibit biological activity. Wasserman et al. [8] showed that experimental myasthenia gravis was produced following immunization against Ab's generated to a potent ACh-like agonist. The animals displayed transient muscle weakness paralleling the Ab titres.

Studies in our laboratory are aimed at generating anti-Id Ab's that recognize the membrane opiate receptor in neural tissue [9, 10]. The opiate receptor has not been highly purified and characterized molecularly; hence generation of Ab's directed against it has not been achieved. As an alternative approach the anti-Id route was utilized as a means of producing anti-receptor Ab.

## GENERATION AND PURIFICATION OF ANTI-MORPHINE ANTIBODIES

New Zealand White rabbits were immunized s.c. with 10 mg of carboxymethylmorphine-BSA in 0.5 ml of phosphate-buffered saline (pH 7.4) emulsified with an equal vol. of complete Freund's adjuvant; booster injections were then given every 3 weeks over 4 months and, 7 days after the last one, the animals were bled from the inner marginal ear vein and serum separated. Active immunization was monitored by following the serum binding capacity of $[^{14}C]$morphine. AM-Ab's were isolated from the γ-globulin fraction by one-step affinity chromatography in which 3-0-carboxymethylmorphine was coupled to AH-Sepharose 4B [11].

To check whether the ligand-binding characteristics of the AM-Ab's paralleled that of the opiate receptor, the relative affinity of AM-Ab's for different opiate ligands was determined in competition studies where varying concentrations of the ligands were used to displace $[^{14}C]$morphine from the Ab's. After affinity-chromatography purification, AM-Ab's bound morphine in a saturable manner with a $K_D$ of 19.5 nM. Preferential binding of agonists over antagonists was seen with a series of ligands that displaced $[^{14}C]$morphine in a stereospecific manner, the following values being found for $IC_{50}$ (the final concentration, nmol/ml, of unlabelled compound that inhibited by 50% the binding of [$^{14}$C]morphine to AM-Ab's): morphine, 0.033; levorphanol, 0.450; dextrophan, 260; naloxone, 700; ketocyclazocine, 35% inhibition at 100 nmol/ml (highest concentration used because of solubility limit); met-enkephalin and beta-endorphin, respectively 6% inhibition at 10,000 nmol/ml and none at 9.7. Even though this binding profile did not precisely parallel that to the opiate receptor, AM-Ab's appeared to function as images of the opiate receptor.

## GENERATION OF ANTI-MORPHINE ANTI-IDIOTYPIC ANTIBODIES

Guinea pigs were immunized against the purified rabbit AM-Ab's, initially 10 mg s.c. in complete Freund's adjuvant and then weekly booster injections in incomplete adjuvant. From 10 to 16 weeks after the start, serum was collected and screened for the ability to bind [$^{14}$C]morphine and immunoprecipitate rabbit Ab's. Two of 10 guinea pigs immunized developed a strong immune response against the rabbit Ab's as determined by direct immunoprecipitation and the interfacial ring test. One of them exhibited transient catatonia, a centrally mediated response characteristic of opiate agonists. Antisera from the actively immunized animals were collected and stored at -20°.

## INTERACTIONS OF ANTI-IDIOTYPIC ANTIBODIES WITH OPIATE RECEPTORS

In the procedure [12] for determining the opiate receptor binding of [$^3$H]naloxone, brains minus cerebellum from 5-10 mice were pooled and homogenized for 20 sec in 30 vol. chilled 50 mM Tris-HCl buffer pH 7.7 with a Polytron PT-10 set at 5. After centrifugation at 49,000 g for 15 min at 4°, the pellet was re-suspended in an equal vol. of buffer and incubated alone at 37° for 30 min. Following re-centrifugation and re-suspension in Tris-NaCl buffer (50 mM Tris, 100 mM NaCl; pH 7.4 at 25°), 2.0 ml of homogenate and 0.1 ml Tris-NaCl or levorphanol (final concn. 1 µM) were incubated for 10 min at 25°. Then 0.1 ml of [$^3$H]naloxone (final concn. 0.3-10 nM) was added and the incubation continued for 20 min. Bound [$^3$H]naloxone was separated from free ligand by filtration (Whatman GF/B, glass) and washed twice with 5.0 ml chilled Tris-NaCl. Specific binding was calculated by subtracting the binding in the presence of levorphanol from that in its absence; values represent fmol bound naloxone/mg protein. Inhibition of naloxone binding by guinea pig antiserum was determined by assaying [$^3$H]naloxone binding after incubating the homogenate for 60 min at 25° with antiserum at final dilutions from 1:1100 to 1:22. Normal guinea pig serum and pre-immune guinea pig serum served as controls in the binding assays.

With the brain preparation, the guinea pig antiserum produced a concentration-dependent inhibition of specific binding of [$^3$H]naloxone to opiate-binding sites, maximally (37-70%) when at 1:22 dilution in the incubation mixture. The ability of the antisera to inhibit binding varied from animal to animal and appeared to undergo transient cycles of activity similar to that reported by Strosberg [5] with anti-alprenolol anti-Id responses. The inability to completely abolish binding may reflect the specificity of the anti-Id Ab for one opiate receptor subtype. Naloxone is a universal ligand which binds several subtypes of receptor. Possibly the anti-Id Ab recognizes only one subtype (most likely mu) and in turn does not inhibit all naloxone binding. To analyze in detail the binding of [$^3$H]naloxone, it was added to the mouse whole-brain preparation already pre-incubated for 60 min at 25° with a 1:22 dilution of anti-Id serum: the observed $K_D$ (nM) and Bmax (fmol/mg protein) were 3.66 and 185.7 respectively, compared with 3.87 and 116.7 in the no-serum control. Evidently in this non-competitive binding by Ab's the affinity of the receptor for radio-ligand was not significantly altered but the total number of binding sites appeared to decrease.

Generally anti-Id Ab's produce competitive inhibition of ligand binding, easily explicable because the Ab is considered to be an immunological image of the natural ligand; but anti-morphine anti-Id Ab's may also bind to components of the receptor outside the active site and so alter binding.

## BIOLOGICAL RESPONSE TO ANTI-IDIOTYPIC ANTIBODIES

The *in vitro* ligand binding studies demonstrate that guinea pig antiserum binds to opiate receptors. In turn, it was of interest to ascertain whether recognition and binding of anti-Id Ab's to opiate receptors would initiate a physiological response as reported with other receptor systems. The biological activity of the anti-serum was determined in two bioassays for opiates: the action on isolated guinea pig ileal longitudinal muscle and the analgesic response to injection of Ab's into the brain of mice.

A 10-20 cm segment of ileum from male Hartley guinea pigs was removed ~10 cm from the ileocaecal junction following washing with warm Krebs-bicarbonate solution. The longitudinal muscle was removed and mounted in a 45 ml tissue bath filled with oxygenated Krebs-bicarbonate containing 1.25 μM chlorpheniramine. The muscle was subjected to 1 g of force, and after equilibration for 1 h at 37° contractions were induced by electrical stimulation at 80 V, 1.1 Hz, 2 msec frequency, and recorded via a myograph transducer (Fig. 1).

Anti-Id antiserum produced a concentration inhibition of the induced contractions similar to that observed with opiate agonist-like compounds, maximally ~50% as with >400 μl the inhibition was no greater. Moreover, the opiate antagonist naloxone did not reverse the inhibition. The anti-Id antiserum produced a significant change in opiate receptor function that was not altered by antagonists.

To ascertain analgesic response, either normal serum or anti-serum was injected intracerebroventricularly into the brain of male mice. The tail-flick latency test was applied: a standard thermal

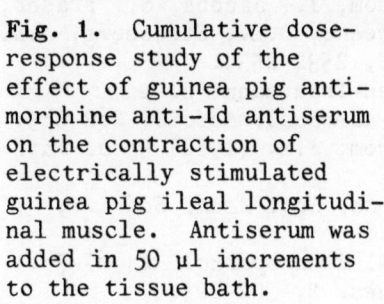

**Fig. 1.** Cumulative dose-
response study of the
effect of guinea pig anti-
morphine anti-Id antiserum
on the contraction of
electrically stimulated
guinea pig ileal longitudinal muscle. Antiserum was
added in 50 µl increments
to the tissue bath.

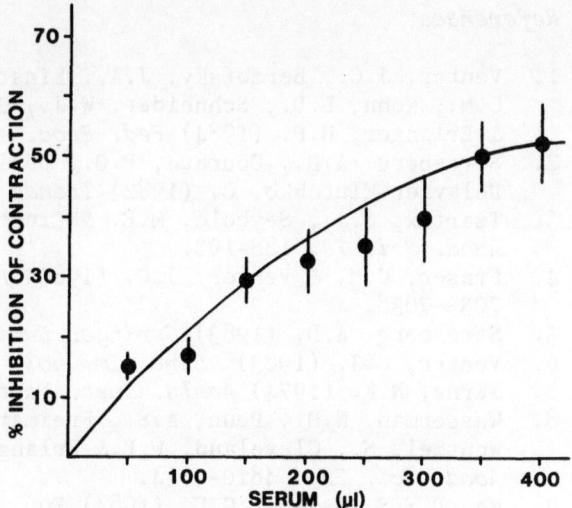

**Table 1.** Analgesic responses of serum-treated mice in the tail-
flick test, at different times after treatment. Responses are
expressed as % analgesia $= 100 \times (T_t - T_c)/(10 - T_c)$ where $T_c$ and $T_t$
represent the response latency period before and after the treatment respectively. Values are means $\pm$ S.E.M. of 5 animals.

| Guinea pig serum | 15 min | 30 min | 45 min | 60 min |
|---|---|---|---|---|
| Normal (5 µl; see text) | 6.7 ±4.7 | 3.4 ±7.5 | 7.1 ±3.8 | 4.1 ±5.4 |
| Antiserum | 8.2 ±2.6 | 12.1 ±3.6 | 10.8 ±3.1 | 4.5 ±2.0 |

stimulus is applied to the mouse's tail and the response time (flick
of the tail) noted. Analgesics increase the time required to respond to the stimulus (antinociception). The antiserum produced a
modest increase in response time as compared to control at 30 and 45
min after administration (Table 1), indicating that the anti-Id Ab
could induce a biological response in an animal.

These studies indicate that an Ab can be generated indirectly
by the anti-Id route which binds to the opiate receptor and can
initiate a biological response. Studies are in progress to develop
monoclonal anti-Id Ab's which are specific for the opiate receptor
and can be used as reagents for its detailed molecular and functional
characterization.

*Acknowledgement*

This work was supported in part by U.S. Public Health Service
grants NS 19584 and DA 02853.

*References*

1.  Venter, J.C., Berzofsky, J.A., Linstrom, J., Jacobs, S., Fraser,
    C.M., Kohn, L.D., Schneider, W.J., Greene, G.L., Strosberg, A.D.
    & Erlanger, B.P. (1984) *Fed. Proc. 43*, 2532-2539.
2.  Strosberg, A.D., Couraud, P.O., Durieu-Trautmann, O. &
    Delavier-Klutchko, C. (1982) *Trends Pharmacol. Sci. 3*, 282-285.
3.  Tzartox, S.J., Seybold, M.E. & Linstrom, J.M. (1982) *Proc. Nat.
    Acad. Sci. 79*, 188-192.
4.  Fraser, C.M. & Venter, J.C. (1980) *Proc. Nat. Acad. Sci. 77*,
    7035-7038.
5.  Strosberg, A.D. (1983) *Springer Semin. Immunopathol. 6*, 67-78.
6.  Venter, C.J. (1983) *Surv. Immunol. Res. 2*, 302-305.
7.  Jerne, N.K. (1974) *Annls. Inst. Pasteur, Paris 125c*, 373-389.
8.  Wasserman, N.H., Penn, A.S., Freimath, P.I., Treptow, N.,
    Wentzel, S., Cleveland, W.L.& Erlanger, B.F. (1982) *Proc. Nat.
    Acad. Sci. 79*, 4810-4814.
9.  Ng, D.S.S. & Isom, G.E. (1984) *Eur. J. Pharmacol. 102*, 187-
    190.
10. Ng, D.S.S. & Isom, G.E. (1985) *Biochem. Pharmacol.*, submitted
    for publication.
11. Wahid, F.A. & Isom, G.E. (1980) *Res. Comm. Subst. Abuse 1*, 451-
    458.
12. Caruso, T.P., Larson, D.L., Portoghese, P.S. & Takemori, A.E.
    (1980) *J. Pharmacol. Exp. Ther. 213*, 539-544.

#B-5

# ANTI-IDIOTYPE ANTIBODIES FOR THE STUDY OF MEMBRANE RECEPTORS:  THE DOUBLE MONOCLONAL ANTIBODY APPROACH

Johan Hoebeke, Jean-Gerard Guillet,
Soulaima Chamat and A. Donny Strosberg

Laboratory of Molecular Immunology
Institut Jacques Monod, Université Paris VII
2 place Jussieu, F75251 - Paris Cedex 05, France

*Rabbit polyclonal anti-idiotypic antibodies (anti-Id Ab's) raised against anti-catecholamine Ab's bind to ß-adrenergic receptors and modulate the hormone-sensitive adenylate cyclase. To gain more insight into the interaction between anti-idiotypes and receptors, the unlimited availability of purified components such as monoclonal antibodies (MAb's) would be of considerable advantage.*

*To this end, MAb's against the ß-adrenergic antagonist alprenolol were produced and characterized for their specificity towards ß-adrenergic agonists and antagonists. One of the MAb's (Ab37A4), with binding properties similar to those of the receptor, was selected for the induction of anti-Id Ab's. They were produced by fusion of NS-1 myeloma cells with splenocytes of mice immunized by i.v. injections of fixed hybridoma cells bearing a MAb specific for ß-adrenergic ligands. Among 23 hybridoma supernatants recognizing the idiotype, 6 were found to inhibit hapten binding and 3 of these recognized ß-adrenergic receptors.*

Since the first report of Sege & Peterson [1] that anti-Id Ab's against anti-insulin Ab's can recognize the insulin receptor and mimic the physiological effects of the hormone, several groups have used the anti-Id route to study hormone or neurotransmitter receptors [e.g. 2]. Most reports described the properties of polyclonal anti-Id Ab's. Such Ab's are more liable to artefactual responses, as described later, and rapidly disappear during the immune response. Their heterogeneity prevents a structural approach to the study of their effects on receptors. For this reason, we undertook the production of monoclonal anti-Id Ab's. A similar approach was taken by Cleveland et al. ([3], & #B-3, this vol.) who selected monoclonal

anti-Id Ab's during the primary response against the nicotinic
acetylcholine agonist BisQ.

Our interest resides in possible structural homologies shared
by the anti-hormone Ab's and the hormone receptors. We therefore
first raised anti-hormone MAb's, to select among them the Ab which
had binding properties most similar to those of the receptor. Monoclonal anti-Id Ab's were then raised against that well-defined Ab.
A similar two-step approach was recently described for the production of monoclonal anti-Id Ab's which recognize the mammalian reovirus receptors [4, 5].

Based on the experience we gained during this work, using the
β-adrenergic receptor as model, we here highlight some experimental
tips and precautions, requisite for the correct assessment and use
of anti-Id Ab's in the study of membrane receptors.

## THE MONOCLONAL IDIOTYPE ANTIBODIES (Id Ab's)

### Choice of the hapten-ligand

The ligand molecules most suited to induce Ab's for induction
of a receptor-specific anti-Id response should be as small and as
rigid as possible, because the larger the molecule the higher the
possibility that antigenic determinants may be present which are
irrelevant to the binding of the molecule to the receptor. This
ambiguity is exemplified by the dopamine- and serotonin-specific
ligand, spiperone, which carries a spirodecanone moiety as a potential hapten which is not involved in the binding on the neurotransmitter receptor ([6]; cf. P.M. Laduron, #A-4 in Vol. 13, this series). What is true for synthetic molecules is *a fortiore* valid for
larger peptide hormones. While these hormones may give a polyclonal
response which carries the correct idiotype, the monoclonal responses should be checked with smaller fragments that are known to be
involved in receptor recognition.

The ligand molecules should carry a reactive group to allow the
coupling to a carrier protein or polysaccharide. If such a reactive
group does not exist, the ligand can be chemically modified but the
modified molecule should be assessed for its affinity towards the
receptor. A large loss in affinity indicates that parts essential
for binding are altered by the modification and, accordingly, that
the Id Ab's will only poorly reflect the binding properties of the
receptor.

For the study of the β-adrenergic receptor, we selected alprenolol as a ligand which possesses the properties we have discussed.
The reactive vinyl moiety on the phenyl ring can be modified without
loss of affinity for the receptor [7] (Fig. 1).

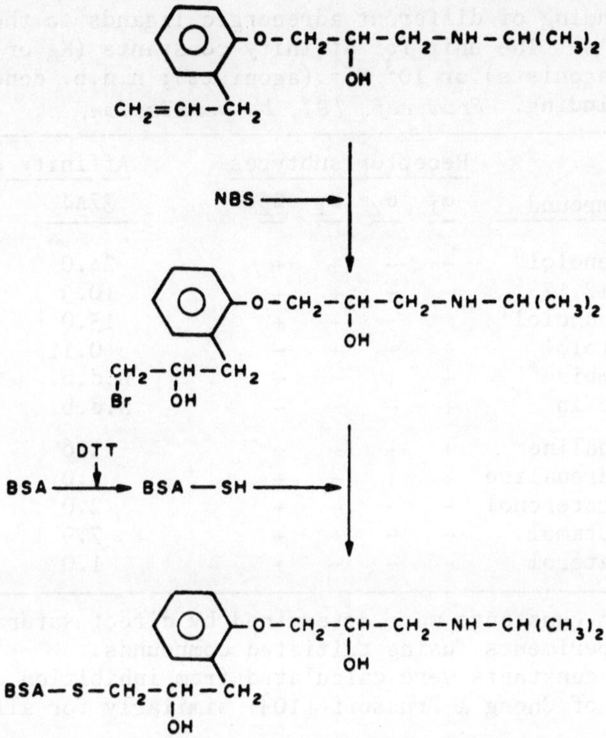

**Fig. 1.** Reaction scheme for the preparation of the hapten–protein conjugate. NBS = *N*-bromosuccinimide; BSA = bovine serum albumin; DTT = dithiothreitol. *From ref. [7], by permission.*

## Choice of the carrier

To raise anti-hapten Ab's a minimal amount of substitution of the carrier molecule is needed. To increase the number of reactive residues, two ways are open. The one we used was to completely unfold the carrier (BSA) in 6 M guanidinium hydrochloride and to reduce existing disulphide bridges for reaction with bromhydrinal-prenolol [7] (Fig. 1). The other possibility is to modify reactive residues in the native protein with multifunctional reagents, e.g. succinylation of the amino groups which increases the number of car-boxyl groups that react, after carbodiimide activation, with primary or secondary amino groups of the ligand [8].

## Characterization of the monoclonal idiotype antibodies

The first screening of hybridoma supernatants is most easily performed by an enzyme immunoassay using the ligand bound to a carrier protein different from that used for immunization [9]. When highly radioactive ligands for the receptor are available, the

Table 1.  Binding of different adrenergic ligands to the 37A4 and the 10E2 MAb's.  The unit for affinity constants ($K_A$ or $K_I$) is $10^6$ $M^{-1}$ (antagonists) or $10^3$ $M^{-1}$ (agonists); n.d.b. denotes no detectable binding. *From ref. [9], by permission.*

|  | Compound | $\alpha_1$ | $\alpha_2$ | $\beta_1$ | $\beta_2$ | 37A4 | 10E2 |
|---|---|---|---|---|---|---|---|
|  |  | | | | | Affinity constants | |
|  |  | Receptor subtypes | | | | | |
| *Antagon-ists* | Alprenolol* | − | − | + | + | 24.0 | 6.6 |
|  | CGP-12,177* | − | − | + | + | 10.3 | 2.4 |
|  | Propranolol† | − | − | + | + | 15.0 | 4.2 |
|  | Practolol† | − | − | + | − | 0.11 | 0.068 |
|  | Yohimbine† | − | + | − | − | n.d.b. | n.d.b. |
|  | Prazosin* | + | − | − | − | n.d.b. | n.d.b. |
| *Agonists* | Adrenaline | + | + | + | + | 1.6 | n.d.b. |
|  | Noradrenaline | + | + | + | + | 1.0 | n.d.b. |
|  | Isoproterenol | − | − | + | + | 2.0 | 13.3 |
|  | Salbutamol | − | − | + | + | 7.9 | 0.5 |
|  | Procaterol | − | − | − | + | 1.0 | n.d.b. |

* Association constants were determined by direct saturation binding experiments  using tritiated compounds.
† Inhibition constants were calculated from inhibition curves by the method of Cheng & Prussoff [10]; similarly for all agonists.

affinity of the Ab's present in the hybridoma supernatants can be checked by a radioimmunoassay in Farr's mode [11].  The choice of an MAb with properties similar to the receptor can be further assessed in an enzyme immunoassay using as inhibitors molecules with minimal structural  resemblance to the hapten but with appreciable receptor affinity [12].  Finally, a series of molecules that are not recognized by the receptor should be tested for their affinity for the Ab. Table 1 summarizes the binding properties of the anti-alprenolol MAb's Ab37A4 and Ab10E2.  Since Ab37A4 recognized all β-adrenergic agonists and especially the natural hormones adrenaline (epinephrine) and noradrenaline, unlike Ab10E2 which recognized only the synthetic agonists, Ab37A4 was chosen for raising monoclonal anti-Id Ab's.

## THE MONOCLONAL ANTI-IDIOTYPE ANTIBODIES (anti-Id Ab's)

### The immunization route

Using syngenic animals for raising monoclonal anti-Id Ab's has the advantage that the response is limited to the anti-idiotype. The precautions needed for selecting out anti-isotype or anti-allo-type responses are thus superfluous.  However, the injection of the Id Ab's in complete Freund's adjuvant could give rise to other arte-facts.  Since *Mycobacterium butyricum*, the inflammatory agent of the adjuvant, is a polyclonal mitogen, it can stimulate B-lymphocytes

Fig. 2.  Screening for hybridomas
secreting anti-Id Ab's.
A.  Fixation of secretion products on the Id-bearing Ab37A4
hybridoma.
B.  Inhibition of Ab37A4 binding
on KLH-Alp by the secretion
products.
The black bars indicate hybridomas secreting Ab's positive in
both tests.
*From ref. [14].*

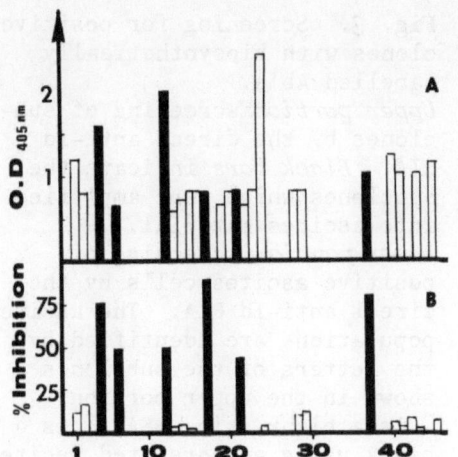

which are unrelated to the anti-Id response, e.g. those secreting
anti-carbohydrate Ab's.  These Ab's might react with membrane glycoproteins or those secreting rheumatoid factors, which will react
with the Fc part of Ab's [13].  To avoid the possibility of such
artefactual responses, we decided on i.v. injection of glutaralde-
hyde-fixed Ab37A4 hybridoma cells, carrying the Ab at their surface
[14].  A similar immunization method was recently used for inducing
mouse monoclonal anti-Id Ab's against human B-lymphomas [15].

Screening for monoclonal anti-Id Ab's

Working in a syngenic system entails some screening difficulties since most assays are based on recognition by a second Ab
(radiolabelled or enzyme-coupled) which cannot distinguish between
the Id and the anti-Id Ab's.  We describe here some ways to circumvent this difficulty.

Adsorption of the hybridoma cells to the solid phase instead of
the Id Ab considerably decreases the recognition of the Fc part of
the Ab by anti-Fc Ab's.  This is probably due to the low accessibility of the Fc region at or near the cell surface.  Using an anti-
Fc Ab at a dilution which is unresponsive to the hybridoma cells,
any positive response will be due to the Fc-region of the adsorbed
anti-Id (Fig. 2A).

Another way to distinguish between the two Ab's (Id and anti-Id)
is to radio-iodinate one of them [16].  We prefer the use of biosynthetically labelled hybridoma products because radio-iodination can
induce changes in antigen recognition [17].  Fig. 3 shows how we
used this technique to subclone hybrids positive for anti-Id Ab
secretion.

**Fig. 3.** Screening for positive clones with biosynthetically labelled Ab's.
*Upper portion:* screening of sub-clones by the direct anti-Id RIA. *Black bars* indicate the subclones which were amplified into ascites material.
*Lower portion:* screening of positive ascites cells by the direct anti-Id RIA. The ascites populations are identified by the letters of the subclones shown in the upper portion.
C is a blank using BSA; F is a blank using an unrelated ascites population. *From ref. [14].*

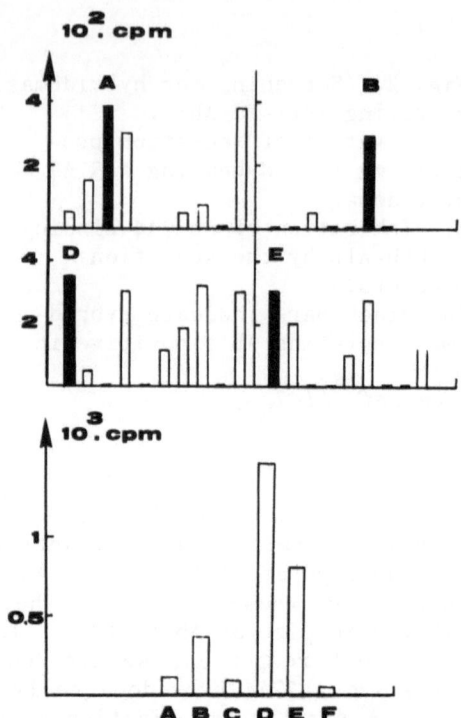

## Screening for receptor recognition

The most stringent criteria must be used to ascertain without doubt that the anti-Id Ab specifically recognizes the receptor. For this reason, we tested the anti-Id Ab's against a variety of plasma-membrane preparations or whole cells carrying the β-adrenergic receptor. Three different cell types from three different species were used: turkey erythrocyte membranes [18], P815 murine mastocytoma cells [19] and the human epidermoid cells A431 [20]. The use of cells from the same species that produced the anti-Id excludes natural Ab's directed against different species. Moreover, a negative control was used (a rabbit B lymphoma), devoid of β-receptors. Non-specific adsorption of Ab's to plasma membranes can thus be detected (Fig. 4).

From the 24 anti-Id Ab's screened in the first test (Fig. 2A), only 3 met the criteria for definition of receptor-recognizing anti-Id Ab's. That all of the three were able to interfere with ligand binding by the Id Ab (Fig. 2B) confirmed the validity of our selection. One of the monoclonal anti-Id Ab's (Ab2B4) has now been further characterized [21]. It immunoprecipitates and immunoblots the receptor of A431 cells; moreover, adenylate cyclase on A431 membranes shows a stimulation, which the β-blocker propranolol inhibits.

**Fig. 4.** Screening for β-receptor
binding capacity of hybridoma
supernatants.
*From top to bottom:*
binding on P815 plasma membranes;
binding on turkey erythrocyte
membranes;
binding on A431 human epidermoid
cells;
binding on rabbit B lymphoma
cells.
*From ref. [14].*

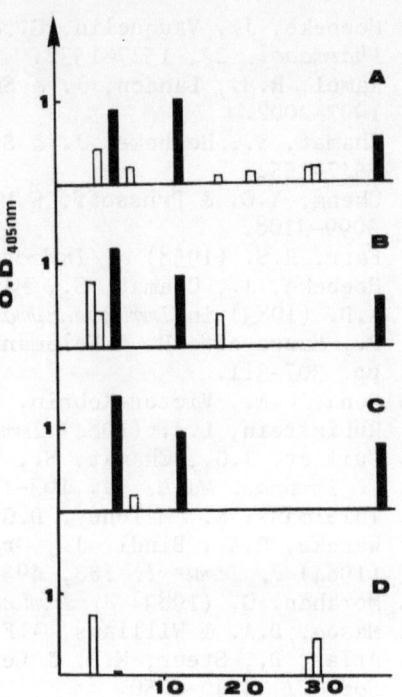

*Acknowledgements*

This work was supported by the Centre National  de la Recherche
Scientifique (A.T.P. Pharmacologie des Neuromediateurs), the Institut National   de la Santé et de la Recherche Medicale (Contrat No.
509929), the Université Paris VII, la Ligue Française contre le
Cancer, and l'Association pour le Developpement de la Recherche sur
le Cancer and la Fondation de la Recherche Medicale.

*References*

1.  Sege, K. & Peterson, P.A. (1978) *Proc. Nat. Acad. Sci. 77*, 2443-
    2447.
2.  Strosberg, A.D. (1983) *Springer Sem. Immunopath. 6*, 67-78.
3.  Cleveland, W.L., Wassermann, N.H., Sarangarajan, R., Penn, A.S.
    & Erlanger, B.F. (1983) *Nature 305*, 56-57.
4.  Noseworthy, J.H., Fields, B.N., Dichter, M.A., Sobotka, C.,
    Pizer, E., Perry, L.L., Nepom, J.T. & Greene, M.I. (1983) *J.
    Immunol. 131*, 2533-2538.
5.  Kauffman, R.S., Noseworthy, J.H., Nepom, J.T., Finberg, R.,
    Fields, B.N. & Greene, M.I. (1983) *J. Immunol. 131*, 2539-2541.
6.  Laduron, P. (1982) in *Advances in Dopamine Research [Adv.
    Biosci., Vol. 37]* (Kohsaka, M., ed.), Pergamon, Oxford, pp. 71-82.

7.  Hoebeke, J., Vauquelin, G. & Strosberg, A.D. (1978) *Biochem. Pharmacol. 27,* 1527-1532.
8.  Kamel, R.S., Landon, J. & Smith, D.S. (1979) *Clin. Chem. 25,* 1997-2002.
9.  Chamat, S., Hoebeke, J. & Strosberg, A.D. (1984) *J. Immunol. 133,* 1547-1552.
10. Cheng, Y.C. & Prussoff, W.M. (1973) *Biochem. Pharmacol. 22,* 3099-3108.
11. Farr, R.S. (1958) *J. Infect. Dis. 103,* 239-262.
12. Hoebeke, J., Chamat, S., Marullo, S., Guillet, J.G. & Strosberg, A.D. (1983) in *Immunoenzymatic Techniques* (Avrameas, S., Druet, P., Masseyeff, R. & Feldmann, G., eds.), Elsevier, Amsterdam, pp. 307-311.
13. Bona, C.A., Victor-Kobrin, C., Manheimer, A.J., Bellon, B. & Rubinstein, L.J. (1984) *Immunol. Rev. 79,* 26-44.
14. Guillet, J.G., Chamat, S., Hoebeke, J. & Strosberg, A.D. (1984) *J. Immunol. Meth. 74,* 163-171.
15. Thielmans, K., Maloney, D.G., Meeker, T., Fujimoto, J., Doss, C., Warnke, R.A., Bindl, J., Gralow, J., Miller, R.A. & Levy, R. (1984) *J. Immunol. 133,* 495-501.
16. Morahan, G. (1983) *J. Immunol. Meth. 57,* 165-170.
17. Mason, D.A. & Williams, A.F. (1980) *Biochem. J. 187,* 1-20.
18. Atlas, D., Steer, M.L. & Levitzki, A. (1974) *Proc. Nat. Acad. Sci. 71,* 4246-4250.
19. Durieu-Trautmann, O., Delavier-Klutchko, C., Hoebeke, J. & Strosberg, A.D. (1985) *Eur. J. Pharmacol.,* in press.
20. Delavier-Klutchko, C., Hoebeke, J. & Strosberg, A.D. (1984) *FEBS Lett. 169,* 152-155.
21. Guillet, J.G., Kaveri, S.V., Durieu, O., Delavier, C., Hoebeke, J. & Strosberg, A.D. (1985) *Proc. Nat. Acad. Sci.,* in press.

*Note added by Editor*

Vol. 13 of this series (1984; *Investigation of Membrane-Located Receptors*, ed. E. Reid et al.) contains hormone and adrenergic receptor material pertinent to the foregoing article, including an outline by A.D. Strosberg of β-adrenergic receptor purification.

#NC(B)

NOTES and COMMENTS related to the foregoing topics

Comments related to particular contributions:

   #B-1, #B-2 & #B-3, p. 145
   #NC(B)-1, p. 146
   #NC(B)-4, p. 147
   #NC(B)-5 to -7, p. 148 (also, for #NC(B)-7, p. 149)

#NC(B)-1

*A Note on*

# CHARACTERIZATION OF IDIOTYPES AND $V_H$ GENES OF VARIOUS MEMBERS OF A48-IDIOTYPIC NETWORK PATHWAY

C. Bona, A. Manheimer and C. Victor-Kobrin

Department of Microbiology
Mount Sinai School of Medicine
1 Gustav Levy Place, Annenberg Building
New York, NY 10023, U.S.A.

The analysis of the immunochemical properties of four members of the A48 idiotypic pathway, viz. A48-$\beta$2-6 fructosan-binding mono-clonal protein ($Ab_1$) and polyclonal syngeneic Ab's – anti-Id ($Ab_2$), anti(anti-Id) ($Ab_3$) and anti[anti(anti-Id)] ($Ab_4$) – showed that $Ab_1$ and $Ab_3$ shared A48 idiotypes while both $Ab_2$ and $Ab_4$ bound to $Ab_1$.

We extended this study by using monoclonal Ab's displaying $\beta$2-6 and $\beta$2-1 fructosan-binding activities, monoclonal $Ab_3$'s obtained from adult animals immunized with $Ab_2$ and monoclonal Ab's specific for A48, UPC10 and MOPC173 monoclonal proteins. The mature $V_H$ gene of these three myelomas derived from the 441-4 germ line gene. Immunochemical analysis of 6 $Ab_1$ monoclonal Ab's exhibiting binding for $\beta$2-6 fructosan or $\beta$2-6 and $\beta$2-1 fructosan shows that they express A48 and UPC10 idiotypes. Only 3 of them express the 173-Id. In Northern blotting, the mRNA from all these hybridomas hybridizes with 441-4 germ-line probe. The study of 10 monoclonal $Ab_3$'s showed that 4 of them bind to both $\beta$2-1 and $\beta$2-6 fructosan, and one binds to only $\beta$2-1 fructosan. All of them express the A48 and UPC10-Id and only two express the 173-Id. The mRNA of 7 hybridizes with the 441-4 germ-line gene probe.

None of 9 monoclonal anti-A48 and anti-UPC10-Id Ab's hybridized with the 441-4 probe, but 5 of them hybridized with J558 probe, suggesting that they belong to this family.

These data clearly suggest that $Ab_1$ and $Ab_3$ represent a single family of Ab's which can be activated through A48 regulatory idio-types and that anti-Id Ab's originate from a different gene family.

*Refs. noted by Senior Editor:-*

Bona, C.A., Goldberg, B., Metzger, D., Urbain, J. & Kunkel, H.G. (1984) *Eur.J.Immunol. 14,* 548-552.- 'Cross-reaction of anti-idiotypic antibodies specific for rabbit and murine anti-A1 allotype antibodies with Fc fragment of human immunoglobulins.'

Bona, C.A., Victor-Kobrin, C., Manheimer, A.J., Bellon, B. & Rubinstein, L.J. (1984) *Immunol. Rev. 79,* 25-44.- 'Regulatory arms of the immune network.'

#NC(B)-2

*A Note on*

# MONOCLONAL INTERNAL IMAGE ANTI-IDIOTYPIC ANTIBODIES OF HEPATITIS B SURFACE ANTIGEN

Yasmin M. Thanavala, Angela Bond, *Richard Tedder, Frank C. Hay and Ivan M. Roitt

Department of Immunology and *Department of Virology
The Middlesex Hospital Medical School
London W1P 9PG, U.K.

Using a monoclonal antibody (Ab) to hepatitis B surface antigen (HBsAg) we have raised 6 monoclonal anti-idiotypic Ab's. If the anti-idiotype is behaving as an 'internal image' of HBsAg, then, like antigen it should bind polyclonal anti-HBsAg sera raised in different species where it is less likely that idiotypes unrelated to antigen binding would be identical. Using polyclonal goat, swine, rabbit and human anti-HBsAg sera in an indirect immunofluorescence technique we have shown that two of the hybridomas stained strongly (Fig. 1). No staining was seen with various control sera or with hybridomas of unrelated specificities.

We have also shown the failure of swine antisera to stain after selective absorption of the hepatitis *a* but not the *d* or *y* specificities, showing a relationship of internal image reactivity to a restricted part of the hepatitis system.

Specificity of staining was also confirmed by experiments involving the elution of goat anti-HB's from a specific (HBsAg) and an irrelevant (ovalbumin) antigen column.

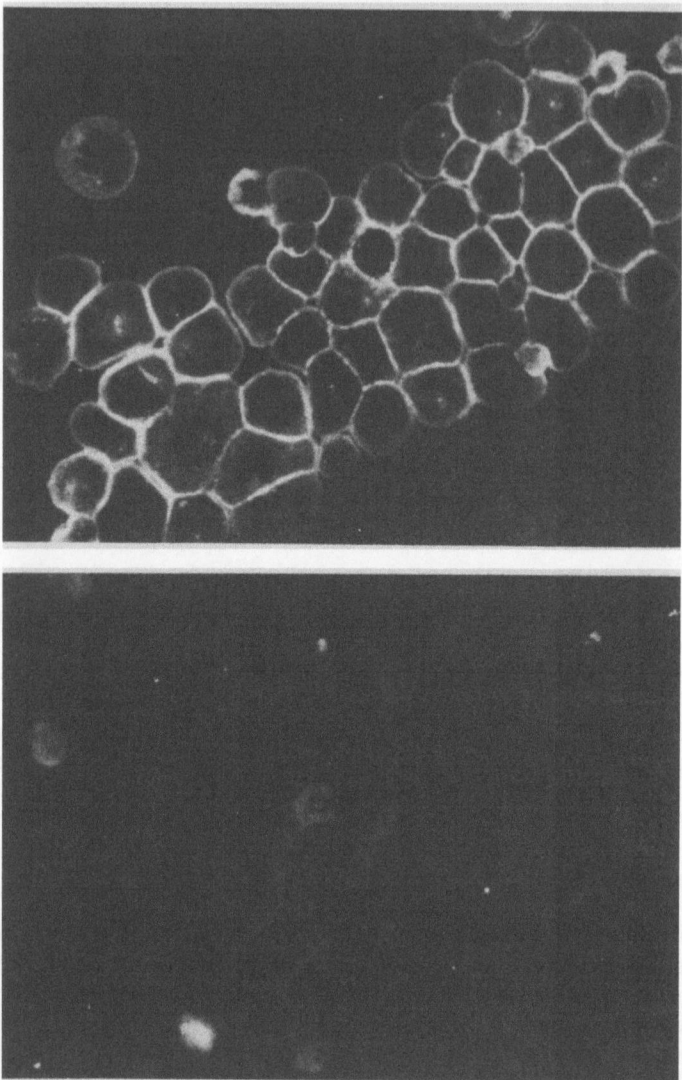

**Fig. 1.** *Top:* Immunofluorescent staining of anti-idiotype hybridoma cells for HBsAg internal image with goat anti-HBsAg serum. Cytocentrifuge smears were prepared (with $1 \times 10^6$ cells/ml), air-dried, fixed for 20 min in ethanol/acetic acid (95:5 by vol.) and stored in cold phosphate-buffered saline till used. *Bottom:* Control slides made from cells not reacting with the immunizing idiotype. Initially the smears were incubated with either normal goat serum or a DE52-purified polyclonal goat anti-HBsAg for 1 h followed by 3 washes in the saline. After a further incubation with fluoresceinated rabbit anti-goat Ig for 20 min, the slides were washed 3 times in the saline and prepared for examination in a fluorescence microscope by epi-illumination.

#NC(B)-3

*A Note on*

## STRUCTURE AND EXPRESSION OF GENES ENCODING
## IgA HEAVY CHAINS

Katherine L. Knight

Department of Microbiology and Immunology
University of Illinois Medical Center
Chicago, IL 60680, U.S.A.

Quantitative analysis of rabbit secretory IgA with anti-allotype antibodies directed against the $V_H$ or $C_\alpha$ regions revealed the presence of recombinant IgA molecules in which the $V_H$ allotype was encoded by one chromosome and the $C_\alpha$ region was encoded by the homologous chromosome. Similar studies with IgG and IgM showed that IgA contained the greatest number of recombinant molecules and that the probable order is V---$C_\mu$---$C_\gamma$---$C_\alpha$ [1] for genes that encode Ig heavy chains. In an effort to establish the order of the heavy-chain genes in rabbit and to determine the gene rearrangements which result in the recombinant Ig heavy chains, we have begun to clone the rabbit heavy-chain genes.

IgA and IgG cDNA clones were isolated from cDNA libraries prepared from RNA of rabbit mammary tissue or spleen, respectively [2, 3]. The IgA-encoding cDNA was cloned into the expression vector pUC8 and the lysates of *E. coli* transformed with this plasmid were examined by radio-binding studies for the expression of molecules bearing α chain antigenic determinants [4]. Several clones encoding the IgA-g subclass were identified; no clones encoding the IgA-f subclass were found. The amino acid sequence of IgA-g heavy chains was deduced from the nucleotide sequence of these cDNA clones, and examination of the cysteine residues has allowed us to predict that secretory component binds IgA at residues 299 and/or 311 of the heavy chains.

The cDNA clones encoding rabbit IgG and IgA, as well as clones encoding mouse $J_H$, $C_\mu$ and $C_\varepsilon$, were used to screen rabbit recombinant phage and cosmid libraries [5, 6] prepared from a rabbit homozygous for the heavy-chain chromosomal region. Overlapping recombinant clones allowed the identification of $J_H$, $C_\mu$, $C_\gamma$ and $C_\varepsilon$ genes on a 90 kb stretch of DNA; these genes are found in a 5'-$J_H$-$C_\mu$-$C_\gamma$-$C_\varepsilon$-3' orientation. In addition, four $C_\alpha$ genes were found; but these have not been linked to the other heavy-chain genes. Southern blot analysis of rabbit sperm DNA indicates that the rabbit genome contains as many as 10 $C_\alpha$ genes. The genomic $C_\alpha$ genes have been cloned in the expression vector pSV2, and mouse myeloma cells have been transfected with them [7]. The transformed cells will be examined for the expression of rabbit $C_\alpha$ chains by radio-binding analysis to determine whether these genes are expressible.

Restriction site polymorphism is found for $J_H$, $C_\mu$, $C_\gamma$ and $C_\alpha$ genes. Cells synthesizing recombinant IgA molecules will be isolated from gut of rabbits heterozygous for the heavy-chain haplotype, and the DNA will be examined for rearrangement of $C_\alpha$ and $J_H$ genes. The restriction site polymorphism of the two haplotypes for the $J_H$ genes should allow us to determine whether the expressed $C_\alpha$ gene is rearranged to $J_H$ on the same chromosome or on the homologous chromosome. Rearrangement to $J_H$ of the homologous chromosome would indicate that the recombinant IgA heavy chains are indeed due to a somatic recombination event.

## Acknowledgement

Our investigations are supported by grants from the National Institutes of Health.

## References

1. Martens, C.L., Gilman-Sachs, A. & Knight, K.L. (1981) in *The Immune System*, Vol. 1 (Steinberg, C.M. & Lefkovitz, I.L., eds.) Karger, Basel, p. 291
2. Martens, C.L., Moore, K.W., Steinmetz, M., Hood, L. & Knight, K.L. (1982) *Proc. Nat. Acad. Sci. 79*, 6018-6022.
3. Efstratiadis, A., Moniatis, T., Kafatos, F.C., Jeffrey, A. & Vournakis, J.N. (1975) *Cell 4*, 367-378.
4. Knight, K.L., Martens, C.L., Stoklosa, C.M. & Schneiderman, R.D. (1984) *Nucleic Acids Res. 12*, 1657-1670.
5. Manistis, T., Hardison, R.C., Lacy, E., Lauer, J., O'Connell, C., Quon, O., Sim, G.K. & Efstratiadis, A. (1978) *Cell 15*, 687-701.
6. Steinmetz, M., Winoto, A., Minard, K. & Hood, L. (1982) *Cell 28*, 489-498.
7. Oi, V.T., Morrison, S.L., Herzenberg, L.A. & Berg, P. (1983) *Proc. Nat.Acad. Sci. 80*, 825-829.

#NC(B)-4

*A Note on*

# HUMAN:MOUSE ANTIBODY-PRODUCING TRANSFECTOMA CELL LINES

### Vernon T. Oi and *Sherie L. Morrison

Becton-Dickinson
Monoclonal Center
2375 Garcia Avenue
Moutain View
CA 94043, U.S.A.

*Department of Microbiology and
the Cancer Center, Institute
for Cancer Research, College
of Physicians and Surgeons
Columbia University
New York, NY 10032, U.S.A.

A new means to generate antibodies (Ab's) with desired antigen-specificities and biological effector functions is to create trans-fectoma Ab-producing cell lines. These cell lines are derived from lymphoid cell lines that have been transfected with novel immuno-globulin genes using DNA-mediated gene transfer techniques. The con-text is our attempts at understanding the regulation of immunoglobu-lin gene expression. A direct outcome of these studies has been a means of generating human Ab molecules with mouse antigen-specifici-ties.

Human:mouse chimaeric Ab genes were generated using recombinant DNA techniques to join together mouse variable-region exons and human immunoblobulin constant-region exons. These chimaeric immuno-globulin light- and heavy-chain genes were ligated into mammalian expression vectors and transfected into lymphoid cell lines so as to generate Ab-producing transfectoma cell lines. We have generated human $IgG_1(\kappa)$ and $IgG_2(\kappa)$ anti-phosphocholine Ab's, where the phos-phocholine binding specificity is derived from the mouse S107 myel-oma Ab. Using recombinant DNA techniques to clone desired variable-region genes and manipulate human immunoglobulin constant-region genes, human Ab's with desired biological functions and antigen-specificities can be generated. This general approach should enable us to further understand the structural and functional correlation of Ab molecules.

---

*Some points noted by the Senior Editor*

At the start of his Forum talk [no publication text] V.T. Oi explained that the present approach whereby a piece of DNA is incor-

porated into a stable cell line, and is eventually expressed, is an
alternative to transferring DNA into a cell and then, usually within
48-72 h, looking for transient expression through measuring what
happens to that piece of DNA either in transcription of RNA off that
gene or in translation of the RNA transcript of the gene. To designate  lymphoid cell lines that are producing immunoglobulins from
genes transfected into the cells, the term 'transfectoma'  cell lines
has been proposed, in contrast with hybridoma cell lines. To transfect a piece of *E. coli*  DNA into myeloma cells, we can use a DNA-
calcium phosphate precipitate ($Ca^{2+}$, then phosphate, added to the
DNA) or, using lysozyme, we can make protoplasts from the bacterial
cells (harbouring the plasmid that contains the immunoglobulin gene
of interest) and do a classical polyethylene glycol fusion between
the protoplasts and the myeloma cells. With an appropriate bacterial
enzyme as a marker, selection can furnish in 2 weeks those lines
that express the gene that one is conserving.

Besides the studies outlined in his Forum abstract (preceding
p.), V.T. Oi has been extending earlier work [ref. below] on 'segmental flexibility' in the Ab molecule, to ascertain what structure is
actually responsible (the end-region?).   Other  studies have shown
that, in general, any human Ab could be constructed (maybe as an aid
to immunotherapy) by utilizing the mouse repertoire furnished by the
many monoclonal Ab's and hybridoma cells lines that already exist.

*Some publications co-authored by V.T. Oi:*

Herzenberg, Leonore A., Hayakawa, K., Hardy, R.R., Tokuhisa, T., Oi,
V.T. & Herzenberg, Leonard A. (1982) *Immunol. Rev. 67*, 5-31.-
*"Molecular, Cellular and Systemic Mechanisms for Regulating IgCH
Secretion".*

Oi, V.T., Vuong, T.M., Hardy, R., Reidler, J., Dangl, J., Herzenberg,
L.A. & Stryer, L. (1984) *Nature 307*, 136-140.- *"Correlation between
segmental flexibility and effector function of antibodies".* It
*appeared that "the effector functions of immunoglobulin isotypes may
be controlled in part by the freedom of movement of their Fab arms".*

Sharon, J., Gefter, M.L., Manser, T., Morrison, S.L., Oi, V.T. &
Ptashne, M. (1984) *Nature 309*, 364-367.- *"Expression of a $V_H C_\kappa$
chimaeric protein in mouse myeloma cells." The study entailed construction of $V_H C_\kappa$-containing plasmids bearing the chimaeric gene
and a selectable marker; transfection into BALB/c mouse myeloma
cells was by protoplast fusion.  After injection of a selected
clone  into BALB/c mice, the presence of the chimaeric protein in
ascites fluid was shown by an RIA which "measured the ability of the
ascites fluid to compete with the $^{131}I$-labelled idiotype-bearing
molecule, 36-65 Fab, for the binding of rabbit anti-Id antiserum".*

#NC(B)-5

*A Note on*

# DIFFERENT WAYS TO MODIFY MONOCLONAL ANTIBODIES

G. Köhler,* B. Baumann, J. McCubrey,
A. Iglesias, A. Traunecker and D. Zhu

Basel Institute for Immunology
487 Grenzacher Strasse
CH-4005 Basel, Switzerland

The availability of hybridoma tissue culture lines secreting monoclonal antibodies (MAb's) of pre-defined specificity opens new possibilities for altering Ab structure *in vitro* and creating molecules with novel properties. We have used three different approaches to modify a given IgM anti-trinitrophenyl (TNP) MAb, secreted by the hybridoma line Sp6.

(1) A TNP-negative subline of Sp6 was isolated, expressing the heavy chain together with a non-specific light chain. This line was fused to mitogen-stimulated mouse spleen cells to generate a set of secondary hybridomas, which were screened for the restoration of the original anti-TNP Ab specificity. Complementation was found to be frequent and achieved with 10 different light chains. In such cases, affinity variations are expected, and the structural restrictions imposed on light chains can be studied when a given specificity has to be maintained together with one heavy chain.

(2) Spontaneously arising variants were selected using a suicide selection technique, by coupling TNP covalently to the surface of the Sp6 cell and incubating them in the presence of complement. Variants with reduced secretion, reduced lytic activity and reduced affinity were thus selected and molecularly analyzed. The mutant phenotypes were found to be due to DNA insertions, deletions and frameshift alterations.

(3) The heavy- and light-chain genes of the Sp6 line were cloned into a pBR322 plasmid carrying a dominant selectable marker. Transformation experiments into B-myeloma lines resulted in stable transformants secreting functional, pentameric IgM. *In vitro* modified genes which encode for a $V_H$-$C_K$ chimaeric protein were expressed in L-chain-producing B cell lines to create idiotype-positive light

---

* now at Max-Planck Inst. für Immunobiologie, D-7800 Freiburg, F.R.G.

chain dimers.    In addition, the heavy- and light-chain-carrying plasmid was introduced into mouse oocytes.    Transgenic mice expressing Sp6 IgM at 10% of the total IgM serum level were obtained.    At this level of expression parental immunoglobulin gene rearrangements were not inhibited.

============

*AMPLIFICATION OF THE FOREGOING FORUM ABSTRACT*

*Editor's excerpts from a paper regarded by G. Köhler as helpful in outlining the methodology (approaches numbered as above):*

Köhler, G., Baumann, B., Iglesias, A., McCubrey, J., Potash, M.J., Traunecker, A. & Zhu, D. (1984) *Med. Oncol. & Tumor Pharmacol.* **1,** 227-233  *(Title as for the Abstract)*

*by kind permission of Pergamon Press*

## MATERIALS AND METHODS

(1) *Cell lines and fusion.-* The lines Sp1, 2, 6 & 7 are described [1].    Antigen negative subclones, containing either the specific heavy or the specific light chain, were isolated.    8-Azaguanine resistant lines were selected which were unable to grow in hypoxanthine, aminopterin and thymine (HAT)-complemented medium.    These lines were fused to 4-day lipopolysaccharide (LPS)-stimulated Balb/c spleen cells.    Supernatants of the resulting hybridoma lines were tested for the restoration of original specificities of the 4 lines. *[Table omitted; some abbreviations hereafter were not in the original text.- Ed.]* ........The derivation of monoclonal rat anti-mouse µ-Ab's and their allocation to the different domains have been described [2]......

(2) *Suicide selection of variants.-* A detailed description of this method and the selection of variant lines are described elsewhere [3-5].

(3) *Cloning and selection.-* The Sp6 µ- and κ-genes were cloned [6] and inserted into a pBR322 plasmid which carried a bacterial neomycin-resistance gene under SV40 promotor control.    A description of the vector and transfection into B-cell lines are given elsewhere [7].

## RESULTS

We have generated variants of mainly one hybridoma line (Sp6) which secretes mouse IgM, κ-Ab's with anti-TNP specificity [1] .........
The IgM can easily be purified by passing culture supernatants on TNP derivatized bovine serum albumin (BSA) coupled to Sepharose and eluting the IgM with 0.1 N HCl.    Antigen negative sublines were established expressing the Sp6 µ-heavy chain (K) (Sp6/HK) or expressing only the Sp6 (L)- and the MOPC21 (K)-light chain (Sp6/LK). Azaguanine-resistant and HAT-sensitive variant lines were isolated. Such lines were used to generate secondary hybridomas.

(1) *Complementation of Ab specificity.* - The antigen non-binding HK- or L-variants of 4 different lines were fused to LPS-stimulated mouse spleen cells and the resulting secondary hybridomas were screened for restoration of the original specificity.  In general, heavy chain donating lines were restored by spleen cell derived light chains at an overall frequency of 1/90 whereas light chain donating lines were complemented at least 30- to 40-fold lower frequency, which was indistinguishable from the frequency of 'background' specificities in the LPS pool........[Re] usage of the complementing light chain variable region genes as analyzed by DNA-restriction mapping.- Ten different κ-chains were able to complement the Sp6 μ-chain for (anti-TNP) activity........ the most frequently isolated light chains used the same V-genes as the original hybridoma lines.

With this method it is therefore possible to generate quite easily a whole array of Ab's of a given specificity.  The fact that the heavy chain remains identical makes such Ab's a valuable tool to study the influence of light chains to the Ab combining sites.

(2) *Suicide selection against cells secreting lytic IgM.* - [We......] analyzed cell lines making altered IgM.  A novel way to select for variants was introduced [3].

The haptens TNP or phosphorylcholine (PC) were covalently linked to the cell membrane of Sp6 or PC700 (IgM, κ, anti-PC).  Cells were then separated from each other in semi-solid medium and incubated in the presence of complement.  The IgM secreted by a normal cell bound to the hapten on its own surface and in the presence of complement the cells lysed.  Thus cells secreting less or less lytic IgM were enriched......[The deletions, largely μ-chain, were] at the DNA level....

All the C-terminal deletions so far sequenced were due to single or double base-pair deletions leading to a string of out-of-frame amino acids and premature termination........ The deletion proteins were used to map a variety of rat MAb's to the 4 mouse μ-constant region domains.  These Ab's were screened for their effects in two standard serological tests which routinely employ anti-IgM: hemolysis and hemagglutination. The anti-idiotypic Ab 20-5 was a strong inhibitor of both......[cf. the inhibition and enhancement seen respectively with anti-κ and 7 anti-μ-Ab's....] ...... A variant (igm21) has been isolated [3] which cannot be distinguished from wild-type IgM by SDS-PAGE or isoelectric focusing, which nevertheless differs by 200-fold in the capacity to lyse TNP-SRC........

(3) *Gene cloning, modification and expression.* - A more direct way to modify Ab's is the cloning of their genes, the *in vitro* manipulation of their DNA and the transfection of the altered genes into cell lines capable of expressing the product.   The light [6] and the heavy chain (unpublished) were cloned and sequenced and a plasmid was constructed which carried  both genes together with a bacterial

**Fig. 1.** Production of chimaeric Ab's. *Top:* The plasmid pKmV$_H$14 carries the kanamycin-resistance gene *(black),* pBR322 sequences *(//////)* and an insert of the Sp6 κ gene from the BamHI to the SalI restriction sites. The Sp6 variable heavy chain gene (VDJ) was inserted into the XbaI and the Hind3 restriction sites in front of the κ enhancer element (E$^κ$). *Bottom:* SDS-PAGE under reducing conditions of products formed after transfection of pKmV$_H$14 plasmid into X63/0 myeloma cells. Stably transformed lines No. 3 & No. 15 show a band with slightly lower mobility than the Sp6 light chain *(right).* Light-chain and pKmV 14-directed production of V$_H$C$_κ$ in (X63/0 − No. 15 × igm 10) hybrids 8 and 21 is observed. Only in these hybrids is the heavy-chain-dependent idiotype 20-5 restored.

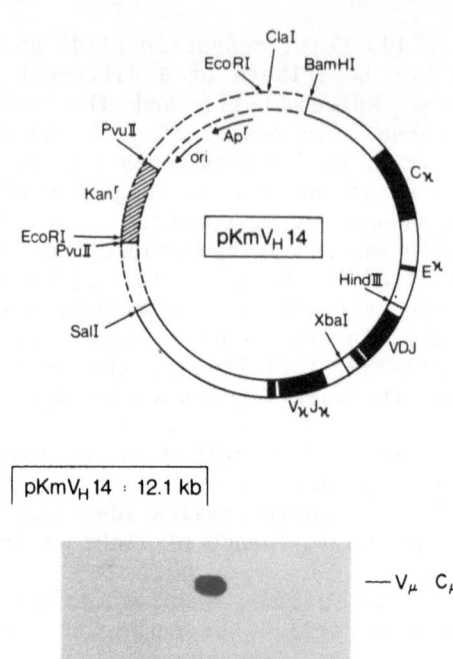

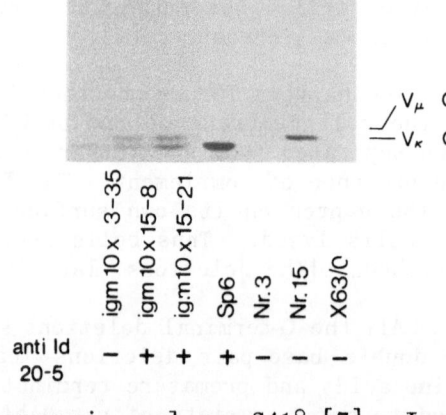

gene conferring resistance to the neomycin analogue G418 [7]. In an initial screening different cell lines were tested for their capacity to become transformed using the bacterial protoplast fusion technique. Furthermore, the amount and quality of the transfected IgM was measured........ The amount was scored in a number of individual transformants using radioimmunoassays; it is the mean amount of IgM produced by one gene copy. Thus the μ- and κ-gene in X63/0 produce about 40-fold less IgM than the original Sp6........ For the purpose of maximal expression of modified genes, however, only the X63/0 line was suitable.

Into this line we transfected the κ-light chain gene of Sp6, which contained the Sp6 μ-heavy chain variable region in its large intron (Fig. 1). We found that the chimaeric product V$_μ$6-C$_κ$6 was the only product made by stably transformed X63/0 lines. Fusing

this line to igm10, a Sp6 light chain producer, we found that the
idiotype, as revealed by $^{125}$I labelled (20-5) MAb-binding, was restored (Fig. 1).    Thus a novel small antigen-binding molecule was
generated which might be of interest when conventional Ab's are too
big (or too sticky) to travel to the site of antigen expression.

## DISCUSSION

[Concerning] medical applications,...... it seems that only rarely
an Ab meets all the requirements which it ideally should have.   For
example, Ab's which are used for removal of cellular components *in
vivo* or *in vitro* would be the least harmful and possibly also the
most effective if they used the host's own defence mechanisms....... It
would be very interesting to see whether genetically engineered high
affinity IgM Ab's, which could use mouse variable heavy and light
chain segments and human constant regions, would be superior to the
conventional MAb's used.    The construction of chimaeric, antigen
binding κ-dimers [V.T. Oi  & colleagues; cf. #NC(B)-4, this vol] has
shown that technically such manipulations are feasible.   This class
of engineered small antigen-binding molecules or those selected from
mutant cells are alternative ways to make mono-Fab-like molecules
[8].    They might be useful when only one binding site is required
(to avoid modulation of cell surface antigens) when a smaller molecule with less non-specific interactions (via the Fc-receptor) and
possibly a higher elimination rate are needed [8].........

Another application of MAb's is their use in one step purification of biological materials [9].   With the possibility of manufac-
turing  protective antigens for active  immunization against [e.g.]
malaria, large scale purification with suitable MAb's may become
important.   Such Ab's should withstand repeated elution procedures,
a property mainly dependent on their variable regions.   Furthermore,
the gentler the elution of an antigen the greater the chance of
maintaining its original 3-dimensional structure.   Low affinity Ab's
may be the  best choice for such needs.........

*References*

1. Köhler, G. & Milstein, C. (1976) *Eur. J. Immunol. 6*, 511-519.
2. Leptin, M., Potash, M.J., Grützmann, R., Heusser, C., Schulman,
   M., Köhler, G. & Melchers, F. (1984) *Eur. J. Immunol. 14*, 534-542.
3. Köhler, G. & Schulman, M. (1980) *Eur. J. Immunol. 10*, 467-476.
4. Köhler, G., Potash, M.J., Lehrach, H. & Schulman, M. (1982) *EMBO
   J. 1*, 555-563.
5. Schulman, M.J., Heusser, C., Filkin, C. & Köhler, G. (1982)
   *Molec. Cell Biol. 2*, 1033-1043.
6. Hawley, R.G., Schulman, M.J., Murialdo, H., Gibson, D.M. & Hozumi,
   N. (1982) *Proc. Nat. Acad. Sci. 79*, 7425-7429.
7. Ochi, A., Hawley, R.G., Hawley, T.,Schulman, M.J., Traunecker, A.
   Köhler, G. & Hozumi, N. (1983) *Proc. Nat. Acad. Sci. 80*, 6351-6355.

8. Yelton, D.E. & Scharff, M.D. (1982) *J. Exp. Med. 156*, 1131-1148.
9. Secher, D.S. & Burke, D.C.(1980) *Nature 285*, 446-450.

# # NC(B)-6

*A Note on*

## RESTRUCTURING ENZYMES AND ANTIBODIES

### Greg Winter and Michael Neuberger

MRC Laboratory of Molecular Biology
Hills Road, Cambridge CB2 2QH, U.K.

It has recently become possible to introduce DNA into lymphoid cell lines [1-3]. By constructing lines which are stably transfected with immunoglobulin gene DNA, it is possible to direct the synthesis and secretion in good yield of antibody (Ab) molecules that display specific desired characteristics. Thus, exon shuffling can be used to direct the production of chimaeric Ab's in which mouse-encoded antigen-specific variable regions are joined to human constant regions. In this way, a line secreting a chimaeric IgE Ab has been established where the $C_\varepsilon$ constant region is of human origin (thus allowing binding to human mast cells) and the variable regions are of mouse origin and form a binding site for the hapten 4-hydroxy-3-nitrophenylacetyl [4]. In other experiments, gene fusions have been performed to make a recombinant Ab in which the Fc portion of the Ab is replaced with an active enzyme moiety (staphylococcal nuclease), thus showing that this is a viable approach for making Ab's exhibiting novel effector functions [5].

However, a detailed analysis of the structural basis of Ab interactions either with antigen or with effector proteins (e.g. complement, Fc receptors, etc.) should be feasible by the use of site-directed mutagenesis. Oligonucleotides can be used to alter one or several nucleotides at a time, and to make insertions and deletions. As a model system for Ab mutagenesis, we can take the directed mutation of the enzyme tyrosyl tRNA synthetase [6]. The crystallographic structure of this enzyme is known, and mutations have been made to delete side-chains which are known to make hydrogen-bonding contacts with the substrate, ATP [7-9]. By this means we find that hydrogen bonds can be worth between 0.5 and 4 kcal/mol to substrate binding [10], and are sometimes even a liability [7, 8]. Thus to understand the interaction of protein and ligand, the construction of an energy map is a first step. Removing unfavourable contacts [8] or making new direct favourable contacts between enzyme and substrate has enabled the enzyme's affinity for ATP to be increased. Furthermore, by distorting the polypeptide backbone, it has been possible to indirectly alter side-chain contacts with substrate, and also to increase affinity [8, 9]. The opportunity to apply these techniques to the Ab molecule has now come.

*References*

1.  Neuberger, M.S. (1983) *EMBO J. 2*, 1370-1378.
2.  Oi, V.T., Morrison, S.L., Herzenberg, L.A. & Berg, P. (1983) *Proc. Nat. Acad. Sci. 80*, 825-829.
3.  Ochi, A., Hawley, R.G., Hawley, M.J., Shulman, A., Traunecker, J., Köhler, G. & Hozumi, N. (1983) *Proc. Nat. Acad. Sci. 80*, 6351-6355.
4.  Neuberger, M.S., Williams, G.T., Mitchell, E.B., Jouhal, S.S., Flanagan, J.G. & Rabbitts, T.H. (1985) *Nature 314*, 268-270.
5.  Neuberger, M.S., Williams, G.T. & Fox, R.O. (1984) *Nature 312*, 604-608.
6.  Winter, G., Fersht, A.R., Wilkinson, A.J., Zoller, M.J. & Smith, M. (1982) *Nature 299*, 756-758.
7.  Wilkinson, A.J., Fersht, A.R., Blow, D.M. & Winter, G. (1983) *Biochemistry 22*, 3581-3586.
8.  Wilkinson, A.J., Fersht, A.R., Blow, D.M., Carter, P. & Winter, G. (1984) *Nature 307*, 187-188.
9.  Carter, P., Winter, G., Wilkinson, A.J. & Fersht, A.R. (1984) *Cell 38*, 835-840.
10. Fersht, A.R., Shi, J.-P., Knill-Jones, J., Lowe, D.M., Wilkinson, A.J., Blow, D.M., Brick, P., Carter, P., Waye, M.M.Y. & Winter, G. (1985) *Nature 314*, 235-238.

#NC(B)-7

*A Note on*

# DIDEOXY mRNA SEQUENCING AS A METHOD TO ANALYZE ANTIBODY DIVERSITY

Gillian M. Griffiths[*], Claudia Berek,
Matti Kaartinen and Cesar Milstein

MRC Laboratory of Molecular Biology
Hills Road, Cambridge CB2 2QH, U.K.

The antigenic specificity of an antibody (Ab) molecule is determined by the amino acid sequence of the light and heavy chain variable regions. We have analyzed the extent to which various possible mechanisms for generating Ab diversity contribute in an immune response to a simple antigen, the hapten 2-phenyloxazolone. In order to identify the relative contributions of the different mechanisms which can generate Ab diversity (e.g. the use of different germ-line V genes, different combinations of V, D and J gene segments and the role of somatic point mutations), nucleotide sequence information was required. We therefore analyzed the V region expression of a series of oxazolone-specific hybridomas by sequencing the mRNA isolated from these cells. The major advantage of this technique is the speed and ease with which the required sequence information can be generated.

The method used was a modification of Sanger's dideoxyDNA sequencing technique [1]. A synthetic oligonucleotide primer is designed which is exactly complementary in sequence to a relatively conserved region of the mRNA template. Using such a primer it is possible to initiate synthesis of a complementary DNA (cDNA) by enzymatic extension with reverse transcriptase in the presence of all 4 deoxynucleotide triphosphates (dNTP's) one of which can be labelled with $^{32}$P. The cDNA transcripts can be terminated in a nucleotide-specific fashion by the incorporation of dideoxy nucleotide analogues which, since they lack a 3'-hydroxy group, will prevent further chain elongation. For example, random incorporation of ddC into the growing cDNA chains will lead to chain terminations at all positions corresponding to G residues in the mRNA template. Since each transcript begins at the same 5' site as defined by the primer then size-fractionation of the products on a gel can be used to reveal the relative

---

[*]now at Pathology Dept., Stanford University School of Medicine, CA 94305, U.S.A.

distances between G residues in the mRNA template. Hence by carrying out cDNA syntheses in the presence of each of the 4 dideoxy nucleotide analogues it is possible to use gel fractionation to deduce the sequence of bases in the mRNA template.

In order to determine the V-region sequences of hybridomas, 3 different primers were produced which were complementary to 15 nucleotides from the 5' ends of the constant regions of each of the κ light and the μ and γ heavy chains. Two further primers were used to obtain full V-region sequences for the majority of hybridomas studied. These primers were chosen to hybridize to relatively conserved regions within the V regions themselves. Using these 5 primers complete V-region sequences could be obtained for both the light and the heavy chains [2]. This approach offered three major advantages over other methods of analyzing V-region diversity in a comparative study of the sort which we wanted to make. Firstly, very few primers were required to sequence very many V regions. Secondly, extensive purification of the mRNA was not required, since the primer itself constitutes an important step in the purification, annealing only to those regions which are complementary in sequence. The major advantage of this technique was the speed and ease with which the sequence information could be generated: from cells to sequence need take only 2 days and involve relatively little work.

Using this technique we sequenced 50 different hybridomas from three different stages of the immune response to the hapten 2-phenyloxazolone. The results revealed that early after primary immunization the Ab response is dominated by the expression of a single pair of heavy- and light-chain V genes with very little evidence of somatic mutation [3]. However, at later stages somatic mutation of these genes seemed to provide Ab's with higher affinities for the antigen [4]. After secondary immunization, however, the expression of other germ-line variable-region genes becomes important in the maturation of this response [5].

*References*

1. Sanger, F., Nicklen, S. & Coulson, A.R. (1977) *Proc. Nat. Acad. Sci. 74*, 5463-5468.
2. Kaartinen, M., Griffiths, G.M., Hamlyn, P.H., Markham, A.F., Karjalainen, K., Pelkonen, J.T., Makela, O. & Milstein, C. (1983) *J. Immunol. 130*, 937-945.
3. Kaartinen, M., Griffiths, G.M., Markham, A.F. & Milstein, C. *Nature 304*, 320-324.
4. Griffiths, G.M., Berek, C., Kaartinen, M. & Milstein, C. (1984) *Nature 312*, 271-275.
5. Berek, C., Griffiths, G.M. & Milstein, C. (1985) *Nature*, submitted.

#NC(B)-8

*A Note on*

# EFFECT OF MULTIVALENT BINDING ON DISSOCIATION OF MONOCLONAL ANTIBODIES FROM CELL SURFACE ANTIGENS

S.M. Hobbs, J.M. Styles and C.J. Dean

Section of Tumour Immunology
Institute of Cancer Research
Clifton Avenue, Belmont
Sutton, Surrey SM2 5PX, U.K.

Clearance of antibodies (Ab's) from cell surfaces has been studied *in vitro* with two IgG monoclonal Ab's (11/160 and 12/FCIII) directed against surface antigens of the rat fibrosarcoma HSN.TC. When Ab-coated HSN.TC cells were incubated in fresh medium at 37°, 12/FCIII was lost rapidly (8% of the initial binding remaining after 2 h), whereas 11/160 was cleared relatively slowly (85% of the initial binding remaining after 10 h). Gel-filtration analysis of radiolabelled Ab shed into the supernatant medium over the first 5 h showed the majority (>80%) of the recovered activity as intact IgG, with the remainder as low-mol. wt. fragments. Loss of radiolabelled monoclonal Ab from the cells was progressively inhibited by treatment of monoclonal-Ab-coated cells with increasing concentrations of antiglobulin before incubation at 37°, and was accompanied by a greater output of degraded fragments.

Taken with our earlier observations on the behaviour of polyclonal alloAb's under similar conditions [1], these results are consistent with a model of Ab-cell surface interactions that involves an equilibrium between Ab in solution, Ab attached to the cell by one binding site, and Ab bound by divalent attachments.

*Reference*

1. Hobbs, S.M., Styles, J.M., Dean, C.J. & Shepherd, P.S. (1983) *Immunology 50*, 565-573.

## Comments on material in #B

*Comments on* #B-1, D.P. Lane et al. - MAb SPECIFICITY
          #B-2, F.A. Simmen, D.R. Headon et al. - SCREENING cDNA's

**D.P. Lane, replying to D.S. Secher.**- PAb 204 indeed recognizes a denaturation-resistant epitope; it is a very unusual Ab, not part of the normal response but raised by immunizing with an Adeno/SV40 hybrid. **Reply to J. Kappler.**- The advantage of inserting the cDNA at the end of the β-galactosidase gene rather than just behind the initial Met is that generally a more stable fusion protein results. **M.M. Davis,** commenting on a reply by **D.R. Headon** who said that the 29 false positives were all different by Southern analysis: one wonders how many possible proteins can fill a given Ig binding site; here you have 30 different possible reactants for the same MAb (including the correct cDNA clone).

*Comments on* #B-3, B.F. Erlanger et al. - ANTI-RECEPTOR Ab's

**C.M. Lewis, to B.F. Erlanger.**- Is there any cross-reaction between anti-BisQ Ab's and sera of patients with myasthenia gravis? **Reply.**- We are interested in this because it would indicate that the patient's serum contained anti-Id Ab's. We have been examining patients for this reaction and we have found it. Unfortunately, 'normal' serum (i.e. from people without myasthenia gravis) also gave positive results. We are at present trying to refine the assay. **Reply to another questioner, who asked:** is it possible that BisQ is a bivalent cholinergic ligand, thus complicating the anti-Id response to this compound? - I do not see how this would cause any complications. In any case, although we have a bias in favour of this kind of reaction, our dose-response curves are 'S'-shaped just as they are with ACh, indicating monovalence. *(See overleaf for a comment.)*

**J. Hoebeke asked:** (1) how do you explain the inhibitory activity of your 'agonist-like' anti-Id Ab in AChR reconstituted vesicles? (2) does your anti-Id Ab inhibit α-bungarotoxin binding to the receptor? **Replies.**- (1) The inhibition of ion flux occurs in the presence of agonists (carbamylcholine). Therefore, the Ab has the opportunity to react with the receptor in the activated (open) state and may block ion flux sterically. (2) α-Bungarotoxin inhibition can be demonstrated by ELISA. However, we have not been able to demonstrate it in the IF reaction with *Torpedo* tissue. We do not know why.

**Remark by E.A. Kabat.**- With their bivalent DNP hapten, Valentine & Green had to have a long chain of ~8 $CH_2$'s to achieve bivalency. This would be longer than the BisQ which then is probably not able to react with two Ab combining sites.

*Comments on* #B-4, G.E. Isom - ANTI-MORPHINE ANTI-Id Ab's
                #B-5, J. Hoebeke - ANTI-Id Ab's FOR MEMBRANE RECEPTORS

**Remark by B.F. Erlanger to G.S. Isom.**- You could establish that you have anti-Id Ab's by purifying your antisera on an anti-morphine Ab column and eluting with morphine or other ligand. **Questions by T. Reilly to J. Hoebeke:** (1) is your anti-Id response predominantly an IgM response? (2) have you determined the affinity of your anti-Id MAb for the β-adrenergic receptor? **Replies.**- (1) We see a high percentage of IgM's among our anti-Id Ab's but don't feel that anti-Id Ab's must be IgM. (2) We have not determined this. **Comment by B.F. Erlanger.**- Our anti-AChR Ab (F8-D5) and the two anti-adenosine receptor MAb's are IgM. **J. Hoebeke, answering M.E. Bardsley.**- Whole brain synaptic membranes were run under denaturing conditions (SDS-PAGE/Western blot) in our efforts to establish whether anti-Id Ab's to β-adrenergic receptor recognized only one molecular species (consisting of only one polypeptide chain) in the brain. This is interesting with respect to the receptor subtypes, specificity of anti-Id and molecular characterization of β-adrenergic receptor.

*Comments on* #NC(B)-3, K.L. Knight - GENES ENCODING IgA CHAINS[*]

**J. Hoebeke asked:** are there functional differences between the two IgA subclasses αf and αg? **Answer.**- We don't know, but we do know that there is an enormous difference in the proteoloytic susceptibility; the f subclass - which is covalently bound - is almost completely resistant to digestion, whereas the g subclass, which is not covalently bound, is very sensitive to digestion. The ratio of f to g is approximately the same in all secretions and in serum, although there is so little in rabbit serum that it is hard to tell.

**Question by E.A. Kabat.** - Are there genes for latent allotypes? Has anyone succeeded in cloning them? **Reply.**- The story with the latent allotypes is that for the κ chain the rabbit has 2 genes, κ1 & κ2. They represent two kinds of gene, one of which predominates and the other is expressed in extremely low quantities but does not in fact encode one latent allotype; for the κ chain evidently there is no extra gene encoding a latent allotype. For the γ chain there are also reports of latent allotypes for the constant region. In all the rabbits we have looked at in our colony there is only a single gene encoding γ; so clearly there is no gene encoding the latent C-γ. But this applies merely to our rabbits, and we have not examined them for latent allotype; so we are now looking at rabbits

---

[*] The 'Note' does not cover all points in the talk that led to the questions, and the replies (based on a tape) are not vouched for.-*Ed*.

from a laboratory where latent C-γ has been described, to see if we can find an extra gene in them. We don't think there is, and don't think there are latent allotypes. For γ it may be a matter merely of somatic mutation to explain latent γ, where a single base change is involved. For κ I can offer no explanation at present.

*Concerning 'latent allotypes' [Tape transcript, not vouched for].*-
For those unfamiliar with latent allotypes, I now briefly describe them. They are unexpected allotypes that have been described for the heavy chain - both the variable region and some of the constant regions (γ and κ chains). Thus, one may find A1 in a supposedly A2 homozygous animal. Why? We are in the process of cloning the variable region, and we have many variable-region genes; so it is very hard at this stage to sort out which is encoding what allotype.- The testing is not simple. My feeling is that latent allotypes for the variable region of the heavy chain may in fact exist, taking account of a report that an attempt to generate anti-Id to injected anti-A2 in an A1 rabbit gave molecules that reacted with anti-A2. Although they were not all identified, at least they were not A1 such as one would expect in an A1 animal; they do seem to have A2-like sequences (investigated at NIH). A possible inference is that there is a latent A2 allotype. An effort is being made in our laboratory to make a cDNA library from one of these animals for confirmation, and ultimately get the germ-line gene.

*Comments on* #NC(B)-4, V.T. Oi - Ab-PRODUCING TRANSFECTOMAS

*Editor's note.*- S.J. Kaufman (Session Chairman) remarked that 'Ig engineering' could be regarded as having started with observations by V.T. Oi and others that it was possible to transfect myeloma cells with an Ig genetic plan and get secreted Ig in rather good yields from these cells - in contrast with many eukaryotic expression systems where the yields are rather poor. In relation to the following discussion points, attention is drawn to the editorial appendage to #NC(B)-4.

**V.T. Oi, replying to D.S. Secher.**- We have not studied by e.m. the angle between Fab's in MAb's of different sub-classes. **Comment by A. Feinstein.**- We have shown that the angle in IgG1 is ~90°. **Question by D.P. Lane.**- I was very impressed by your depiction of the human C-region/mouse V-region transfectoma in mice or an ascites tumour. Did you have any problems with rejection of the tumour since it is expressing foreign epitope? **V.T. Oi's reply.**- No. We did not expect the experiments to work, and in fact some mice died before we realized they were developing tumours. The amounts of Ab produced are not so great as with the best hybridomas, but they are comparable to those with less good hybridomas.

*Comments on* #NC(B)-5, G. Köhler et al. - MODIFICATION OF MAb's
           #NC(B)-6, G. Winter - RESTRUCTURING ENZYMES AND Ab's
           #NC(B)-7  G.M. Griffiths et al. - Ab DIVERSITY ANALYSIS

**Remark by M.M. Davis.**- Concerning the observation that the TNP occurred exclusively in the B-cell hybridomas made from the transgenic mice, one possible explanation is that the frequency of a good heavy-chain rearrangement (V-D-J) is sufficiently low (~ 10%) that many B-cells which would ordinarily not be able to produce a good Ig heavy chain are now just expressing the α-DNP construct. **G. Köhler agreed:** perhaps the additional DNA rearrangements we see in these hybrids are non-functional, but we haven't checked yet. **To J. Kappler, who asked:** is the idiotype in the transgenic mouse expressed on the surface of a majority of B-cells? - Yes; ~50% of the B-cells stain, for TNP and for the idiotype. We have not investigated whether this idiotype is functional (as judged by TNP activating a majority of these B-cells). **Further answer:** when we immunize the transgenic mice with TNP-red blood cells and measure TNP activity, little change is found, nor in red-cell activity; but possibly the TNP never reaches the cells to stimulate them because of the abundance of Ab's.

**Question by C.A. Morrison to G. Winter.**- In your *Staph.* nuclease-Fab hybrid, surely you can identify which band is which at ~45 K on your SDS gel by Western blot? **Reply.**- Probably both bands are *Staph.* nuclease-Fab hybrids (perhaps partly processed), so a Western blot would not be of much use.

**Remarks by S.M. Hobbs on observations by G.M. Griffiths** [who furnished only a partial text of her talk].- Hybrids generated at 7 days were all $\gamma_1$, whereas at 14 days you found some μ and $\gamma_2$ hybrids. Surely this is the opposite to the situation normally found for the isotype distribution (μ first, followed by γ) of the serum Ab response? **Reply.**- The isotype distribution of hybrids probably doesn't reflect serum Ab, in view of artefacts caused by hybrid selection criteria etc. Other fusions have generated μ hybrids in this system at 7 days.

**D.P. Lane, to G.M. Griffiths.**- How do you reconcile your results on the secondary response recruiting at 30% frequency H-chain and L-chain germ-line V genes not seen in the primary response with immunological memory? **Reply.**- This is unclear. The sample size may be insufficient. Alternatively, clones may be stimulated later in the primary response. **Lane.**- Maybe B cells do not carry the memory! **Questions by A. Feinstein.**- (1) Concerning interpretation of the relationship of your findings to the question of what is in the B memory, when do you think the somatic mutation or selection of new germ-line genes occurs? **Reply.**- We think they occur between the primary and the secondary response. (2) But couldn't they have existed yet not become dominant during the primary response, and then been selected in the secondary response?  **Reply.-** We think not.

*Comment on #*NC(B)-8, S.M. Hobbs - DISSOCIATING MAb's FROM ANTIGENS

**S.J.  Kaufman.-** We have found that cross-linking of  membrane
antigens with primary Ab's or primary and secondary Ab's  also  alters
the turnover of the cell-surface antigens, and that cross-linking
in many cases drives an association between the antigen and the
Triton-insoluble framework of the cell.   The latter association may
very well alter the  stability and fate of cell-surface antigens.
Perhaps it would be worth checking this in your cells as well.

_____

*#NC(B)-7: Amplification of some methodological approaches, excerpted
by Senior Editor from ref. [3] (also [2]) in #NC(B)-7*

Kaartinen, M., Griffiths, G.M., Markham, A.F. & Milstein, C. (1983)
*- mRNA sequences define an unusually restricted IgG response to 2-
phenyloxazolone and its early diversification. Nature 304, 320-324.*

"Synthetic oligonucleotide primers [e.g.  for γ-chain, μ-chain,
heavy-chain oxazolone-idiotype] designed to pair with defined bases
within segments of mRNAs were used to initiate reverse transcription.
Using dideoxynucleotides, specific stops in the cDNA can be generated and the nucleotide sequence [merely 'likely assignments' in the
case of some difficult residues] determined by gel methods" [as
previously described from the Cambridge laboratory: Hamlyn, P.H., et
al. (1978) *Cell 15*, 1067-1075]. cDNA synthesis was effected with an
incubation system which included, besides the transcriptase and
poly(A)$^+$-enriched RNA, deoxy- ($^{32}$P-labelled in one case) and dideoxy-
nucleoside triphosphates and the primer complementary to a specific
site  on the mRNA, e.g. bases 22-37 for the γ-chain constant region.

The 15 monoclonal antibodies to the oxazolone hapten were
obtained 7 days after immunization, when the distribution of avidi-
ties is narrower than after 14 days [Kaartinen, M., et al. (1983)
*J. Immunol. 130*, 937-945].  The antibodies were mostly of the IgG1
class.

*A methodological ref. noted by Senior Editor:*

Westerwoudt, R.J., McFarlane, B.M., McSorley, C.G., McFarlane, I.G.
& Williams, R. (1985) - *Improved Fusion methods. IV. Technical
Aspects.  J. Immunol. Meths. 77*, 181-196.  This paper in the series
concerns an RIA for detecting circulating Ab's reacting with the
hepatic asialoglycoprotein receptor protein.

*Colloquia whose relevance is not confined to sect. #B:-*

(1) (1984) *Biochem. Soc. Trans. 12*, 737-756.- 'Effector functions of
the constant region of immunoglobulin molecules.'
(2) (1985) *Biochem. Soc. Trans. 13*, 1-18, & 94-124.- 'Monoclonal an-

tibodies: their application in exploring cells and molecules.'

Topics in (1) include: C1 activation; lymphocyte, K cell and phagocyte Fc receptors; monocyte binding domains on IgG; roles of the immune system and mucus in protection of mucous membranes.

A.F. Williams opens (2) by considering what MAb's see at cell surfaces.    T.A. Springer et al. deal with the LFA-1, Mac-1 leucocyte adhesion glycoprotein family and its deficiency in a heritable human disease.    Then (p. 6) O. Acuto et al. consider molecular analysis of the human T-cell antigen receptor by use of MAb's, and (p.10) J.R. Birch et al. of Celltech Ltd. survey the production of MAb's in large-scale cell culture. Other MAb topics relate to X-31 influenza virus haemagglutinin (J.J. Skehel et al., p. 12), ACh receptor structure studies (J. Lindstrom et al., p. 14), and microtubule studies entailing microinjection of tubulin Ab's into living cells (J. Wehland & K. Weber, p. 16).

Topics in the the subsidiary contributions to (2) include avian desmin,   insulin  receptor,  granulocyte  differentiation  antigens, intestinal enterocyte and brush-border proteins,  ribonucleoprotein-Sm complex (to which Ab's occur in systemic lupus erythematosus), alkaline phosphatases, IgG effector sites, platelet  glycoproteins, T-cell markers on chronic  lymphocytic leukaemia lymphocytes,  myotube lysis in relation to myasthenia gravis, ACh receptor studies, apoprotein E, and carcinoma cell lines.

A colloquium on MAb's in the study of differentiation,  including a contribution by R.M.E. Parkhouse on nematode antigens, may be in press; ref. otherwise as for (2).

*Reminder to those who have access to Vol. 2, this series:-*

In 'Preparative Techniques' (1973; available from E. Reid as its Editor), O. Cromwell deals with conjugation and other pertinent techniques in his review of 'Protein Purification by Immunoadsorption'.

Section #C

CYTOSKELETAL, MYOCYTE AND NEURONAL STUDIES
WITH ANTIBODIES

#C-1

# IMMUNOLOGICAL APPROACHES AND TECHNIQUES EMPLOYED IN TUBULIN AND MICROTUBULE RESEARCH

Christopher R. Birkett and Keith Gull

Biological Laboratory, University of Kent
Canterbury, Kent CT2 7NJ, U.K.

*It was through the use of indirect immunofluorescence (IF) using monospecific polyclonal antisera that it was first possible to visualize in their entirety the arrays of microtubules within cells. Now, by using specific monoclonal antibodies (MAb's), it is possible to distinguish subsets of microtubule populations within these arrays. Some MAb's bind only to species-specific tubulins and are even capable of defining individual electrophoretic tubulin isotypes within the full spectrum of tubulin polypeptides produced by a single organism. Because it is possible to exactly determine the nature of the binding site of a MAb, it becomes a probe for molecular structure, the sensitivity and specificity of which has not previously been available. As the library of MAb's expands it should therefore be possible to investigate the specific protein constituents of microtubules and the nature of their interaction with many other cellular organelles. Such work is still in its comparative infancy, but already there are examples of how this approach will be able to benefit our understanding of the microtubules within the cell.*

Microtubules are found in all eukaryotic cells. As implied by the name, they are tubular structures with a diameter of 25 nm and of variable length; the wall is composed of a helical arrangement of protofilaments generally totalling 13. There are various structures in cells which incorporate microtubules. Besides the spindle and flagellum, which are of widespread occurrence, arrays of microtubules are to be found in the cytoplasm of most cells, and protozoans in particular exhibit numerous examples of highly specialized microtubule-containing structures. By the application of certain drugs which interfere specifically with microtubules, it can be demonstrated that there is an actual requirement for microtubules in the functioning of these structures, e.g. during mitosis, flagellar beating, organelle translocation and vesicular secretion. Although the generation of a motive force is implied directly in each of these examples, the actual role played by microtubules, either alone or in concert with other cellular components, is still largely speculative.

The most abundant protein in microtubules is tubulin.   It has mol. wt. 110,000, and is a heterodimeric protein comprised of equimolar amounts of two structurally related polypeptide subunits, α- and β-tubulin (each ~55,000 mol. wt.).   There are various non-tubulin proteins which are stoichiometrically associated with tubulin in microtubules, so-called microtubule-associated proteins (MAP's), and others such as dynein in the flagellum which interact with specific microtubule structures.   Microtubule biochemistry and cell biology have had several monographs and reviews [1-4] devoted to them.

This article does not aim to be a complete review of the immunological techniques that have been implemented in the course of microtubule research. Accordingly, there is no mention of techniques such as quantitative immunoassay (RIA, ELISA, etc.) or immunoaffinity purification, for which there are published examples and standard methodological approaches have been employed. The focus is therefore on those techniques that have been particularly informative in microtubule laboratories in the past and which continue to pass from strength to strength – IF, immunoprecipitation, immunoblotting – and on the more recent applications of MAb's in particular, which can be used to probe the microtubule proteins with unique sensitivity.

## IMMUNOFLUORESCENCE (IF) TECHNIQUES

IF techniques have played a major part in elucidating the nature of the microtubule network found within cells. Being of 25 nm diam., individual microtubules are not resolved under the light microscope. Transmission electron microscopy (TEM) requires ultrathin sections and so produces a selective impression of a cell's microtubules. Three-dimensional information from TEM about structures containing microtubules can be achieved by skilful serial sectioning, or more recently by the use of high-voltage electron microscopy, studying mounts of complete cell cytoskeletons.   However, owing to the high level of specialization of these two techniques, such investigations have been rather few.   A major advantage of fluorescence microscopy is that it can resolve the complete microtubular content of an individual cell.

The basic indirect-IF technique involves fixation of the cell, permeabilization and reaction with the anti-tubulin Ab's (say a mouse monoclonal) [5].   The cells are then washed and  reacted with a second Ab that can recognize the first Ab and itself carries a fluorescent tag, say a goat anti-mouse Ab labelled with fluorescein isothiocyanate (FITC) or rhodamine.   The material is then mounted and viewed under good quality optics in a microscope fitted with UV epi-fluorescence illumination. Related procedures using a second Ab tagged with peroxidase [6] or colloidal gold [7] merely require a good optical microscope.

Fixation and permeabilization procedures may affect the ease with which visualization of microtubules can be achieved (reviewed by Weber & Osborn [8, 9]). Where there are no specific examples in the literature for a particular cell type, it is worth comparing several fixation methods directly on the cell in question. One problem associated with fixation is that of antigenic masking demonstrated recently by Franke with anti-vimentin antisera [10].

IF microscopy has now been used to visualize microtubules in many vertebrate cells in culture [9, 11-13] and in eukaryotric microbes such as *Physarum* [14], *Dictyostelium* [15] and yeasts [16]. The technique has also been extended to examine the microtubules of mammalian cells in tissues [17]. The use of second Ab's tagged with electron-opaque labels has allowed the general technique to be applied at the electron-microscopic level [18, 19]. Double-labelling experiments using Ab's also directed against other cell components have been employed to visualize inter-relationships between microtubules and, for example, other components of the cytoskeleton such as intermediate filaments or actin microfilaments [20, 21]. The separation of spindle microtubules and chromatin during mitosis can be shown dramatically by DNA stains such as 'DAPI' (4,6-amidino-2-phenylindole) or propidium iodide together with microtubule IF [22].

Obtaining anti-tubulin antisera of high enough quality for IF experiments has sometimes been a problem. Tubulins from neuronal sources have been purified sufficiently well to theoretically allow the production of good antisera, but the molecule is rather poorly antigenic, probably owing to its evolutionary structure conservation [23]. Concentration of antisera by purification on a tubulin affinity column has proved to be a necessity for improving titre, sometimes ~50- to 500-fold.

The advent of MAb's has had a marked effect on the ease with which indirect-IF staining may be performed. A suitable ascites antiserum preparation can now be diluted at least 500-fold.

The evolutionary conservation of the tubulin molecule [23] means that MAb's or polyclonal antisera raised against a particular tubulin will often cross-react well with tubulins from different sources [24]. However, this cannot always be taken for granted. In a recent study using *Dictyostelium* the monoclonal anti-tubulin MAb Tu 1 [25] failed to 'illuminate' microtubules [26]. Interestingly, 4 polyclonal antisera also failed similarly; IF was achieved with the anti-yeast α-tubulin YL 1/2 [27], whose sister Ab YOL 1/34 was likewise successfully used in another laboratory [28]. However, in mammalian axons and in *Trypanasoma rhodesiense* YL 1/2 has itself been the cause of differential cytochemical and IF staining. YOL 1/34 stains microtubules along the full length of rat axonal tissue and also within the cerebellum, whilst YL 1/2 staining is restricted to the cerebellar glia [29]. In the case of the trypanosome cell, YL 1/2 stains only flagellar microtubules, the large number of pellicular and cytoplasmic microtubules remaining unstained. In contrast, YOL 1/34 stained the complete complement of microtubules [30].

Such examples of differential IF staining may be seen as anomalies to the operator whose intention is to visualize the full complement of microtubules in a particular cell or tissue. In reality they can yield information on the nature of tubulin and its organization in microtubules; but in practice to obtain the complete picture of microtubule organization in a cell it seems advantageous to assimilate a variety of monoclonals or polyclonal antisera, or at least in the first instance to use one of the more widely acknowledged generally staining monoclonals such as YOL 1/34 [27] or DM 1B [31]. The use of 'discriminating anti-tubulin monoclonals' in the analysis of tubulin polypeptide isotypes is discussed later.

Tissue-cultured cells have figured prominently in microtubule IF investigations. By virtue of the very thin cytoplasmic layer in a well-spread cell, microtubules extending into the cytoplasm are easily visualized against a very low level of background fluorescence. Resolution of individual microtubules is feasible. The classic display is of a large number of individual microtubules radiating out to the cell periphery and often originating from a focal point lying close to the nucleus, the microtubule-organizing centre [8, 12].

The specificity of the staining pattern can be addressed by classical controls and the fact that microtubules are often susceptible to depolymerization by anti-microtubule drugs such as colchicine or nocodazole [32], or physical agents such as cold. Further, treatment with vinblastine produces paracrystalline arrangemnts of tubulin in the cytoplasm which can be visualized by IF microscopy [8].

The IF image is usually photographed using a high-speed low-grain film such as Ilford XP-1 or Agfapan Vario-XL chromogenic films. Photographic recording of microtubule IF images has been simplified by the introduction of anti-fade mounting media. Specimens mounted in polyvinyl alcohol- or glycerol-based buffers (pH 8.5) containing 1 mg/ml *p*-phenylenediamine [9] or n-propyl gallate [33] no longer fade rapidly during microscopic inspection, and can be stored successfully at 4° in the dark for several weeks.

## MICRO-INJECTION OF ANTI-TUBULIN ANTIBODIES INTO CELLS

Being a fairly recently applied technique, the full implications of the micro-injection of anti-tubulin MAb's into live cells have not yet been realized. Wehland et al. recently injected the monoclonal anti-α-tubulin YL 1/2 into Swiss 3T3 cells [34], PtK$_2$ and A 549 (human lung carcinoma) cells [35], and reported that with increasing amounts of Ab an increasing level of microtubule aggregation and bundling occurs. This is apparently not due strictly to polyvalent Ab cross-linking since the distance between the bundled microtubules is ~20 nm, some 8–10 nm greater than bivalent IgG cross-linking. Injecting ~10% of the cell volume of a solution of 2 mg/ml IgG decorated the microtubules without altering saltatory movement and organelle translocation. Higher Ab concentrations inhibited these cellular functions and caused vimentin intermediate filaments to collapse onto the nucleus. Even higher concentrations produced mitotic arrest and dispersion of Golgi bodies.

When the injected Ab was labelled with the fluorochrome rhodamine and the cell was viewed with video-intensification microscopy, fluorescently tagged microtubules were stable for several hours, allowing rearrangements of microtubules to be followed. This experimental evidence is complementary to biochemical and ultrastructural data relating to microtubular participation in intracellular movement and organization of Golgi elements, and of close interaction with intermediate filaments.

A separate study of microtubule/intermediate filament interaction has also employed micro-injection of anti-tubulin monoclonals [31]. Using four specific MAb's and a polyclonal antiserum, two MAb's (DM 1A and DM 3B3) and the serum resulted, when injected, in the collapse of the vimentin filaments from a normal reticulate pattern into aggregates around the nucleus. The intermediate filament collapse takes 90–120 min for completion. The Ab's did not promote a major depolymerization of the microtubules.

From both laboratories [31, 34, 35] it is suggested that this observed collapse of vimentin filaments is caused by the blockading of specific binding sites which are present on the surface of the tubulin molecule and are normally required to maintain the spatial orientation of vimentin fibres. Clearly the use of micro-injected MAb's has great potential, particularly in the field of interactive investigations of the type reported here.


## IDENTIFICATION OF TUBULIN POLYPEPTIDES

Purification methods for tubulin rely at least partially on its ability to polymerize into microtubules *in vitro* [36–38], and amongst other requirements this depends on there being a certain critical concentration of tubulin present. Mammalian brain extracts contain tubulin as 10–12% of the total soluble protein, and brain has been the tissue of choice for tubulin purification for many years. Microtubule research has now spread to many diverse organisms, not least for the reason  that there are excellent experimental models to be found amongst plants and lower eukaryotes. Microtubule purification from these non-mammalian sources has proved considerably more diffi-cult, the major  problem being the very much lower levels of tubulin encountered (reportedly 0.05-0.1% of total soluble protein) [39-42]. Since tubulin is only a minor component in many heterogeneous tissue extracts it can be a difficult job to identify the polypeptides, particularly on 2-D gels.

If the sample is of mammalian or avian origin then preliminary identification can be made on gels on the basis of co-electrophoresed samples of the  respective  neurotubulins.   This is, however, an unsuitable approach for lower eukaryotic and plant proteins which can show marked variation in isoelectric and SDS electrophoretic mobilities [43].   In this case identification can be made on an immunological basis, given the availability of cross-reacting tubulin antisera.   The well-proven technique of immunoprecipitation has now been complemented by the recently introduced methodology of immuno-blotting (Western blotting) [other relevant arts. include #D-1.- *Ed.*].

## IDENTIFICATION BY IMMUNOPRECIPITATION

Immunoprecipitation is of particular value when radiolabelled antigen is used, so allowing a clear distinction to be made between antigen and precipitants. Use of SDS to denature the protein mixture prior to the addition of the antiserum and the presence of a non-ionic detergent during the precipitation are conducive to clear results [44, 45].   To avoid the need for titrating a second, preci-pitating antiserum, it is common to use a specific adsorbent to bring about the final precipitation, either species-specific anti-immunoglobulins or Protein A bound to Sepharose, or whole *Staphylo-coccus aureus* cells which have been heat-killed and formalin-fixed [46, 47]. Immunoprecipitation is of particular value for identifying specific polypeptides in mixtures produced by the *in vitro* transla-tion of mRNA.   Such translation mixes may contain many hundreds of labelled polypeptides, the electrophoretic nature of which may differ from the same native polypeptides should the latter be subject to post-translational modification *in vivo*.

## IDENTIFICATION BY IMMUNOBLOTTING

Methodology for the recently introduced immunoblotting technique [48, 49] is already diverse, but invariably entails transfer of the gel-electrophoresed proteins, usually electrophoretically, onto a support medium to which they bind.   The advantage of creating a replica on a macroporous support is that the immunoglobulin (Ig) probes can more readily gain access to the proteins on the blot, where protein is on the surface rather than within the pores of a polyacrylamide gel.   Removal of unbound Ig is likewise facilitated.

Antigen-specific Ig's are then located with the aid of a second, Ig-specific antiserum or *S. aureus* Protein A labelled with either horse-radish peroxidase or $^{125}I$ [50, 51]. Visualization is accomplished by incubation with a chromogenic substrate such as 4-chloro-1-naph-thol or, with $^{125}I$, by autoradiography.

Choice of a suitable support matrix is dictated by the ultimate use for which it is intended. Diazotized papers [52] bind proteins covalently, have a high binding capacity and allow Ab's to be removed under denaturing conditions so that subsequently the blot can be re-probed. Charge-modified nylon sheets only bind proteins electro-statically but with sufficient strength to allow multiple re-probing of blots [53]. Unlike the diazo-papers, the protein-binding capacity of the nylon sheet does not decay rapidly nor does the nylon require activation immediately prior to use. Pure nitrocellulose membrane is the most commonly used blot-matrix due to its ease of use and relative inexpensiveness. Binding of proteins is thought to be electrostatic but of a lower strength than with charge-modified nylon; consequently the conditions required to remove Ab from anti-gen also remove the blotted proteins from the nitrocellulose. However, nitrocellulose blots can be re-probed with antisera dif-fering in specificity without removing signals from previous probes.

A further feature of nitrocellulose blots is the ability to stain the membrane for protein. The commonly used stains amido-black, Coomassie Blue, etc., bind directly to both diazotized paper and charge-modified nylon. All blotting techniques incorporate a method for preventing the binding, directly to the membrane, of the Ab's used to probe the blot. Generally this involves incubating the matrix directly after the transfer in a concentrated solution of a non-antigenic protein, commonly BSA, Hb or gelatin. With nitrocel-lulose membrane, however, similar protection can be achieved by con-ducting the immuno-probing in the presence of Tween 20 (0.05% v/v) [54, 55]. This modification allows the same nitrocellulose blot to be post-stained with amido-black or using the more sensitive India-ink method [55]. When using a protein to block nitrocellulose immuno-blots a separate gel must be blotted and subjected to staining.

One particularly good example of the usefulness of Western blot analysis of tubulin is the identification of *Dictyostelium* tubulins. 2-D gels showed no polypeptides migrating with the normal 2-D coor-dinates of most tubulins (pI's 5.3-6.0, mol. wts. 53,000-58,000). Immunoblot screening of such gels with the α-tubulin probes YOL 1/34 and YL 1/2 and a polyclonal anti-tubulin antiserum as a comparative β-tubulin probe demonstrated that the *Dictyostelium* subunits have isoelectric points between 6.2 and 6.7, more basic than anticipated, and 'hidden' amongst many other gel spots [28].

## TUBULIN POLYPEPTIDE ISOTYPES

Tubulin is ubiquitous in all eukaryotes, and in all examples cited so far there are distinct α- and β-subunits. Sequence determinaton has been carried out on the pig brain species [56] and indicates ~40% sequence homology between the α- and β-subunits. It seems likely that α- and β-tubulin polypeptides may have had a common ancestral gene. In most organisms there are multiple genes for both α- and β-tubulin and many identifiable polypeptide variants of tubulin subunits can exist, 17 isotypes being separated by isoelectric focusing of brain tubulin [57]. The proportion of this variation that can be attributed to expression from different genes, or by post-transcriptional and post-translational modification, is unknown. Tissue-specific and cell type-specific tubulins have also been identified [44, 58, 59]. There is also an organelle-specific tubulin variant which is found in flagella of *Chlamydomonas reinhardtii* [60], *Crithidia fasciculata* [61] and *Physarum polycephalum* [44]. The so-called α3-tubulin is the result of a post-translational modification of existing α-cytoplasmic tubulin.

Questions that are now being levelled in tubulin cell biology include whether or not the multiple isotypes do represent multi-function subsets: how are they distributed within the microtubules within the cell? Alongside biochemical and molecular genetic surveys, investigations using MAb's can be highly informative.

## INVESTIGATIVE IMMUNOLOGICAL APPROACH

MAb's are renowned for their exquisite specificity and their very defined epitope binding requirements. To date the most closely investigated anti-tubulin is the Ab YL 1/2. The epitope which it recognizes is part of the carboxy-terminal end of the α-tubulin molecule [34]. At its synthesis tubulin is endowed with a tyrosine residue at its carboxy-terminus; but in brain tissues particularly, specific enzymatic cleavage removes this terminal residue. The tyrosine can be added back *in vivo* enzymatically. YOL 1/2 binding is reliant on the carboxy-terminal residue having an aliphatic side group [62]; consequently it binds only to the tyrosylated subspecies of α-tubulin. Other specific residues extending back from the terminus have been identified as important for optimal binding.

Since YL 1/2 binds to the surface of native microtubules it is clear that the carboxyl terminus of the α subunit must exist on the outside of the molecule. The above-mentioned apparently anomalous cytochemical staining of rat cerebellum [29] indicates that tubulin de-tyrosylation is selectively performed in the axons of developing rat brain. Whether or not the differential staining of trypanosome pellicular and flagellar microtubules [30] is due to tyrosylation/de-tyrosylation remains to be determined.

Certainly the flagellum-specific α-tubulin is not modified by tyrosylation alone, and reportedly the addition of [³H]acetate in *Chlamydomonas* points to acetylation being an important feature in the production of α3 [63]. Whether tyrosylation is also an issue is unknown. A second pair of anti-α-tubulin monoclonals 3F3 and 16D3 also differentiate between the trypanosome flagellum - stained only by 3F3 - and pellicle [64]. The nature of the differentiation by these Ab's is also unknown.

Another monoclonal anti-α-tubulin which seems to differentiate a subset of microtubules is the Ab 1-6.1 [65]. This does not illuminate the full complement of cytoplasmic microtubules in IF with mouse and human fibroblasts that can be detected when a polyclonal antiserum· is used to double-stain the same cell [66]. It does, however, bind to the very small subset of colcemid-resistant microtubules. The implication is that Ab 1-6.1-positive microtubules contain a colchicine-resistant α-tubulin and that where a microtubule contains a high proportion of this subunit, it is more resistant to the drug-induced depolymerization. As yet no correlation of this Ab has been made with isotype specificity, but it will be of interest to see whether subset definition is found.

We have generated a MAb to *Physarum* tubulin, KMP-1, which is demonstrably specific for the slime mould α-tubulin [67]. In Western blot analysis of total protein homogenates of an evolutionally diverse series of organisms, KMP-1 binds only to *Physarum* tubulin. A sister MAb KMX-1, which is β-specific, demonstrates the normal cross-species reactivity normally experienced with anti-tubulin Ab's (Fig. 1). KMP-1 also clearly demonstrates α-tubulin isotype specificity. *Physarum* α1-tubulin is found in the myxamoeba, but peptides of similar electrophoretic mobility are expressed on the plasmodium which also expresses a novel α2 isotype [41, 44]. KMP-1 binds to the amoebal α1-tubulin but recognizes only a subset of the plasmodial α1 polypeptides and does not recognize the α2-tubulin isotype. Thus the KMP-1 MAb represents an Ab that is able to discriminate not only between the α-tubulins of evolutionarily distinct organisms but even between the α-tubulin isotypes expressed within one cell. Ab's of this type should prove extremely useful in studies of tubulin polypeptide evolution and of the significance of expression of multiple tubulins within one cell.

## IN CONCLUSION

All of the techniques described here are standard practice in any laboratory which uses Ab's. What is important in the case of the anti-tubulins is the way in which recent developments have facilitated the use of immunological reagents. Any researcher whose standard immunogen is naturally highly antigenic cannot realize the frustration involved in obtaining enough affinity-purified anti-tubulin Ig to perform a protracted series of IF experiments. The

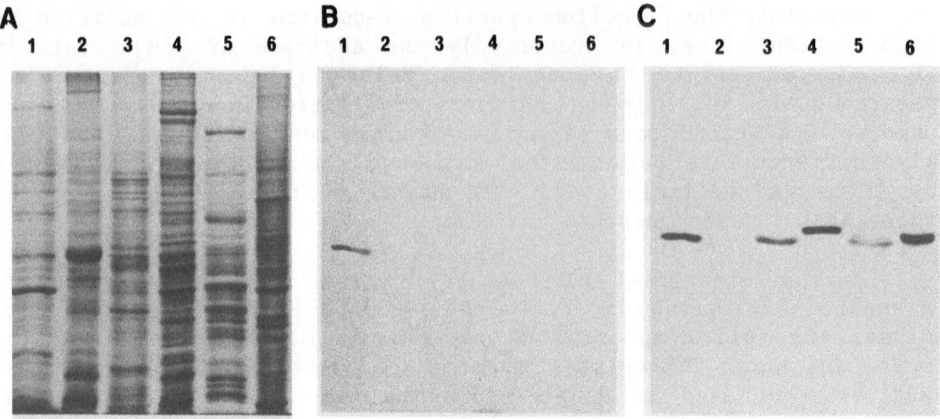

Fig. 1. Slab polyacrylamide gel (**A**), and two replica immunoblots which have been probed with the MAb's KMP-1 (**B**)  [anti-α-tubulin], and KMX-1 (**C**) [anti-β-tubulin].  The gel tracks correspond to homo-genates of (1), the slime mould *Physarum polycephalum* myxamoebae; (2) *Chlamydomonas reinhardtii* flagellates;  (3) *Phaseolus vulgaris* root tips;  (4) the nematode *Caenorhabditis elegans*; (5) rat liver 100,000 *g* supernatant; (6) Vero cells (monkey kidney-cell line). The blotting procedure utilized detergent rather than protein to block the membrane; the second Ab was a peroxidase-conjugate using 4-chloro-1-naphthol as the substrate.

developments of recent years have brought about major  improvements in this area alone. Firstly the design of more fruitful immunization protocols, with cross-linking, proteolytic cleavage of the antigen, etc., has resulted in polyclonal antisera of greater titre; secondly MAb's have, more recently, become available.  Even if one's sole intent is to visualize microtubules *in situ*, monoclonals have allowed greater freedom in experimental design, have increased  resolution in some systems and have opened that avenue of investigation to a number of laboratories where it was impracticable before.

The greatest asset of MAb's, however, is their specificity of binding, which promises to be a most important tool in the tubulin field, as in so many others, and through which they can fulfil a much needed role of  intracellular labelling in the field of cell biology.  The greatest input of data will come from those Ab's where it is possible to define the binding site, either definitively as in the case of YL 1/2, or relatively, say by peptide mapping, as in the case of DM 1A and DM 3B3 and their sister Ab's.

Because the tubulins collectively belong to a multi-protein family in most organisms, there is considerable interest in whether this may confer tissue-specific or even organelle-specific usage.

Interest is all the greater because modification or mutation of tubulin can alter its drug-binding capability.  Since anti-microtubule drugs are used in medical, veterinary and agrochemical spheres there is an ongoing interest both in designing species-specific antagonists and in the extent to which a microtubule poison may cross-react between species.  A more complete understanding of the drug-binding sites will perhaps draw on the results to be obtained from anti-Id Ab's.

As with the polypeptides the tubulin genes are a multi-membered family in many organisms, and there is an obvious interest in the control of expression of such.    In certain cases it should be possible by using isotype-specific Ab's to relate polypeptide-isotype expression to gene expression.  The field of molecular genetics has already provided a number of tubulin gene sequences [68-70] from which both highly conserved and highly specific putative oligopeptides can be identified.  Synthesis of these peptides and subsequently raising MAb's to them is a major way forward in immunology today.  It may have taken longer for immunological techniques to become widely established in microtubule laboratories;   but now they are so useful that their  continued use is assured and likely to be expanded widely.

*Acknowledgements*

Work in this laboratory was supported by grants from the Science & Engineering Research Council, the Agricultural & Food Research Council and the Leukaemia Research Fund.

*References*

1.  Dustin, P. (1978) *Microtubules*, Springer, New York, 452 pp.
2.  Roberts, K. & Hyams, J., eds. (1979) *Microtubules*, Academic Press, London, 595 pp.
3.  Gozes, I. & Littauer, U.Z. (1982) *Scand. J. Immunol. 15*, 299-316.
4.  McKeithan, T.W. & Rosenbaum, J.L. (1984) in *Cell and Muscle Motility*, *Vol. 5* (Shay, J.W., ed.), Plenum Publishing Co., New York, pp. 255-288.
5.  Weber, K., Rathke, P.C. & Osborne, M. (1978) *Proc. Nat. Acad. Sci. 75*,  1820-1824.
6.  Gu, J., de May, J., Moermans, M. & Polak, J. (1981) *Regulatory Peptides 1*,  365-374.
7.  de May, J., Moermans, M., Geuens, G., Nuydens, R. & de Brabander, M. (1981)  *Cell Biol. Int. Rep. 5*, 889-899.
8.  Weber, K. (1976) in  *Cell Motility, book A*  (Goldman, R., Pollard, T. & Rosenbaum, J., eds.), Cold Spring Harbor Press, New York, pp. 403-417.
9.  Osborn, M. & Weber, K. (1982) in *Methods in Cell Biology, Vol. 24* (Wilson, L., ed.), Academic Press, New York, pp. 98-132.
10. Franke, W.W., Schmid, E., Wellsteed, J., Grund, C., Gigi, O. & Geiger, G. (1983) *J. Cell Biol. 97*, 1255-1260.

11. Havercroft, J.C., Quinlan, R.A. & Gull, K. (1981) *J. Cell Sci.* *49*, 195-204.
12. Karsenti, E., Guilbert, B., Bornens, M., Avrameas, S., Whalen, R. & Pantaloni, D. (1978) *J. Histochem. Cytochem. 26*, 934-947.
13. Virtanen, I., Lehto, V.-P. & Lehtonen, E. (1980) *Eur. J. Cell Biol. 23*, 80-84.
14. Havercroft, J.C. & Gull, K. (1983) *Eur. J. Cell Biol. 32*, 67-74.
15. Cappuccinelli, P., Rubino, S., Fighetti, M. & Unger, E. (1982) in *Microtubules in Microorganisms* (Cappuccinelli, P. & Morris, R.N., eds.), Dekker, New York, pp. 71-98.
16. Kilmartin, J.V. & Adams, A.E.M. (1984) *J. Cell Biol. 98*, 922-933.
17. Wolosewick, J. & de May, J. (1981) *Biol. Cellul. 44*, 85-88.
18. Bendayan, M. (1984) *J. Electron Micros. Technique 1*, 243-270.
19. Armbruster, B.L., Garavito, R.M. & Kellenberger, E. (1983) *J. Histochem. Cytochem. 31*, 1380-1384.
20. Blose, S.H. (1981) *Cell Motility 1*, 417-431.
21. Fujiwara, K. & Pollard, T. (1978) *J. Cell Biol. 77*, 182-195.
22. Brinkley, B.R. & Cox, S.M. (1978) *Stain Technol. 53*, 345-349.
23. Little, M., Krauhs, E. & Ponstingl, H. (1981) *Biosystems 14*, 239-251.
24. Piperno, G. & Luck, D.J.L. (1977) *J. Biol. Chem. 252*, 383-391.
25. Viklicky, V., Draber, P., Hasek, J. & Bartek, J. (1982) *Cell Biol. Int. Rep. 6*, 725-731.
26. Roos, U.-P., de Brabander, M. & de Mey, J. (1984) *Exp. Cell Res. 151*, 183-193.
27. Kilmartin, J.V., Wright, B. & Milstein, C. (1982) *J. Cell Biol. 93*, 576-582.
28. White, E., Tolbert, E.M. & Katz, E.R. (1983) *J. Cell Biol. 97*, 1011-1019.
29. Cumming, R., Burgoyne, R.D. & Lytton, N.A. (1983) *Eur. J. Cell Biol. 31*, 241-248.
30. Cumming, R. & Williamson, J. (1984) *Cell Biol. Int. Rep. 8*, 2.
31. Blose, S.H., Meltzer, D.I. & Ferasmisco, J.R. (1984) *J. Cell Biol. 98*, 847-858.
32. de Brabander, M. Geuens, G., Nuydens, R., Willebrords, R. & de Mey, J. (1980) in *Microtubules and Microtubule Inhibitors 1980* (de Brabander, M. & de Mey, J., eds.), Elsevier/North-Holland, Amsterdam, pp. 255-268.
33. Giloh, H. & Sedat, J.W. (1982) *Science 217R*, 1252-1255.
34. Wehland, J., Willingham, M.C. & Sandoval, I.V. (1983) *J. Cell Biol. 97*, 1467-1475.
35. Wehland, J. & Willingham, M.C. (1983) *J. Cell Biol. 97*, 1476-1490.
36. Shelanski, M.L., Gasken, F. & Cantor, C.R. (1973) *Proc. Nat. Acad. Sci. 70*, 765-768.
37. Roobol, A., Pogson, C.I. & Gull, K. (1980) *Exp. Cell Res. 130*, 203-215.

38. Kilmartin, J.V. (1981) *Biochemistry 20*, 3629-3633.
39. Barnes, L.D. & Robertson, G. (1979) *Arch. Biochem. Biophys. 196*, 511-524.
40. Morejohn, L.C. & Foskett, D.E. (1982) *Nature 297*, 426-428.
41. Roobol, A., Wilcox, M., Paul, E.C.A., & Gull, K. (1984) *Eur. J. Cell Biol. 33*, 24-28.
42. Dawson, P.J., Gutteridge, W.E. & Gull, K. (1983) *Mol. Biochem. Parasitol. 7*, 267-275.
43. Clayton, L., Quinlan, R.A., Pogson, C.I. & Gull, K. (1980) *FEBS Lett. 115*, 301-305.
44. Burland, T.G., Gull, K., Sched , T., Boston, R.S. & Dove, W.F. (1983) *J. Cell Biol. 97*, 1852-1859.
45. Mose-Larsen, P., Bravo, R., Fey, S.J., Small, J.V. & Celis, J.E. (1982) *Cell 31*, 681-692.
46. Blose, S.H. & Meltzer, D.I. (1981) *Exp. Cell Res. 135*, 299-309.
47. Surolia, A., Pain, D. & Khan, M.I. (1982) *Trends Biochem. Sci. 7*, 74-76.
48. Towbin, H., Staehlin, T. & Gordon, J. (1979) *Proc. Nat. Acad. Sci. 76*, 4350-4354.
49. Burnette, W.N. (1981) *Anal. Biochem. 112*, 195-203.
50. Symmington, J., Green, M. & Brackman, K. (1981) *Proc. Nat. Acad. Sci. 78*, 177-181.
51. Gershoni, J.M., Hawrot, E. & Lentz, T.L. (1983) *Proc. Nat. Acad. Sci. 76*, 4973-4977.
52. Bittner, M., Kupferer, P. & Morris, C.F. (1980) *Anal. Biochem. 102*, 459-471.
53. Gershoni, J.M. & Palade, G.E. (1983) *Anal. Biochem. 131*, 1-15.
54. de Blas, A.L. & Cherwinski, H.M. (1983) *Anal. Biochem. 133*, 214-219.
55. Hancock, K. & Tsang, V.C.W. (1983) *Anal. Biochem. 133*, 157-162.
56. Krauhs, E., Little, M., Kempf, T., Hofer-Warbinek, R., Ade, W. & Ponstingl, H. (1981) *Proc. Nat. Acad. Sci. 78*, 4156-4160.
57. George, H.J., Misra, L., Field, D.J. & Lee, J.C. (1981) *Biochemistry 20*, 2402-2409.
58. Bibring, T., Baxandall, J., Denslow, S. & Walker, B. (1976) *J. Cell Biol. 69*, 301-312.
59. Kemphues, K.L., Kaufman, T.C., Raff, R.A. & Raff, E.C. (1982) *Cell 31*, 655-670.
60. Lefebvre, P.A., Silflow, C.D., Weiben, E.D. & Rosenbaum, J.L. (1980) *Cell 20*, 469-477.
61. Russell, D.G., Miller, D. & Gull, K. (1984) *Mol. Cell. Biol. 4*, 779-790.
62. Wehland, J., Schroder, H. & Weber, K. (1984) *EMBO J. 3*, 1295-1300.
63. L'Hernault, S.W. & Rosenbaum, J.L. (1983) *J. Cell Biol. 97*, 256-263.
64. Gallo, J.-M. & Anderton, B.H. (1983) *EMBO J. 2*, 479-483.
65. Asai, D.J., Brokaw, C.J., Thompson, W.C. & Wilson, L. (1982) *Cell Motility 2*, 599-614.
66. Thompson, W.C., Asai, D.J. & Carney, D.H. (1984) *J. Cell Biol. 98*, 1017-1025.

67. Birkett, C.R., Johnson, L. & Gull, K. (1985) *FEBS Lett.*,
    submitted.
68. Velenzuela, P., Quiroga, M., Zaldivar, J., Rutter, W.J.,
    Kirschner, M.W. & Cleveland, D.W. (1981) *Nature 289*, 650-655.
69. Neff, N.F., Thomas, J.H., Grisafi, P. & Botstein, D. (1983)
    *Cell 33*,  211-219.
70. Lee, M.G., Lewis, S.A., Wilde, C.D. & Cowan, N.J. (1983) *Cell
    33*, 477-487.

#C-2

# CELL–SURFACE EVENTS DURING MYOGENESIS:
# IMMUNOFLUORESCENCE ANALYSIS USING MONOCLONAL ANTIBODIES

**R.F. Foster and S.J. Kaufman**

Department of Microbiology
University of Illinois
131 Burrill Hall, 407 South Goodwin
Urbana, IL 61801, U.S.A.

*We have raised monoclonal antibodies (MAb's) reactive with $L_8E_{63}$ myoblasts to study the events on the cell surface that accompany skeletal myogenesis. The dynamic state of the myoblast membrane has been described using indirect immunofluorescence with microphotometry, and developmentally defective mutants examined for antigenic expression relative to $L_8E_{63}$ myoblasts. One antibody (Ab), H36, is highly selective for skeletal muscle and is not bound by non-fusing mutant myoblasts. Modulation of differentiation alters expression of H36 antigen.*

During skeletal muscle differentiation, mononucleate myoblasts replicate, undergo distinct morphological and biochemical changes, and fuse to form multinucleate fibres (for recent monographs see [1, 2]). Cultured myoblasts faithfully mimic *in vivo* development and have been employed as a model in studies of myogenesis. As they undergo differentiation, myoblast cultures contain cells at many different stages of development. While approaches such as biochemical analysis, radioisotopic labelling and radioimmunoassays can provide information about the developing cell culture as a whole, they do not distinguish between cells at stages that occur concurrently and they may fail entirely to reflect transient alterations which, if rapid, will be represented by only a small number of cells in the population. Techniques applicable in using MAb's to study the myoblast surface include immunofluorescence (IF) microscopy [3-5]. Its particular value is that, combined with phase-contrast microscopy, it allows determinations on individual cells at identifiable developmental stages and provides information about spatial distribution of the antigens. IF also permits identification of myogenic and non-myogenic cells within primary cultures of muscle tissue.

## PRODUCTION OF MONOCLONAL ANTIBODIES

Following Kennett's method [3], Sp2/0-Ag14 cells were fused with spleen cells from Balb/cV mice immunized with $L_8E_{63}$ rat myoblasts or myotubes.  Hybridoma culture fluids were screened by indirect IF for Ab reactive with the surface of live $E_{63}$ myoblasts and myotubes. Positive hybridomas were cloned in 0.24% agarose containing 50% conditioned medium.  Culture fluid was collected from Ab-producing clones in log phase of growth, supplemented to 0.02% with sodium azide and stored at -20°.

## PREPARATION OF SAMPLES FOR IMMUNOFLUORESCENCE (IF) MICROSCOPY

Ab reactions were carried out on live cells.  Samples were generally prepared at 5°.  At this temperature antigen distribution was not perceptibly different from that on cells fixed with fresh 1% formaldehyde;  the live-cell assay was chosen because some determinants appeared to be partially extracted by formaldehyde.

Cells were grown on acetone-washed, gelatin-coated 18 mm  No. 1 cover-slips in 60 mm plastic dishes.  $E_{63}$ cells were seeded at $2.75 \times 10^5$ cells per dish.  After at least 48 h of growth the monolayers were washed 3 times with serum-free medium, and 30-40 µl of hybridoma culture fluid buffered with 10 mM Hepes, pH 7.3, was placed on each cover-slip.  Culture fluid from Sp2/0-Ag14 was used as a negative control.  After 30 min the cells were rinsed briefly and washed 4 times, each 10 min, with Dulbecco's phosphate-buffered saline (DPBS).  The cells were then incubated with fluorescein-conjugated purified rabbit anti-mouse immunoglobulin (FITC-RAMIg). Prior to the assay, to minimize non-specific binding, commercial FITC-RAMIg (Miles Labs.)  was mixed with 5 vol. of the same horse serum as used in the cell growth-medium, for 60 min at 5°, then brought to a working dilution of 1:30 with DPBS.  To each cover-slip 40 µl of this dilution was applied.  After 30 min, the cells were washed as described above, then fixed with 95% (v/v) ethanol, rinsed twice with DPBS buffer and finally inverted over a  mounting medium (glycerol with buffer; see Fig. 1 legend) containing 10  mg/ml  p-phenylenediamine (PPD) to retard photobleaching during examination [6].    Cover-slips were sealed with Flotexx (Lerner Labs.) and stored overnight at 5°.

## MICROSCOPY AND PHOTOMETRY

Cell preparations were examined with a Zeiss Universal microscope equipped  for fluorescence excitation with epi-illumination optics and HB 100 mercury lamp. Filters used for FITC were BP 485/20 with FT510 beam splitter for excitation, and BP 520-560 barrier filter.  Fluorescence intensity was measured either with a Jena photomultiplier and 110 EMI photometer or with a Zeiss MPM-03 computer-controlled photometer.  Computer-operated shutters limit  excitation to the period of measurement (for data given here, 410 msec). An automatic field stop restricts the sample area excited, minimizing photobleaching. With the Jena photometer, the presence of PPD in the sample retarded bleaching sufficiently to allow satisfactory measurements to be made (Fig. 1). Using fluorescein and samples of FITC-RAMIg similar in thickness to that of cell preparations, the response of both photometers was determined to be linear within the range of intensities measured.

## DEVELOPMENTAL SPECIFICITY OF MONOCLONAL ANTIBODIES

The expression of 24 antigenic determinants was observed in detail on $E_{63}$ cells grown for 2-8 days, enabling us to group the determinants into 6 broad classes according to their expression in relation to development (Fig. 2). The estimated relative intensities indicated reflect the fluorescence of the whole cell surface, taking into account, for example, local accumulations at the cell margin and between neighbouring cells.

We have described development-related changes in topographic distribution of many antigens [7]. We applied microphotometry with IF to demonstrate more objectively the different profiles of expression depicted schematically in Fig. 2. Absolute quantitation of antigens by IF is problematic at best, due to the difficulty of

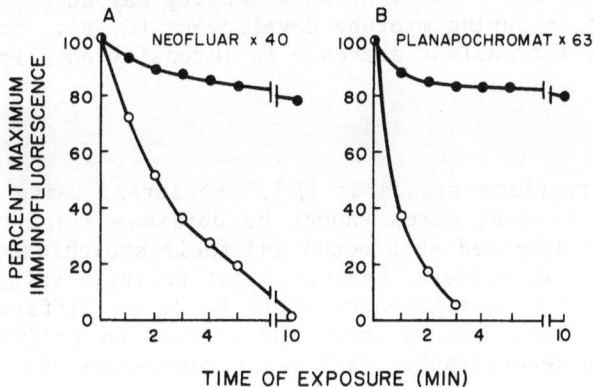

TIME OF EXPOSURE (MIN)

**Fig. 1.** Effect of PPD on photobleaching. E63 myoblasts were grown on glass cover-slips, reacted with H58 MAb and then with FITC-RAMIg. The cells were then fixed with 95% ethanol and mounted in either PBS/glycerol, 1:9 (o——o) or in PBS/glycerol with PPD to 0.01 M (•——•); pH 8.5 in both cases. Photometric measurements were made with a Jena photometer mounted on a Zeiss Universal microscope with 40X neofluar (**A**) and 63X planachromat (**B**) objectives. Storage at 4° for 7 days did not diminish the effectiveness of PPD in inhibiting photobleaching (data not shown).

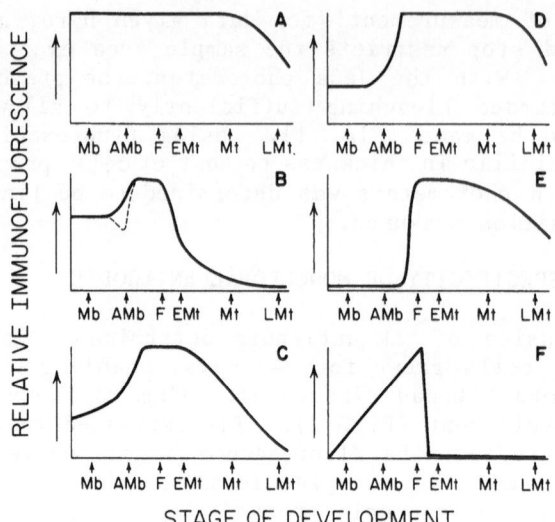

**Fig. 2.** Schematic representation of relative amounts of antigens on developing myoblasts. MAb's reactive with surface antigens were used to study changes on the myoblast membrane. Indirect IF analysis of replicating [Mb], aligned [A] and fusing [F] myoblasts, and early [EMt], mature [Mt] and late [LMt] myotubes was compared. IF intensity may be similar throughout development (**A**), may be greater on mononucleate cells than on myotubes (**B, F**) or may increase with development (**D, E**). A decline in intensity may accompany fusion (**B**) or may occur during myotube development (**C, E**). Some Class B antigens show a transient decrease in intensity on aligned cells (**B, ----**).

obtaining appropriate standards [8]. Moreover, using indirect IF, the number of binding sites cannot be obtained since the number of fluorochrome-conjugated Ab's bound and their stoichiometry with each MAb is unknown. We believe, however, that relative values do reflect the quantity of an antigen accessible to Ab at different stages of development. Thus, Fig. 3 shows the expression profile of antigen H36, a Class D determinant. Each point represents the mean of 30 or more readings on individual cells. The field measured fell within the boundary of the cell and does not reflect IF at the periphery. The increase in H36 IF observed on aligned cells (Fig. 2D) occurs largely at the cell margin and thus is not reflected in Fig. 3.

## CELL SPECIFICITY OF H36

Antigen H36 is of particular interest because its expression is is restricted to muscle cells [9]. It is present on primary skeletal rat myoblasts and rat cardiac myocytes besides $E_{63}$ cells, but it

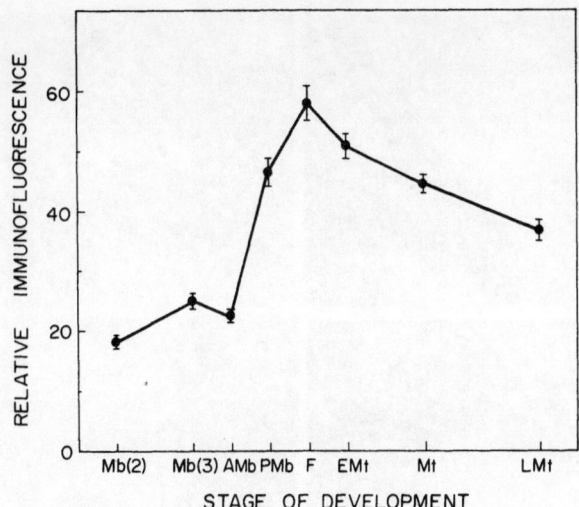

**Fig. 3.** Quantitation of antigen H36 by IF photometry. $E_{63}$ myo-blasts were grown for 2-8 days and processed simultaneously for indirect IF; intensity was measured as for Fig. **1A**. Each point is the mean of at least 30 measurements, ± s.e. (Negative control values were subtracted.) Cells were selected by phase-contrast microscopy: 2-, 3- and 4-day myoblasts, Mb(2) etc.; aligned myo-blasts, AMb; fusing cells, F; early myotubes, EMt; myotubes, Mt; and late myotubes, LMt. Pre-fusion myoblasts, PMb, were identified by their characteristic fluorescence. Estimated diam. of area measured: 20-22 μm.

is lacking on rat-muscle fibroblasts and cardiac endothelial cells (Fig. 4) and on primary cultures and cell lines of other tissue origins, including smooth muscle. The morphologically distinct cell types exhibiting fluorescence, as shown in Fig. 4, were readily dis-cerned in growing cultures, and their myogenic nature confirmed: in the case of skeletal muscle, by subsequent fusion to form myotubes, and in the case of cardiac myocytes by rhythmic contraction.

## MODULATION OF EXPRESSION OF H36

Photometry allows detailed, objective analysis of variations in IF intensity, and helps overcome the difficulties inherent in recor-ding and comparing results based on subjective visual observation. In addition, computer-assisted photometry permits one to store, pro-cess and display values in a visual format; this facilitates, for example, comparison between two populations of cells each of which exhibits widely varying intensities.

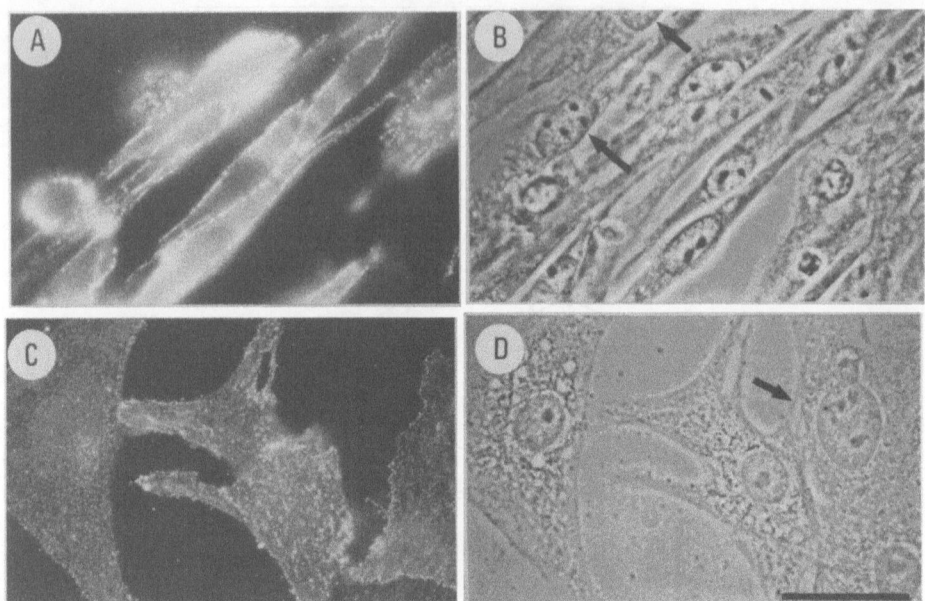

Fig. 4. Specificity of H36 antigen. Primary cultures of newborn-
rat hind limb (**A, B**) and heart ventricle (**C, D**) were stained for
indirect IF using H36 Ab. H36 is observed on myoblasts (**A**) but not
on fibroblasts seen by phase-contrast microscopy (**B**, *arrows*).
Cardiac myocytes also exhibit IF (**C**), but endothelial cells in the
same culture do not stain (**D**, *arrow*). 63x planapochromat. Bar = 30 μm.


H36 increases prior to fusion and is abundant on myotubes. Conditions that modulate fusion, e.g. low-Ca medium or the presence of
30 μM BrdUrd, alter the expression of H36 [9]. When $E_{63}$ cells were
grown in medium containing 90 μM $Ca^{2+}$, myotubes did not form and
myoblasts did not exhibit the increased expression of H36 that precedes fusion in 1.8 mM $Ca^{2+}$ (Fig. 5). This is reflected by the more
limited range of intensities recorded in Fig. 5C.

Fig. 5 (*opposite*). Effect of reduced $Ca^{2+}$ on expression of H36
antigen. $E_{63}$ cells were grown for 4 days in medium containing
1.8 mM $Ca^{2+}$ (**A**), or were grown similarly and then for 2 days in
fresh medium containing 1.8 mM (**B**) or 90 μM (**C**) $Ca^{2+}$. Live cells
were prepared simultaneously for IF with MAb H36. IF intensities
of mononucleate cells were measured, and the distribution of the
values plotted for each population. Cultures grown with 1.8 mM
$Ca^{2+}$ exhibit more bright cells at 6 days when myotubes are forming
(**B**) than at 4 days (**A**). This increased expression of H36 does not
occur in cultures switched to 90 μM $Ca^{2+}$ (**C**), nor do myotubes form.
Negative control values (not subtracted) were <1.5%.

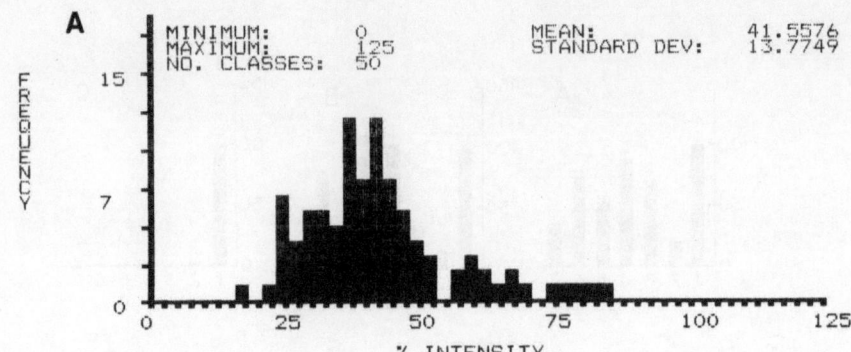

DESCRIPTION OF CON4    FILE
H36IF

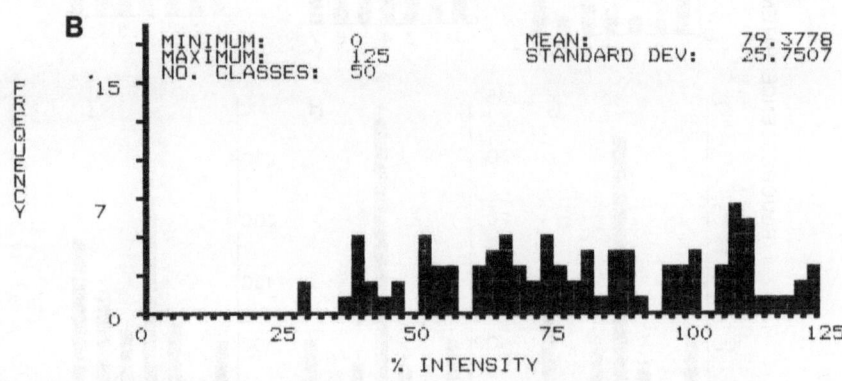

DESCRIPTION OF CON6    FILE
H36IF

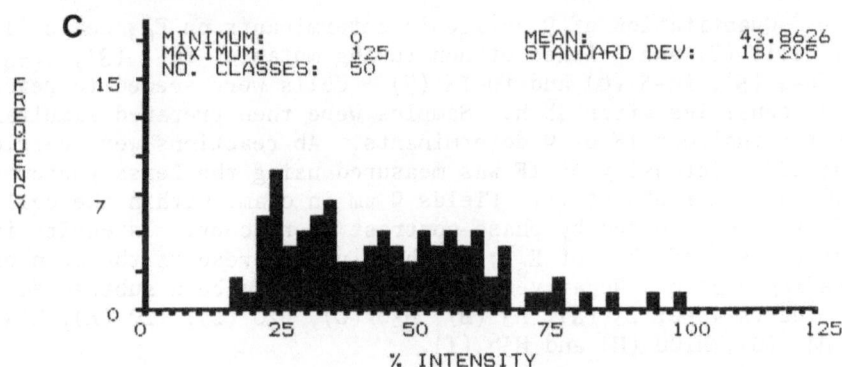

DESCRIPTION OF LOC6    FILE
H36IF

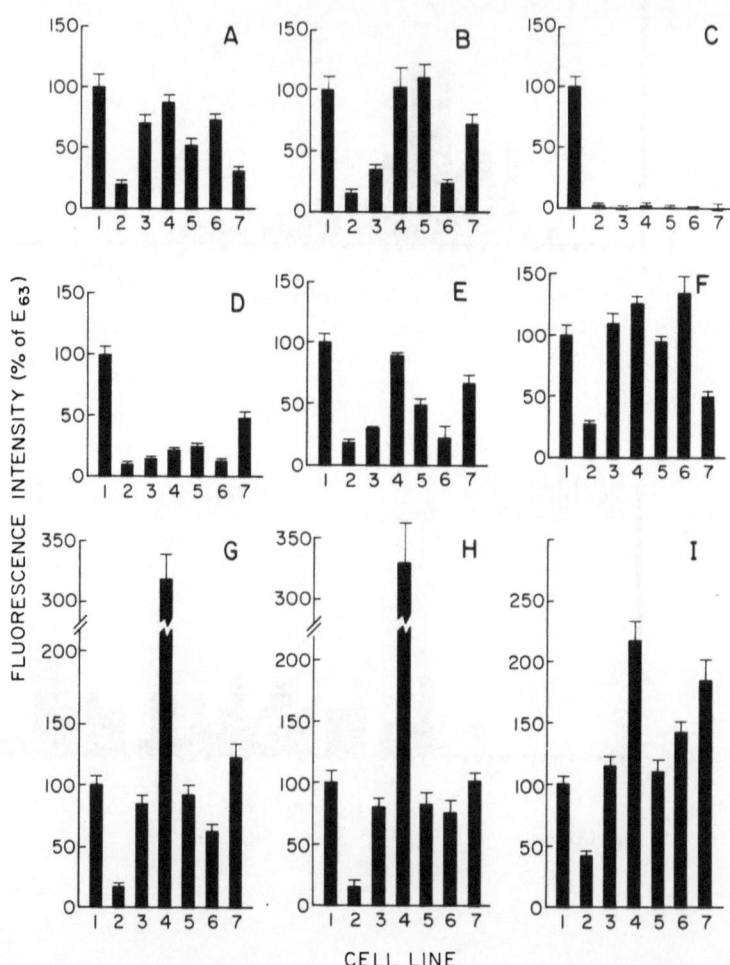

**Fig. 6.** Quantitation of 9 antigenic determinants on $E_{63}$ cells (**1**), Rat-1 cells (**2**) and 5 lines of non-fusing mutants: fu-1 (**3**), fu-3 (**4**), fu-4 (**5**), fu-5 (**6**) and fu-15 (**7**). Cells were seeded to reach similar densities after 48 h. Samples were then prepared simultaneously for indirect IF of 9 determinants. Ab reactions were carried out at 24°. Intensity of IF was measured using the Zeiss photometer and 40× neofluar objective. Fields 9 μm in diam. within the cell boundary were selected by phase-contrast microscopy. Intensity is expressed as % of that of $E_{63}$; each column represents the mean of 30 readings, ±s.e. Negative control values have been subtracted. Determinants were: I3 (**A**), H3 (**B**), H36 (**C**), D20 (**D**), I20 (**E**), H73 (**F**), H48 (**G**), H100 (**H**) and H58 (**I**).

## ANTIGENIC EXPRESSION OF NON–FUSING CELLS

We have examined the expression of H36 and other antigens on developmentally defective, non–fusing (fu-) mutants of $L_8$ isolated in this laboratory [10] and on Rat-1 cells, a line derived from rat fibroblasts.   IF intensity relative to $E_{63}$ cells was measured by photometry (Fig. 6).  H36 was lacking on Rat-1 and on all fu- lines; all determinants were less abundant on Rat-1.   The fu- lines varied in antigenic expression. All fu- lines exhibit, to a varying degree, properties characteristic of neoplastic transformation [10].

## FURTHER APPLICATIONS OF MICROPHOTOMETRY

Along with other applications of MAb's and the ongoing characterization of the antigenic determinants with which they react,  we will continue to use IF to study development-related changes on the cell surface. Further exploration of applications of photometry will allow us to document the expression of determinants, not only on areas representative of the whole cell  but also,  using smaller photometer apertures and more automated methods of data collection [11], to record local changes in topography. Multiple-labelling of the cell surface with MAb's conjugated with different fluorochromes will facilitate observation of the location and re-distribution of specific membrane components relative to one another in the course of development.   By 'mapping' the cell surface in this way for a number  of  determinants,  we  expect  to  demonstrate  further   the dynamic state of the myoblast membrane during myogenesis.

*Acknowledgement*

This work was supported by grant USDHHS GM 28842.

*References*

1.  Pearson, M.L. & Epstein, H.F., eds. (1982) *Muscle Development: Molecular and Cellular Control*, Cold Spring Harbor Laboratory, New York, 581 pp.
2.  Eppenberger, H.M. & Perriard, J.-C., eds. (1984) *Developmental Processes in Normal and Diseased Muscle (Exp. Biol. Med. 9)*, 293 pp.
3.  Lee, H.U. & Kaufman, S.J. (1981) *Devel. Biol. 81*, 81-95.
4.  Kaufman, S.J., Ehrbar, D.M. & Doran, T.I. (1982) *as for 1.*, pp. 281-289.
5.  Kaufman, S.J. & Foster, R.F. (1984) *as for 2.*, 57-62.
6.  Johnson, G.D., Davidson, R.S., McNamee, K.C., Russell, G., Goodwin, D. & Holborrow, E.J. (1982) *J. Immunol. Meths. 55*, 231-242.
7.  Kaufman, S.J. & Foster, R.F. (1985) *Dev. Biol.*, in press.

8.  Haaijman, J.J. (1983) in *Immunohistochemistry* (Cuello, A.C., ed.), Wiley, Chichester, pp. 47-85.
9.  Kaufman, S.J., Foster, R.R., Haye, K.R. & Faiman, L.E. (1985) *J. Cell Biol.*, in press.
10. Kaufman, S.J., Parks, C.M., Bohn, J. & Faiman, L.E. (1980) *Exp. Cell Res. 125*, 333-349.
11. Ploem, J.S. (1982) in *Immunofluorescence Technology* (Wick, G., Traill, K.N. & Schauenstein, K., eds.), Elsevier Medical, Amsterdam, pp. 73-94.

#C-3

# DIFFERENTIAL EXPRESSION OF CELL-SURFACE ANTIGENS ON MUSCLE SATELLITE CELLS AND MYOBLASTS

Frank S. Walsh, Stephen E. Moore and Ramesh Nayak

Molecular Neurobiology Laboratory
Institute of Neurology
Queen Square, London WC1 3BG, U.K.

*The differentiation of mononucleate myoblast cells into post-mitotic multinucleated myotubes is accompanied by the activation and repression of many gene families. We have produced a panel of antibodies (Ab's) that define muscle differentiation antigens in an attempt to produce antigenic markers for each of the main muscle-cell lineages. To determine whether the myoblast lineage can be subdivided further, four Ab's that react with myoblasts were studied in detail: two monoclonal antibodies (MAb's), viz. 24.1D5 and 5.1H11, and rabbit anti-Thy-1 and rabbit anti-N-CAM. The chosen test systems were cell cultures of human muscle that contain myoblasts, fibroblasts and myotubes and cryostat sections of control and diseased human muscle.*

*All four Ab's reacted with myoblasts in culture, but only anti-Thy-1 reacted with fibroblasts, whereas MAb 5.1H11 and similarly anti-N-CAM reacted with both myoblasts and myotubes but not fibroblasts. The most specific reagent found was MAb 24.1D5 which reacts only with myoblasts. Cryostat sections of normal and diseased muscle were used to characterize these reactivities further. Only MAb 24.1D5 showed activity towards adult skeletal muscle, restricted to the small satellite cells that reside beneath the basal lamina of myofibres. Myopathic muscle that contained histochemically identifiable regenerating fibres was reactive with MAb 5.1H11 and anti-N-CAM only. Muscle satellite cells may be the source of myoblasts for regeneration in muscle injury. The present demonstration that satellite cells are antigenically distinct from myoblasts is novel. Future work will determine how and when satellite cells activate the synthesis of myoblast antigens when placed in cell culture.*

The development of skeletal muscle both *in vitro* and *in vivo* has been shown to involve the formation of multinucleated myotubes from

mononucleated myoblast cells [1]. Myoblast cells, in cultures initiated from enzymatically dissociated skeletal muscle, are thought to
be derived from satellite cells which reside within the muscle basement mebrane but outside of the sarcolemma [2]. Growing muscle and
slow, in comparison with fast, muscle have numerous satellite cells
associated with them [3-5]. These observations indirectly support
the proposition that satellite cells represent a reserve pool of
myogenic stem cells. However, there is more direct evidence of the
myogenic competence of satellite cells. Isolated rat myofibres
having intact basement membranes and satellite cells were shown to
be capable of regenerating after traumatization *in vitro*; furthermore, traumatization of myofibres not associated with satellite
cells resulted not in regeneration but in degeneration of the fibre.
In addition, killing of myonuclei with Marcaine or by cutting the
fibre did not prevent the appearance of mononucleated myogenic cells;
but cytosine arabinoside treatment ablated fibre regeneration [6, 7].
These results not only demonstrate the myogenic potential of satellite cells but also indicate a regenerative role in response to
trauma and muscle fibre necrosis.

The nature of the satellite cell with regard to its placement
in the muscle lineage remains unclear. It has been suggested that
satellite cells are dormant myoblasts as they rarely divide and show
little evidence of active protein synthesis. Only when these cells
are activated by injury or disease do they resemble myoblasts; this
observation lends credence to the notion that satellite cells may
represent an earlier and distinct stage in the myogenic lineage.

The consensus of published information points to the satellite
cell being a myogenic reserve cell that is the source of the regenerative response of skeletal muscle [3, 6, 8]. There is, however, a
possible exception to this proposition. Muscle regeneration after
limb amputation in certain amphibian species has been suggested to
occur through dedifferentiation of stump tissues [9-12]. The morphological characteristics of this dedifferentiation have been described in detail, and it is thought that myonuclei of muscle fibres
'bud out' to form mononucleated cells that are competent myoblasts.

The evidence for myogenic capabilities of the resultant mononucleated cytoplasmic compartment is, however, circumstantial [13]
and relies heavily on the inability to detect satellite cells in the
limbs of these amphibians [14]. Satellite cells have, however, been
demonstrated in several amphibian species which have the capacity for
limb regeneration [14-17]. These observations do not preclude the
existence of a myogenic reserve cell population outside the basement
membrane, and lack of satellite cells reflects only an anatomical
absence which need not result in a functional absence. Immunochemical studies [18] on muscle regeneration within the regenerating
amphibian limb were interpreted as supporting the dedifferentiation
hypothesis, but did not address the possibility of myogenic reserve

cells existing outside of the basal lamina. Consequently, the de-differentiation *versus* the myogenic reserve-cell question remains equivocal. Nevertheless the satellite cell appears to be responsible for muscle regeneration in higher vertebrates.

Although the literature describing morphological and functional characteristics of satellite cells is large, biochemical correlates of satellite cell function are few [19].

We report here the use of a panel of Ab's that react with antigens that are developmentally regulated in skeletal muscle, and show that such reagents are of value in determining the lineage relationship between satellite cells and muscle.

## REACTIVITY OF ANTIBODIES WITH MUSCLE CELLS

We have used a number of Ab's in immunochemical studies of muscle gene expression [20, 21]. Some are of little or no value in studying myoblast antigens as the antigens are clearly not present on early muscle cells. However, from amongst several reagents of interest we have selected four that react with antigens expressed on myoblasts, viz. MAb's 5.1H11 and 24.1D5 and rabbit Ab's to the Thy-1 antigen and N-CAM. The reactivities of 5.1H11 and anti-Thy-1 have been described [22, 23]. In muscle-cell cultures myoblasts and myotubes but not fibroblasts react with MAb 5.1H11. With anti-Thy-1 myoblasts and fibroblasts are positive while myotubes are negative. MAb 24.1D5 was produced from a cell fusion of mouse splenocytes immunized against a plasma-membrane fraction isolated from cell cultures of myoblasts and the P3X63Ag8 mouse myeloma line.

Fig. 1 illustrates the reactivity of MAb 24.1D5 with cultured human-muscle myoblasts. These were cloned from a sample of skeletal muscle, can fuse to form myotubes, and can therefore be regarded as a functionally homogeneous population of myoblasts. All the cells in this type of culture express Thy-1 antigen [22] and co-label with MAb 24.1D5 (Fig. 1), showing that MAb 24.1D5 reacts with an antigen expressed by myoblasts. In contrast, skin fibroblast cultures were negative with MAb 24.1D5 (not shown). As myoblasts fuse to form myotubes the reactivity with 24.1D5 antigen is lost. This is shown in Fig. 2 where a myotube cell culture is double-stained with MAb 24.1D5 and Ab to the M subunit of the enzyme creatine kinase which is expressed only by myotube cells. No double-labelled cells were found, showing that MAb 24.1D5 does not react with myotubes. In contrast, there were many MAb 24.1D5-positive myoblasts and M-creatine kinase-positive myotubes.

Ab to the neural cell adhesion molecule N-CAM has been extensively used to study its structure and function in the nervous system [24, 25], but published data are sparse concerning its distribution in skeletal muscle. In one study, however, anti-N-CAM was found

**Fig. 1.** Double indirect
immunofluorescence
analysis of cloned
skeletal muscle myoblast
cells with (**a**) rabbit
anti-Thy-1 Ab, and
(**b**) the same field
labelled with MAb
24.1D5.
Mag. × 62.

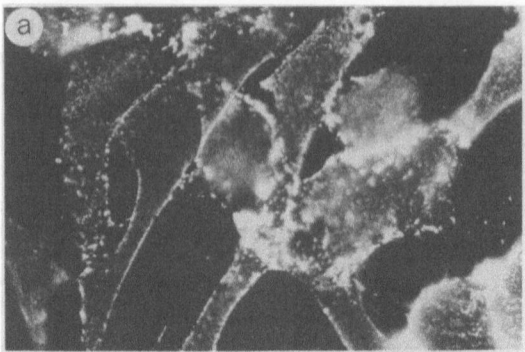

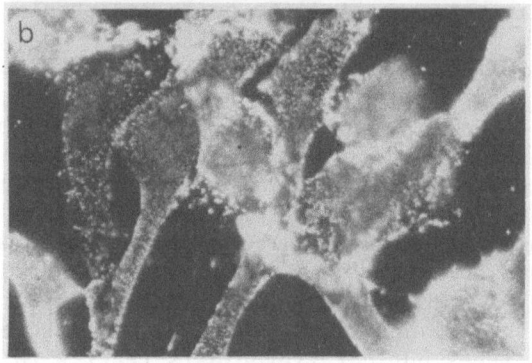

to block the initial adhesion of spinal neurons to skeletal muscle
myotubes, suggesting that N-CAM may be involved in the early events
in nerve-skeletal muscle synapse formation [26]. Double-staining
with anti-N-CAM and Thy-1 of a human myotube cell culture (Fig. 3)
shows that all mononucleate cells in this culture stain with anti-
Thy-1, consistent with previous observations, while anti-N-CAM reacts
with a subpopulation. The N-CAM-positive population was shown to be
myoblasts insofar as anti-N-CAM did not react with fibroblasts. It
was also seen in this culture that Thy-1-negative myotubes expressed
N-CAM. Here a direct overlap was found between these two reagents,
showing that N-CAM is expressed by both myoblasts and myotubes (Fig.3).

Evidently all 4 Ab's react with myoblasts. The least specific
is anti-Thy-1 that reacts with myoblasts in the muscle lineage and
fibroblasts in the non-muscle lineage. The other 3 Ab's reacted only
with cells of the muscle lineage, and MAb 24.1D5 was most specific
as it reacted only with myoblasts. Myoblast antigens appear, then,
to be regulated according to different temporal programmes during
development. In an attempt to analyze the expression of these antigens further, we investigated their expression on cryostat sections
of normal and diseased skeletal muscle.

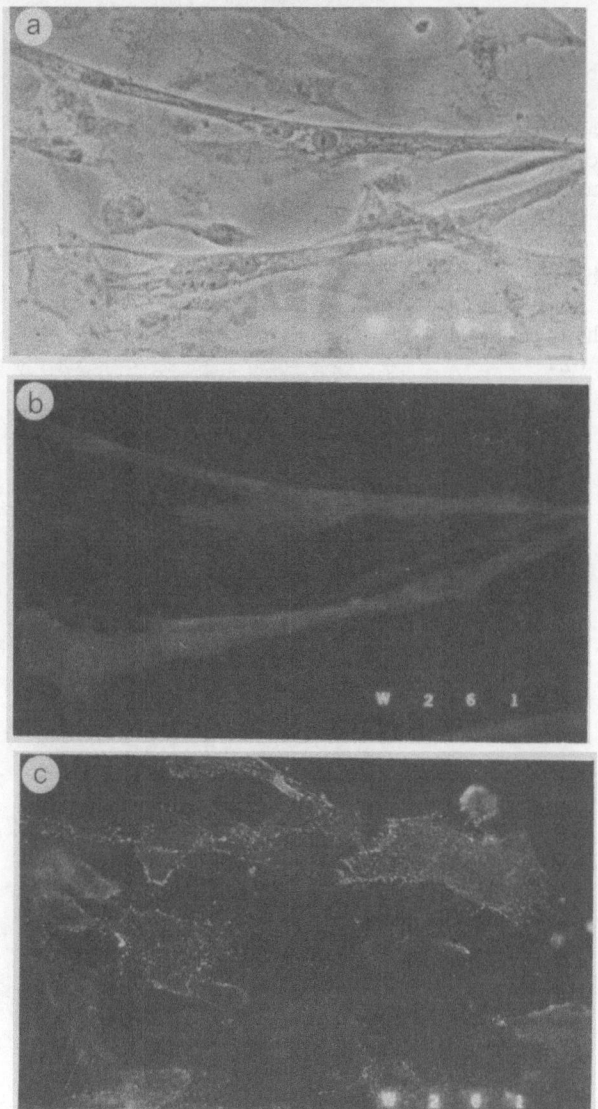

**Fig. 2.** Double indirect fluorescence analysis of a post-fusion
human skeletal muscle culture. (**a**) Phase-contrast micrograph.
(**b**) The same culture labelled with Ab to M-creatine kinase.
(**c**) The same field as (**a**) labelled with MAb 24.1D5. Mag. x 62.

## ANTIGEN EXPRESSION ON CRYOSTAT SECTIONS OF SKELETAL MUSCLE

Cryostat sections of adult human skeletal muscle were subjected
to indirect immunofluorescence analysis with the 4 Ab's used above.
None of these Ab's reacted with the sarcolemma of mature muscle
fibres, indicating that although all 4 react with muscle cells at

Fig. 3. Double indirect
immunofluorescence
analysis of a human
muscle cell culture.
(a) Phase-contrast micro-
graph of the culture.
(b) The same field
labelled with anti-Thy-1
Ab. (c) The same field
labelled with anti-N-CAM
Ab. Mag. x 62.

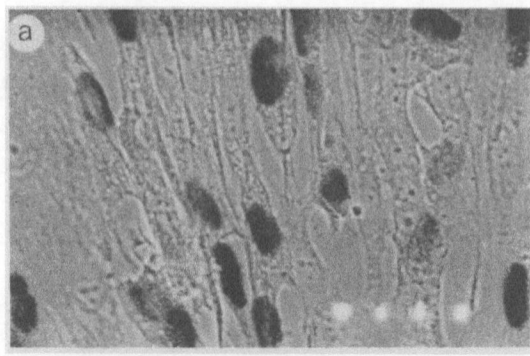

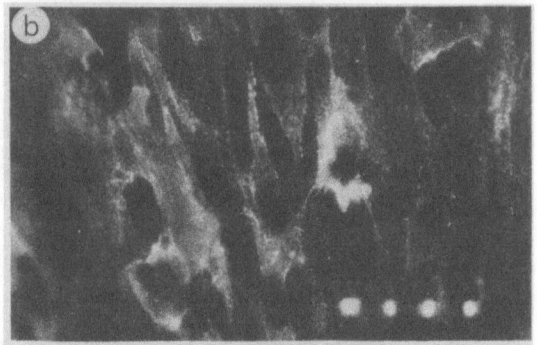

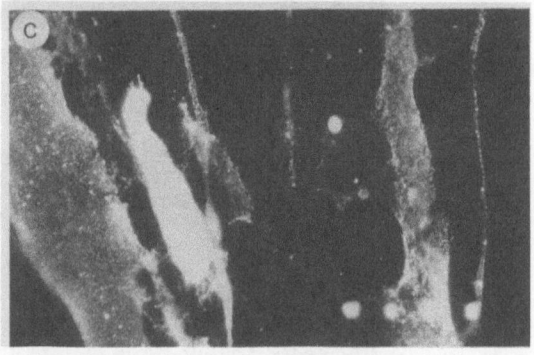

certain stages in their life-cycle, they are all down-regulated by
the time myofibres are terminally differentiated *in vivo*. The only
Ab to show any reactivity with adult muscle sections was MAb 24.1D5.
Reactivity on cross-sections was with small cells intimately associated with the sarcolemma, ~15 µm in diameter (Fig. 4); in longitudinal sections these cells were 60 µm long. They correspond closely
in size and location to muscle satellite cells, suggesting that MAb
24.1D5 reacts with an antigen expressed by satellite cells. Fig. 4
(a & b) also shows that 24.1D5 is found associated with nerve bundles. Here reactivity is seemingly associated with the Schwann cell
membrane as MAb 24.1D5 reacts *in vitro* with these cells but not neurons (not shown). To test for reactivities with muscle cells at

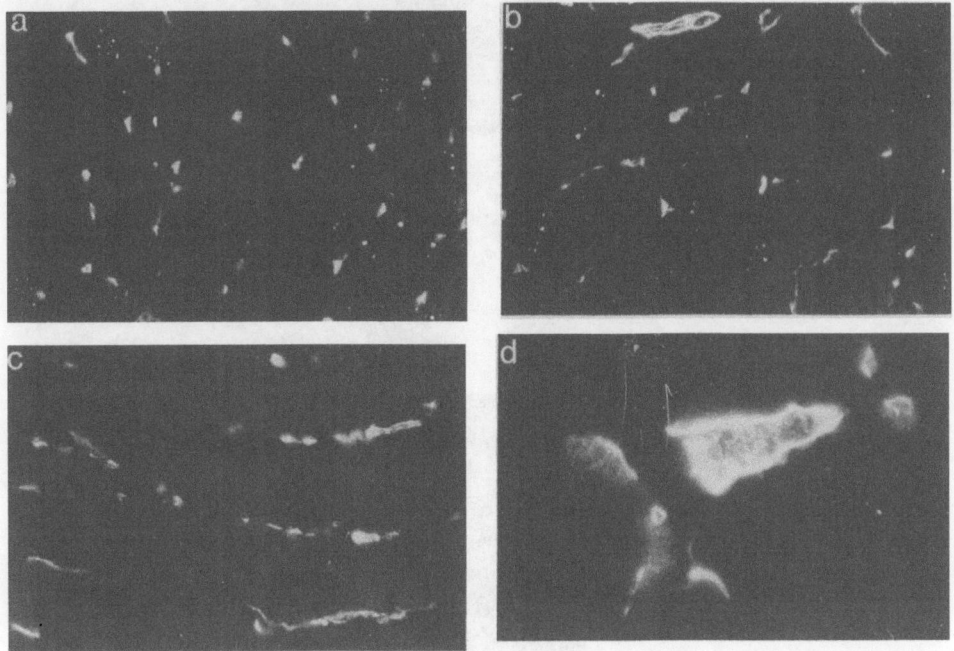

Fig. 4. Indirect immunofluorescence staining of cryostat sections
of baboon intercostal muscle with MAb 24.1D5. (a) Cross-section
showing staining of satellite cells. (b) A sister field showing
satellite cell staining and reactivity with nerve. (c) Longitudinal section showing satellite cells. (d) High-power micrograph
of cross-section showing staining of nerve branch.    Mag.: (a)-(c):
x 62; (d): x 158.

intermediate stages of development, samples of diseased muscle were
examined. Two types of disease state were chosen, viz. diseases such
as polymyositis and muscular dystrophy, where there are often many
regenerating muscle fibres, and diseases like motor neuron disease
which are characterized by areas of denervation.   Two of the Ab's
were found to have identical patternsa of reactivity and will not be
distinguished hereafter.   MAb 5.1H11 reacts with regenerating myofibres on skeletal muscle [23],  and anti-N-CAM was found to have an
identical pattern of reaction.   Fig. 5 shows a micrograph of muscle
from a polymyositis patient in which only the regenerating fibres,
identified by haematoxylin and eosin (H & E) staining, are positively
stained with MAb 5.1H11 and anti-N-CAM.  No reactivity was found on
denervated muscle for either Ab.

    Having demonstrated that MAb 5.1H11 and anti-N-CAM, which react
with myotubes in culture, were reactive with regenerating myofibres,
it was important to determine whether the two 'myoblast-specific'
reagents MAb 24.1D5 and anti-Thy-1 reacted with such fibres.  Fig. 6

**Fig. 5.** Double indirect
immunofluorescence
staining of serial
cross-section of human
skeletal muscle from a
patient with polymyos-
itis.  (a) H & E stain.
(b) MAb 5.1H11 stain.
(c) Anti-N-CAM stain.
Mag. X 62.

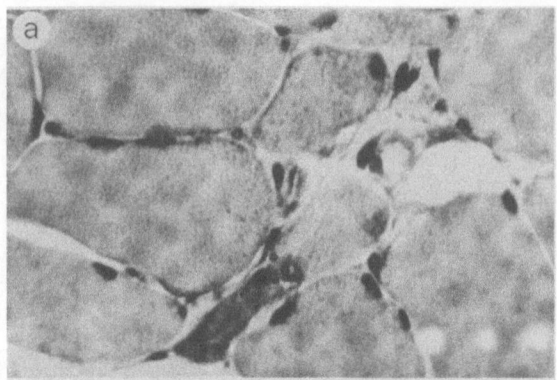

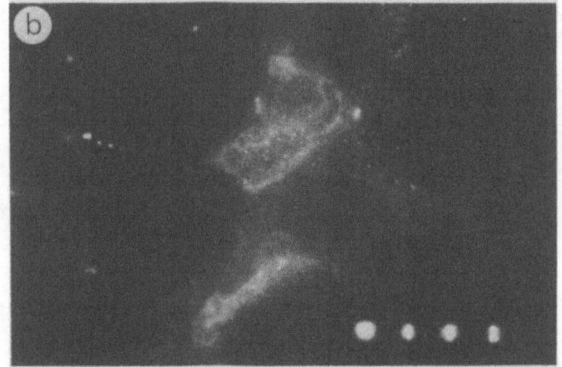

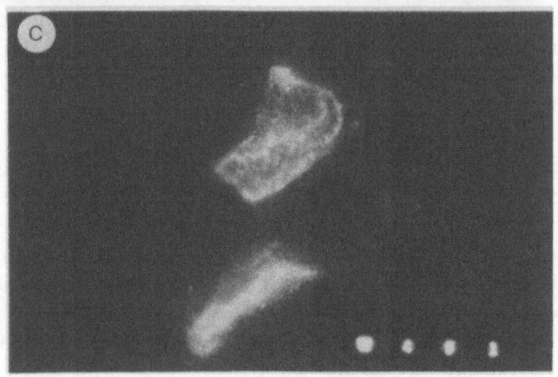

is a composite micrograph of sections of regenerating muscle: it
shows that MAb 5.1H11 reacts with regenerating myofibres, although
MAb 24.1D5 and anti-Thy-1 do not do so.  Fig. 6 also shows a dramatic
distinction in specificity between the latter two Ab's: MAb 24.1D5
reacts with satellite cells (3, near the regenerating fibre) whereas
anti-Thy-1 does not react with  them or the regenerating myofibre
(but on this  section shows some reactivity for blood vessels which
are Thy-1-positive).  Only rarely have we found regenerating fibres
positive for both MAb 5.1H11 and anti-Thy-1; nor does anti-Thy-1 react

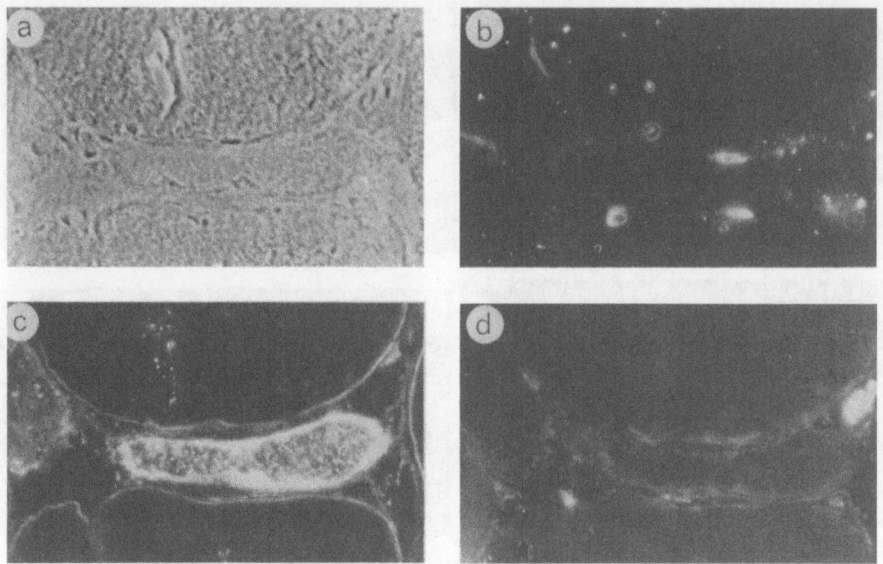

Fig. 6.  Indirect immunofluorescence staining of serial cross-sections of human skeletal muscle from a patient with polymyositis. (a) H & E stain.  (b) MAb 24.1D5 stain.  (c) MAb 5.1H11 stain. (d) Anti-Thy-1 stain.  Mag. x 62.

with denervated muscle although MAb 241D5 shows considerable reactivity, localized to satellite cells. Fig. 7 also shows that fast myosin (detected by reactivity with MAb 29.1D12 [27]) is present in a number of the fibres but is not expressed by satellite cells.

## CONCLUSIONS

Table 1 summarizes the results. While the cell culture results show that all 4 Ab's react with myoblasts *in vitro*, evidently 3 of the 4 do not react with the presumptive myoblast counterpart in mature muscle *in vivo*, namely the satellite cell. One MAb, 24.1D5, was found to react with satellite cells, quite plausibly as the expression of the 24.1D5 antigen in terms of both temporal regulation and cellular specificity *in vitro* is diagnostic of a myoblast-specific antigen, which one might therefore expect to be expressed ubiquitously on myoblastoid mononucleated cells such as the satellite cell. Evidently the patterns of gene expression in a satellite cell and a myoblast are different, and the two cells cannot be considered identical. This observation is consistent with the operational definition of differentiation offered by Jacob & Monod [28] in which they propose: "two cells are differentiated with respect to each other if, while they harbor the same genome, the patterns of proteins which they synthesize is different".

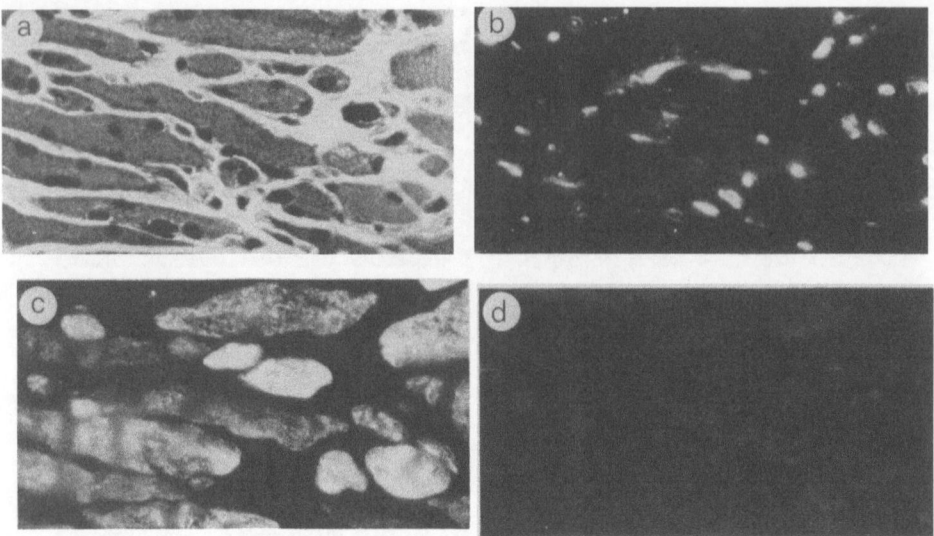

**Fig. 7.** Indirect immunofluorescence staining of serial cross-sections of human skeletal muscle from a patient with motor neuron disease. (**a**) H & E stain. (**b**) MAb 24.1D5 stain. (**c**) MAb 29.1D12 stain. (**d**) Negative anti-Thy-1 stain. Mag. x 62.

This study presents the first convincing report of biochemical differences between satellite cells and myoblasts. As 3 out of 4 reactivities were expressed by myoblasts but not by satellite cells, there must be activation of a number of gene families during the satellite cell→myoblast transition. However, the fact that MAb 24.1D5 reacts with satellite cells and myoblasts shows that both cell types share certain specialized gene products. The only other reactivity that has been convincingly demonstrated on satellite cells and myoblasts is the C3/1 antigen [19]; but it is also present on myotubes and is clearly much less specific than MAb 24.1D5.

The demonstration of immunochemical non-identity between satellite cells and myoblasts (with regard to the Thy-1 antigen and N-CAM and MAb 5.1H11 reactivity) indicates that the notion of the satellite cell being merely a resting myoblast is false. The expression of 24.1D5 antigen and the potential for myotube formation by both satellite cells and myoblasts do, however, indicate that the satellite cell is a member of the myogenic lineage. The combinational differential expression of 24.1D5 antigen and Thy-1 antigen in these two cell types strongly indicates that the satellite cell is a distinct stage in muscle development that may be defined by both immunochemical and anatomical criteria.

**Table 1.**  Summary of reactivities of antibodies reactive with human skeletal muscle

| Antibody reagent | Cell culture analysis | | | Immunohistochemical analysis | |
| --- | --- | --- | --- | --- | --- |
| | Myoblast | Myotube | Fibroblast | Regenerating myofibres | Satellite cells |
| Thy-1 | + | - | + | - | - |
| 24.1D5 | + | - | - | - | + |
| N-CAM | + | + | - | + | - |
| 5.1H11 | + | + | - | + | - |

The four Ab reagents here described have allowed the demonstration of differential gene activation between satellite cells and myoblasts, and indicated several important experimental directions to be followed. It is not known how the transitional expression of these antigens is linked to the cell cycle, and at what stage of the cycle these transitions occur. Recently, however, it has been reported that antigenic differentiation during the myoblast → myotube transition appears to occur during the G1 phase of the cell cycle [29, 30]. While these immunochemical studies are of explicit value, precise temporal analysis of cell-cycle control of gene activation may require analysis at the RNA level.

The effect of myotrophic factors on the cell cycle and biochemical differentiation is likewise obscure. Although several trophic factors have been identified which have effects on cell cycling and morphological differentiation of muscle cells [31-35], the effect of these factors on cell cycling and biochemical differentiation has not been reported. The availability of Ab's such as those now described should enable this gap to be filled.

*References*

1.  Hauschka, S.D. (1974) *Dev. Biol. 37*, 329-344.
2.  Mauro, A. (1961) *J. Biophys. Biochem. Cytol. 9*, 493-495.
3.  Moss, F.P. & Leblond, C.P. (1971) *Anat. Rec. 170*, 421-436.
4.  Schultz, E. (1976) *Am. J. Anat. 147*, 49-70.
5.  Gibson, M.C. & Schultz, E. (1982) *Anat. Rec. 202*, 329-337.
6.  Bischoff, R. (1979) in *Muscle Regeneration* (Mauro, A., ed.), Raven Press, New York, pp. 13-29.
7.  Bischoff, R. (1980) in *Plasticity of Muscle* (Pette, D., ed.), Walter de Gruyter, Berlin, pp. 119-129.
8.  Lipton, B.H. & Schultz, E. (1979) *Science 205*, 1292-1294.
9.  Thorton, C.S. (1938) *J. Morph. 62*, 19-47.
10. Thorton, C.S. (1968) *Adv. Morphogenesis 7*, 205-249.
11. Hay, E.D. (1959) *Dev. Biol. 1*, 555-585.

12. Lentz, T.L. (1969) *Am. J. Anat. 124*, 447-479.
13. Schrag, J.A. & Cameron, J.A. (1983) *J. Embryol. Exp. Morph. 77*, 255-271.
14. Popiela, H. (1976) *J. Exp. Zool. 198*, 57-64.
15. Carlson, B.M. (1973) *Am. J. Anat. 137*, 119-150.
16. Carlson, B.M. & Rogers, S. (1976) *Folio Morph. 24*, 359-361.
17. Flood, L. (1971) *J. Ultrastruct. Res. 36*, 523-524.
18. Kintner, C.R. & Brockes, J.P. (1984) *Nature 308*, 67-69.
19. Wakshull, E., Bayne, E.K., Chiquet, M. & Fambrough, D.M. (1983) *Dev. Biol. 100*, 464-477.
20. Hurko, O., Quinn, C.A., Brown, S.M., Gordon, R.D., Woodroofe, N.M. & Walsh, F.S. (1984) in *Developmental Processes in Normal and Diseased Muscle* (Eppenberger, H.M. & Perriard, J.-C., eds.), S. Karger, Basel, pp. 41-49.
21. Walsh, F.S., Moore, S.E., Woodroofe, N.M., Hurko, O., Nayak, R., Brown, S.M. & Dickson, J.G. (1984) in *Developmental Processes in Normal and Diseased Muscle* (Eppenberger, H.M. & Perriard, J.-C., eds.), S. Karger, Basel, pp. 49-56.
22. Walsh, F.S. & Ritter, M.A. (1981) *Nature 289*, 60-64.
23. Hurko, O. & Walsh, F.S. (1983) *Neurology 33*, 737-743.
24. Edelman, G.M. (1983) *Science 219*, 450-457.
25. Rutishauser, U. (1984) *Nature 310*, 549-554.
26. Rutishauser, U., Grumet, M. & Edelman, G.M. (1983) *J. Cell Biol. 97*, 145-172.
27. Moore, S.E., Hurko, O. & Walsh, F.S. (1984) *J. Neuroimmunol. 7*, 137-149.
28. Jacob, F. & Monod, J. (1983) in *Cytodifferentiation and Macromolecular Synthesis* (Locke, M., ed.), Academic Press, New York, pp. 30-64.
29. Nyugen, H.T., Medford, R.M. & Nadal-Ginard, B. (1983) *Cell 34*, 381-393.
30. Leed, H.U., Kaufman, S.J. & Coleman, J.R. (1984) *Exp. Cell Res. 152* 331-347. [Cf. #C-3, this vol.- *Ed.*]
31. Markelonis, G., Oh, T.H., Eldefrawi, M.E. & Guth, L. (1982) *Dev. Biol. 89*, 353-361.
32. Linkhart, T.A., Clegg, C.H. & Hauschka, S.D. (1981) *Dev. Biol. 86*, 19-30.
33. Florini, J.R. & Roberts, S.B. (1979) *In Vitro 15*, 983-992.
34. Allen, R.E., Dodson, M.V. & Luiten, L.S. (1984) *Exp. Cell Res. 152*, 154-160.
35. Matsuda, R., Spector, D.H., Micou-Eastwood, J. & Strohman, R.C. (1984) *Dev. Biol. 103*, 276-284.

#C-4

# CHOLINERGIC-SPECIFIC NERVE TERMINAL ANTIGENS

V.P. Whittaker, E. Borroni, P. Ferretti and G.P. Richardson

Abteilung Neurochemie
Max-Planck-Institut für Biophysikalische Chemie
Postfach 2841, D-3400 Göttingen, W. Germany (FRG)

*In mammals, transmission of information at synapses depends on >20 chemical transmitters, on which classification of neurones may be based. While the transmitter release mechanism may be common to all presynaptic nerve terminals, the synthesis, storage and eventual disposal of each transmitter require an unknown number of specific proteins. In development, moreover, surface markers specific for the different types of synapse may be needed to regulate synaptogenesis. The ability to recognize transmitter-specific components would be invaluable in developing new morphological methods and for isolating subpopulations of detached nerve terminals (synaptosomes) and their constituent organelles (synaptic vesicles; plasma membranes, p.m.) derived from neurones of defined transmitter type.*

*In our immunological approach, with* **Torpedo marmorata** *electric organ as a rich source of purely cholinergic nerve terminals, means were devised for isolating cholinergic synaptic vesicles and presynaptic p.m. in high purity and then raising antisera to components of these membranes. One antiserum is to a proteoglycan which is a specific component of the vesicle membrane; the other is to a family of gangliosides specifically present in the presynaptic p.m. The former recognizes a similar immunoreactive substance in cholinergic motor nerve and retinal nerve terminals in the rat but is less selective in the mammalian autonomic and central nervous systems, whereas the latter appears to recognize a similar, cholinergic-specific immunoreactive component in all central and peripheral cholinergic terminals, and offers a basis both for a new immunofluorescence (IF) histochemical method for cholinergic nerve terminals and for novel isolation of a cholinergic subpopulation of terminals by immunoaffinity chromatography. The anti-proteoglycan antiserum allows measurement of the content of the vesicles independently of their acetylcholine (ACh) and ATP content and has enabled differences between reserve and re-cycling vesicles to be detected.*

In M.C. Raff's view [cf. #NC(C)-1], there may be thousands or even tens of thousands of different types of neurone in the mammalian CNS.   This may well be true if account is taken of all parameters – molecular, morphological and functional. A simpler way of classifying neurones is by the type of transmitter substance that they utilize to communicate with neighbouring nerve- or effector-cells.   Some 20 or more putative chemical transmitters have been identified, and others may await discovery. A neurone utilizes one – perhaps sometimes two  – such transmitters, and it would be very convenient to have immunochemical methods for recognizing neurones by their transmitter type based on antisera to transmitter-related antigens.   This would be especially useful for transmitters which are not themselves anti- genic or which do not lend themselves to other forms of cytochemical investigation. Such antisera could form the basis of new histochemi- cal methods for tracing pathways or connections involving specific transmitters and, if the transmitter-related antigen were a surface antigen, they could also be used to isolate subpopulations of de- tached presynaptic nerve terminals (synaptosomes) containing only one type of nerve terminal from the  mixed population of synaptosomes derived from neurones of many transmitter types.   A contribution that follows, by P.J. Richardson (#C-5), describes successful work of this kind.

In our approach we have worked with a model system containing only one transmitter and one type of synapse, namely the cholinergic electromotor synapse of the electric ray, *Torpedo marmorata*.   The electric organ's electrocytes  develop embryologically from myotubes [1, 2].   The under (ventral) surfaces of the electrocytes are almost completely covered with nerve terminals which form synapses resem- bling – not unexpectedly – neuromuscular junctions, except that the synaptic vesicles – the storage sites of the transmitter ACh within the terminals – are somewhat larger (90 nm in diameter) than those at the neuromuscular junction (50 nm).   The tissue is 500–1000 times richer in synapses than muscle and is thus an ideal starting  point for the isolation of presynaptic p.m. [3] and synaptic vesicles [4, 5].   Methods have been worked out for obtaining these structures in high purity and yield.

**VESICLE–SPECIFIC PROTEOGLYCAN: A STABLE VESICLE MARKER**

Rabbit antiserum prepared against highly purified cholinergic synaptic vesicles derived from *Torpedo* electromotor nerve terminals yields antibodies (Ab's) directed against a proteoglycan vesicle antigen [6], whose glycosaminoglycan moiety is of the heparan sulph- ate type [7]; it is recognized by the antiserum only if conjugated with protein.   Immunochemical analysis shows the crude vesicle anti- serum to react with several antigens besides the proteoglycan, nota- bly acetylcholinesterase (AChE), a highly antigenic component present not in synaptic vesicles but in contaminating p.m. and/or adsorbed soluble cytoplasmic protein.   Other antigens are present in dorsal

(non-innervated) electrocyte membranes.  This finding illustrates
a problem with polyclonal antisera, viz. their tendency to form Ab's
to highly antigenic contaminants that may be insignificant biochemically but not immunochemically.  Fortunately, in this case the unwanted Ab's can be removed by adsorption with preparations of
electric organ membranes minus vesicles.

Several criteria have been adopted to judge the specificity of
the adsorbed antiserum.  Of these the most important are:  co-purification of the antigen with the organelle assumed to carry it;
recognition of a single component in a 'blot' test; immunocytochemical recognition of the antigen in the expected site.  For the anti-
vesicle specific proteoglycan antiserum these criteria were shown to
be satisfied (Figs. 1a, 1e & 2a–d).

Localization of the antigen to the interior of the vesicle has
been demonstrated by the failure of antiserum to precipitate intact
vesicles; however, lysed vesicles are precipitated [8].

## Cross–reactivity with other cholinergic synapses

Antisera directed against *Torpedo* electromotor vesicle-specific
proteoglycan cross-react with *Torpedo* and rat motor nerve  terminals
[9].    Their ability to react specifically with other – in particular, central – types  of cholinergic endings has not been fully
evaluated, but preliminary work with retina is promising [10].

## THE LIFE–CYCLE OF VESICLES, AND THEIR RE–CYCLING

The cell bodies of the electromotor neurones are exceptionally
rich in rough and smooth endoplasmic reticulum (e.r.), Golgi membranes, polysomes and vesicles, and one would expect synaptic vesicles
to be synthesized there and transported to the terminal by  fast
axonal transport, as has been demonstrated in other systems.  The
anti-vesicle specific proteoglycan antiserum may be used to detect
the formation and transport of synaptic vesicles (Fig. 3a–d) by
indirect immunofluorescence (IF) cytochemistry [11].  As vesicle-
specific proteoglycan is internal to the vesicle,  it must be rendered accessible to the antiserum either by breakdown of the imper-
meability of the plasma and vesicle membranes by the embedding
procedure or detergent treatment (Fig. 3, a & b), or by exteriorization of the vesicle core through stimulus–induced exocytosis (Fig. 3d);
the latter process occurs only at the nerve terminals.  The results
(Fig. 3) indicate that vesicles are assembled in the cell body,
accumulate in the axon hillock and are then transported to the terminal via the axon.  On stimulation they re-cycle: the stimulus-
induced exteriorization of the vesicle interior is reversible [12].

In relation to re-cycling, Fig. 1b shows the distribution, in a
zonal density gradient, of vesicular ACh and vesicle-specific pro-

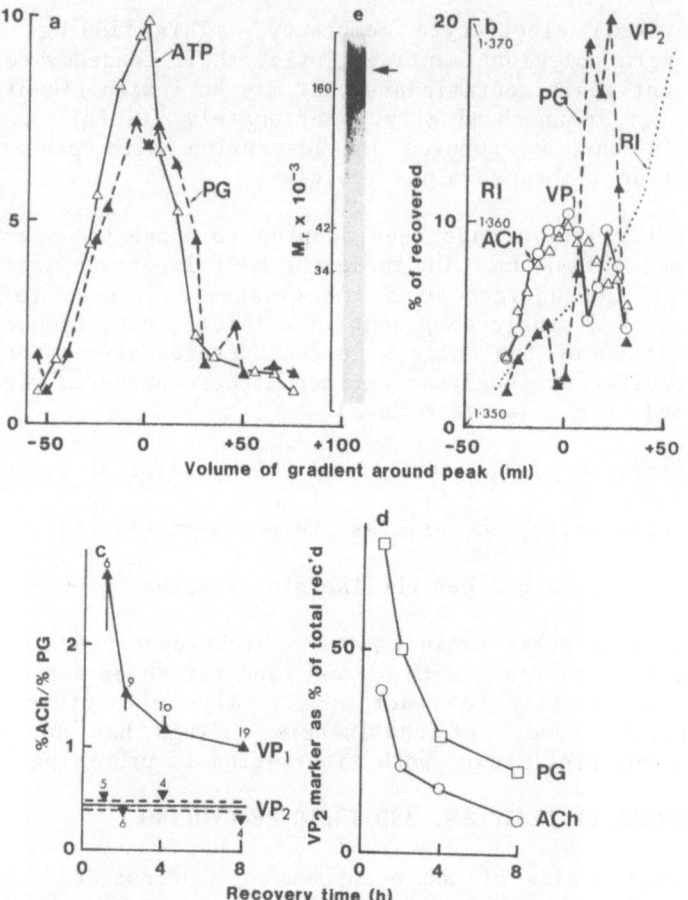

**Fig. 1.** (**a**) Co-purification of the vesicle-specific proteoglycan antigen (▲, PG) with cholinergic vesicles (marker: vesicular ATP) in a density gradient with a zonal rotor [6]. Distributions have been normalized. Assays on 5 fractions on which both ATP and PG were measured gave, for % ATP *vs.* %PG, r = 0.90. The apparently higher ATP/PG ratios between 0 and −50 ml do not, in fact, differ significantly from the apparently lower ones between +5 and +75 ml. (**b, c, d**) Recovery of biophysical characteristics of reserve (VP$_1$) vesicles by re-cycling (VP$_2$) vesicles. (**b**) Assays for ACh (o), ATP (△) & PG (▲) in a zonal gradient after recovery for 1 h. Note separation of fraction into regions of high and low ACh/PG ratios corresponding to the VP$_1$ & VP$_2$ populations. Blocks were excised from electric organs stimulated via the electric lobe and after varying periods of perfusion were frozen, crushed and the vesicles extracted [13]. (**c**) ACh/PG ratio *vs.* time, for VP$_1$ (▲) & VP$_2$ (▼) fractions. Points and bars are means ±S.E.M. for VP$_1$ (6-9 fractions) & VP$_2$ (4-6; for all points, —— & ---- denote mean ±S.E.M.). (**d**) Proportion of VP$_2$ vesicles *vs.* time, measured by PG (⊔) and ACh (o). The vesicles

[continued opposite

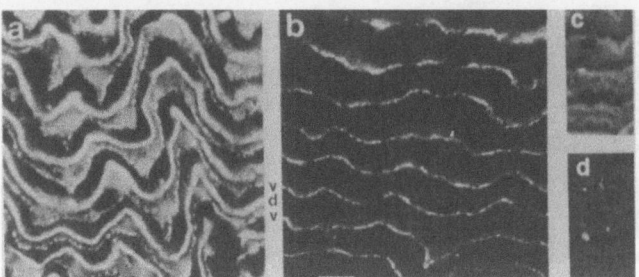

**Fig. 2.** Immunofluorescence (IF) of anti-vesicle antisera to vesicles on 5 µm paraffin sections of *Torpedo* electric organ.
(**a**) Unadsorbed antiserum; (**b**) adsorbed antiserum; (**c**) control serum;
(**d**) vesicle-adsorbed serum. Ventral (v) and dorsal (d) membranes
are indicated. Scale bar = 10 µm.

teoglycan for a vesicle-rich tissue extract derived from perfused
blocks of electric organ recovering from a previous *in vivo* stimulation (0.1 Hz for 3.3 h) through the electric lobe (V.P. Whittaker,
G.H.C. Dowe & J.H. Walker, unpublished). In contrast with Fig. 1a,
ACh and proteoglycan are bimodally distributed. Such a distribution
has been shown earlier [13] to result from the separation of re-
cycling (VP$_2$) vesicles from the reserve (VP$_1$) population. Re-
cycling vesicles are denser, smaller, and only partially filled,
i.e. have a smaller content of osmotically active solutes and free
water than reserve vesicles [14, 15], which constitute almost the
entire population in unstimulated tissue.

Evidently (Fig. 1a) the re-cycling vesicles have a smaller ACh/
proteoglycan ratio than reserve vesicles, just as one would expect
if the proteoglycan is a stable vesicle marker and VP$_2$ vesicles are
only partially filled. As recovery takes place (Fig. 1c) the proportion of re-cycling vesicles becomes progressively smaller and the
mean normalized ACh/normalized proteoglycan ratio for the VP$_1$ population progressively approximates to the value for the normalized
ratio for a single population, namely unity. This is due to the
gradual acquisition by re-cycling vesicles of additional ACh and,
accordingly, of the biophysical characteristics of the resting popu-

*Fig. 1 legend, continued*
are only partially filled (inferred from the discrepancy between the
two curves). (**e**) Immune blot of vesicle proteins separated by SDS-
PAGE and electrophoresed onto a nitrocellulose sheet. The adsorbed
anti-synaptic vesicle antigen recognizes a single constituent of
mol. wt. ~200,000 (*arrow*) which has been identified as a heparan
sulphate-type proteoglycan. Staining was with peroxidase-conjugated
second Ab, diaminobenzidine and hydrogen peroxide.

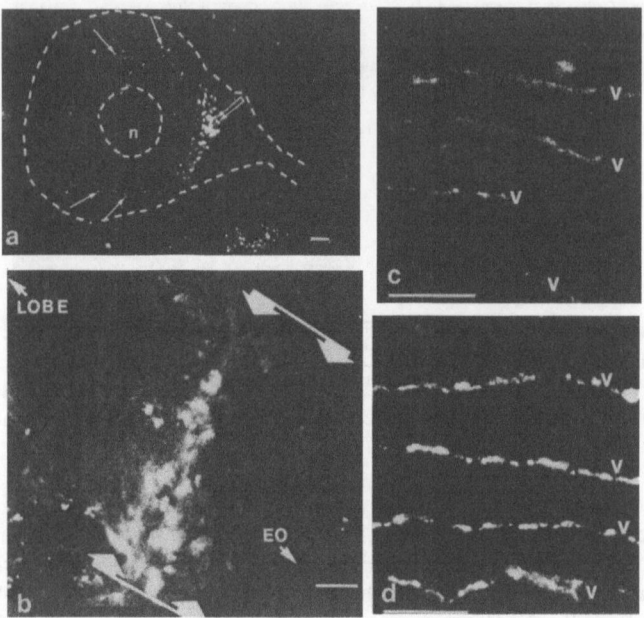

Fig. 3. Vesicle formation and transport detected by the anti-vesicle specific proteoglycan antiserum [11]. (**a**) Punctate (vesicular) staining in the electrometer cell body and accumulation of vesicles in axon hillock awaiting transportation; the superimposed outline of the cell is derived from phase-contrast examination of the section before IF cytochemistry. (**b**) Accumulation of vesicle-specific fluorescence in the axons of the electromotor neurone above a ligature. (**c**, **d**) IF of (**c**) resting, (**d**) stimulated electromotor nerve terminals, showing exteriorization of vesicle proteoglycan by stimulation; this effect is reversible [12]. *Methods:* (**a**) Paraplast embedding, (**b**) Triton X-100 treatment, (**c**) frozen section with no embedding or Triton X-100 treatment.

lation from which they were derived by stimulation-induced recycling. In Fig. 1d we see the diminishing proportion of $VP_2$ vesicles in the total population as recovery continues; the difference in the estimates of this proportion according to whether proteoglycan or ACh is taken as the vesicle marker is, again, a consequence of the partial filling of the vesicles.

**THE APPEARANCE OF VESICLE-SPECIFIC PROTEOGLYCAN IN DEVELOPMENT**

In developing electric organ, vesicle proteoglycan makes its appearance *pari passu* with functional synaptogenesis (Fig. 4) [16]. It also disappears *pari passu* with other presynaptic markers when the electromotor innervation is caused to degenerate by nerve section (E. Borroni & P. Ferretti, submitted for publication).

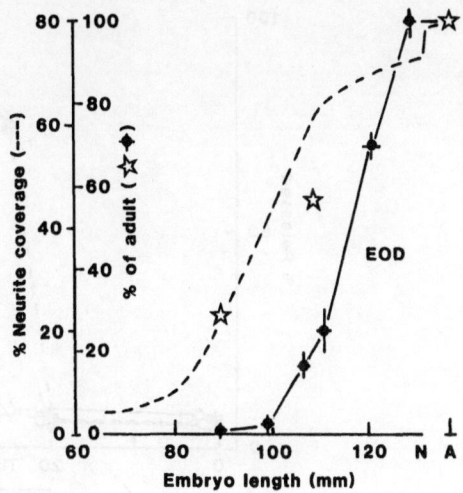

**Fig. 4.** Appearance of vesicle-specific proteoglycan (☆) *pari passu* with neurite coverage of the ventral surface of the electrocyte (----) and somewhat ahead of the acquisition of functional synapses as measured by the magnitude of the electric organ discharge (EOD, ◆) evoked by single stimuli applied through the nerve [16] (— indicates range of lengths pooled, | is S.E.M. of EOD's of this range). Values for newborns (N) and adults (A) are inserted for comparison.

## Chol-1, A CHOLINERGIC CELL-SURFACE MARKER

In other work [9, 17-19] we identified a family of cholinergic surface antigens, consisting of gangliosides, on the presynaptic p.m. of cholinergic nerve terminals. Antisera (anti-TSM) were prepared by injecting *Torpedo* electromotor presynaptic p.m. preparations into sheep – a fortunate choice, as it turned out, because gangliosides are more immunoreactive in sheep than in rodents or rabbits. Anti-TSM antiserum selectively induces complement-mediated lysis of the cholinergic subpopulation of mammalian brain cortical synaptosomes (Fig. 5). This indicates the presence on the external surface of the cholinergic synaptosomes of an antigen ('Chol-1') common to *Torpedo* and mammals.

The inhibition of cholinergic synaptosome lysis by Chol-1 has been used to assay the antigen. In *Torpedo* electric organ Chol-1 antigen has been identified as a class of polysialogangliosides [18, 20], and in mammals two components (bands) were active in inhibiting the selective lysis [21]. The specific population of cytotoxic Ab's has now been purified and used to identify cholinergic neurones in different areas of rat brain (as first presented at the Forum).

## Purification of anti-Chol-1 antibodies

Fig. 6a shows the binding of anti-TSM antiserum and pre-immune serum to guinea-pig brain cortical synaptosomes. The binding of the latter does not cause complement-mediated lysis, indicating the presence of a non-cytotoxic population of Ab's directed against non-relevant antigens. Fig. 6b shows the net binding of the anti-TSM serum to synaptosomes after correcting for binding of the pre-immune serum components.

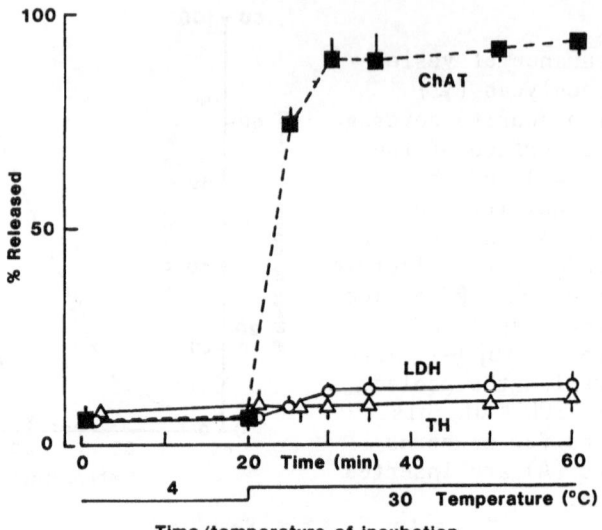

**Fig. 5.** Selective, complement-mediated lysis of the cholinergic subpopulation of guinea-pig brain cortical synaptosomes. The 85% release into the supernatant of the soluble cholinergic synaptosomal marker choline acetyltransferase (ChAT) contrasts with the absence of any release of the soluble adrenergic marker, tyrosine hydroxylase (TH). The small proportion (6%) of the soluble cytoplasmic marker lactate dehydrogenase (LDH) indicates that cholinergic synaptosomes constitute ~6% of the total population in cortex.

To eliminate non-cytotoxic Ab's, the antiserum was affinity-purified on *Torpedo* electric organ gangliosides covalently bound to glass beads. After purification the antiserum showed an 8-fold increase in its capacity to induce selective lysis of cholinergic synaptosomes, indicating the effective purification of the specific population of Ab's (Table 1).

Fig. 7 shows the effect of the affinity-purification of anti-TSM antiserum on its immunoreactivity with synaptosomal p.m. proteins. The crude antiserum recognizes two membrane proteins that are not still detectable after affinity-purification.

Fig. 8 shows that guinea-pig forebrain and *Torpedo* electric organ gangliosides differ generally in TLC migration pattern. When *Torpedo* gangliosides are stained with crude anti-TSM serum (Fig. 9), many give immunopositive reactions, whereas only two from brain are manifest. The affinity-purified serum recognizes the same gangliosides in *Torpedo* electric organ as the crude anti-TSM serum, but only one band migrating close to GT1b is detectable in guinea-pig gangliosides. No staining with the pre-immune serum was observed

*[Text continues on p. 201*

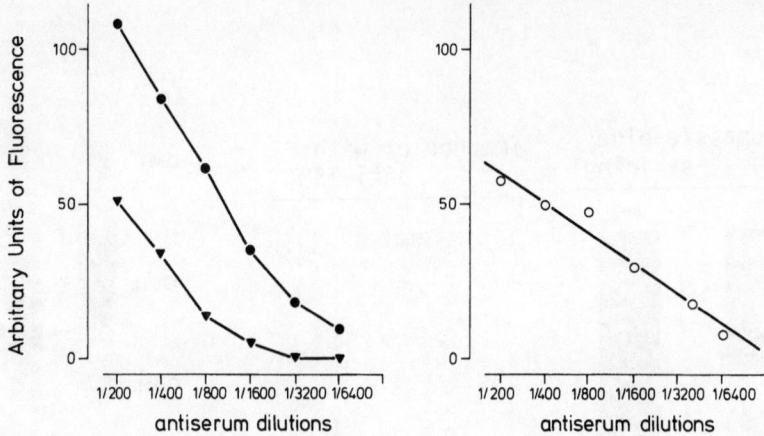

Fig. 6. (a) Binding of anti-TSM serum (●) and pre-immune serum (▼)
to guinea-pig synaptosomes (fraction $P_2$). (b) Net binding of anti-
TSM serum (○) calculated as the difference between the 2 (a) curves.
Serum-treated synaptosomes were incubated with FITC-labelled rabbit
anti-sheep IgG Ab's and the amount of fluorescence bound was measured in an MPFL Perkin-Elmer spectrofluorimeter at 495/520 nm after
solubilizing pelleted synaptosomes in 1% (v/v) Triton X-100.

Table 1. Affinity-purification of anti-Chol-1 Ab's on *Torpedo*
electric organ gangliosides covalently bound [22] to alkyl glass
beads. In the purification, adsorbed Ab's were eluted after immunoadsorption onto columns of the coated beads. For choline acetyltransferase (ChAT), the values represent mean ±S.D. of 6 experiments: the ChAT released is expressed as % of the total ChAT of a
guinea-pig cortical synaptosome preparation and signifies selective complement-induced lysis of the cholinergic subpopulation of
synaptosomes mediated by anti-Chol-1 Ab's. The last column gives
the ratio of this % value to μg IgG in 1.5 ml (= 15 mg of tissue).

| Serum | IgG amount, μg | ChAT released, % | % release/μg IgG |
|---|---|---|---|
| Anti-TSM, crude | 420 | 32.4 ±2.1 | 0.08 |
| Affinity-puri-<br>fied anti-TSM | 48 | 30.6 ±2.1 | 0.64 |
| | 87 | 52.2 ±1.1 | 0.60 |

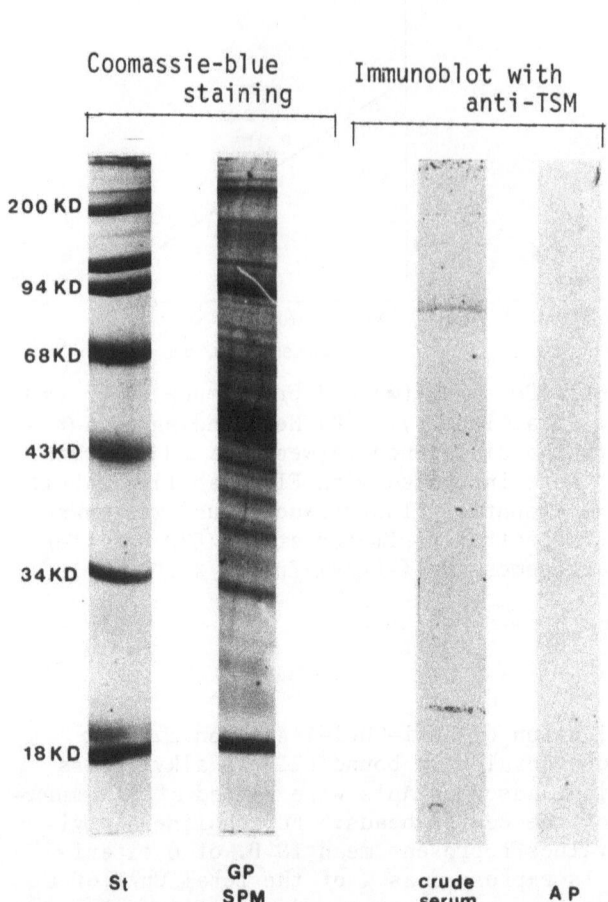

Fig. 7. SDS-PAGE of guinea-pig p.m. proteins (GP SPM): *left*, stained; *right*, immunoblotted with crude and affinity-purified (AP) anti-TSM antiserum (by electrophoretic transfer of proteins onto nitrocellulose paper). KD values represent mol. wt. x $10^{-3}$.

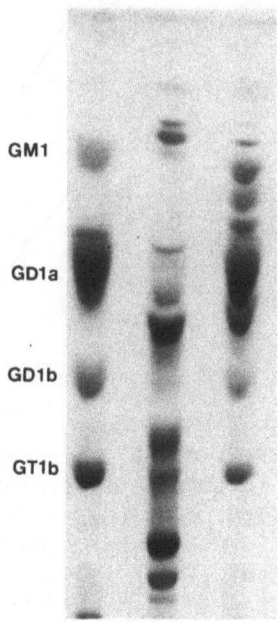

Fig. 8. TLC patterns of *Torpedo* electric organ (EO) and guinea-pig forebrain (GP) gangliosides compared with a standard mixture (St) of ox-brain gangliosides (Seromed, Munich).

After extraction and purification ([23]; resorcinol method [24] for estimating sialic acid content), the gangliosides were separated on Kieselgel 60 HPTLC plates (Merck, Darmstadt) using chloroform/methanol/2% aq. $CaCl_2$ (50:42:11 by vol.) with 2-5 µg sialic acid per lane. Visualization was with *p*-dimethylaminobenzaldehyde.

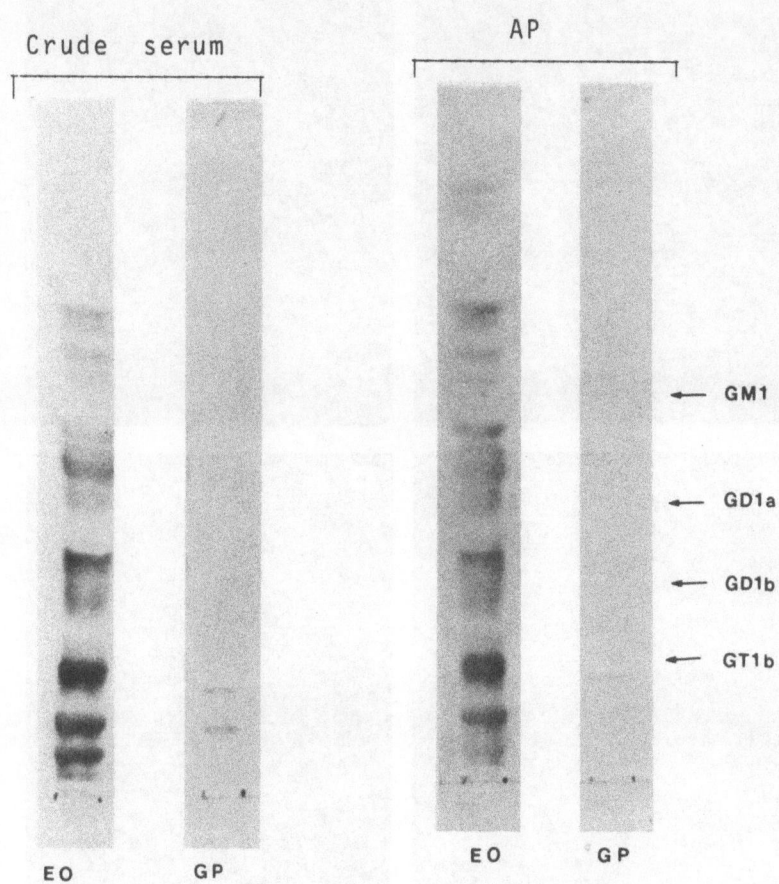

Fig. 9.   Immunostaining on TLC of electric organ (EO) or guinea-pig
forebrain (GP) gangliosides, using a method modified from [25] and
*(two left lanes)* crude anti-TSM serum or *(two right lanes)* affinity-
purified (AP) serum.   *Arrows* indicate the $R_F$'s of standard ox-
brain gangliosides.

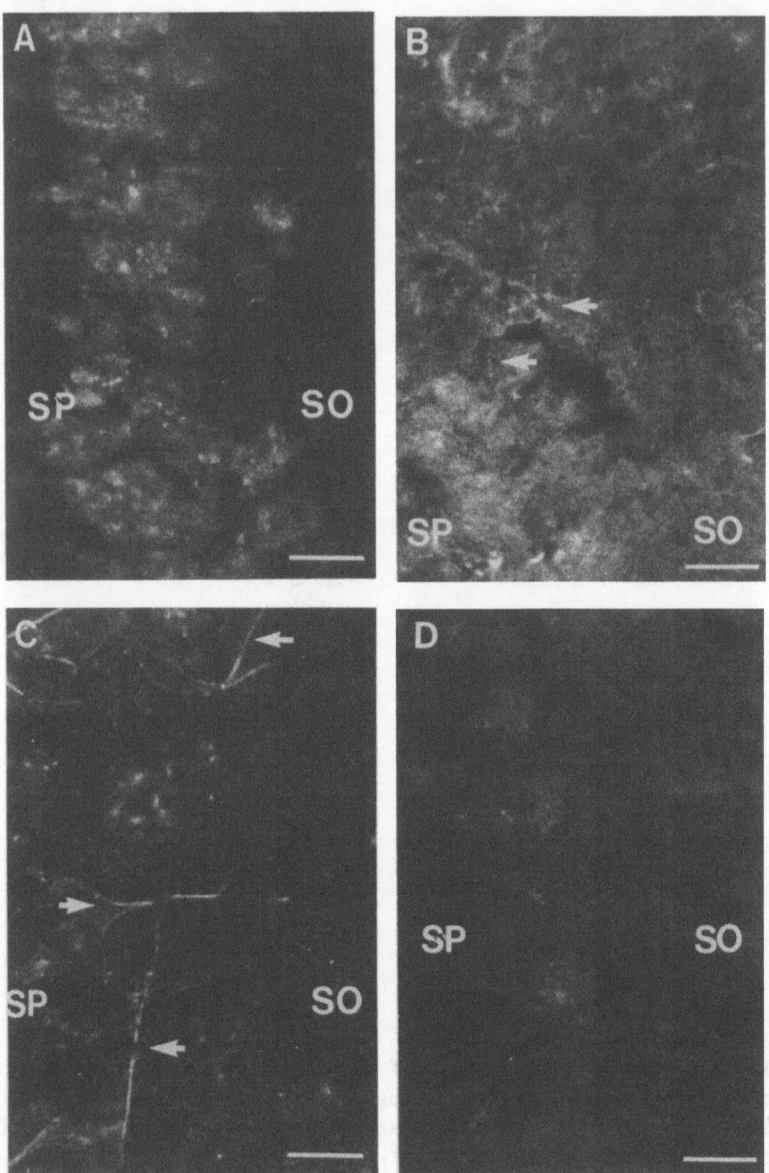

Fig. 10.  IF staining in cryostat sections of the CA 1 area of the
rat hippocampus (cf. Fig. 12).  (**A**) Pre-immune serum, (**B**) crude
anti-TSM antiserum, (**C**) affinity-purified antiserum, (**D**) FITC-
labelled anti-sheep IgG alone.  For SP, SO and other explanatory
points, see text.  Scale bar: 10 μm.

(result not shown).  The affinity-purified  serum is thus a highly
specific detector of a minor ganglioside in guinea-pig brain which
is a specific surface marker for cholinergic nerve terminals;  this
ganglioside,  or  another  immunochemically  identical,  is  one  of  a
family on the cholinergic electromotor terminals of *Torpedo*.    It is
not clear at present whether these gangliosides all contain a common
determinant which has been preserved in evolution or whether electromotor nerve terminals contain a series of unrelated gangliosides
each of which is transmitter-related and antigenic.

## Anti-chol-1 as an immunofluorescence (IF) cytochemical reagent

The affinity-purified anti-Chol-1 antiserum has been used in IF
cytochemical studies of various rat brain areas to test its possible
application  for  detecting  mammalian  central  cholinergic  neurones.
In extension of previous work [17],  unfixed brain cryostat sections
were fixed in acetone at -25°, air-dried and stained by conventional
IF techniques.

Fig. 10 shows the patterns of IF staining obtained in the CA 1
area of the rat hippocampus (delineated in Fig. 12) with (A) pre-
immune serum, (B) crude anti-TSM serum, (C) affinity-purified serum,
(D) FITC-labelled anti-sheep IgG alone.    Punctate staining of cell
bodies in the stratum pyrimidale (SP) is observed with all antisera,
due partly to autofluorescence and partly to non-specific Ab-binding.
Fibres in the stratum oriens (SO) as well as in SP are  stained with
both antisera ( B and C ), but the affinity-purification reduces the
diffuse uniform background staining and allows clear visualization
of terminals in the stratum oriens (C).

Fig. 11 presents IF micrographs of areas from coronal rat-brain
sections (Fig. 12) stained with affinity-purified anti-TSM antiserum
or, as controls (A', B' ), with FITC-labelled anti-sheep IgG alone.
Staining of terminals is shown for cerebral cortex (A), the nucleus
lateralis thalami (B) and the nucleus medialis thalami, pars lateralis (C).  Neurone staining is shown for the nucleus caudatus putamen
(D) and the tractus diagonalis (E).  Fibres were stained (F) in the
gyrus dentatus (GD) that abuts onto the fimbria (Fi).

These preliminary results clearly show that ·the affinity-purified anti-Chol-1 antiserum is a highly promising reagent for detecting cholinergic neuronal elements in mamalian brain.    Further work
will involve a detailed comparison with antisera specific for the
classical cholinergic marker, ChAT.

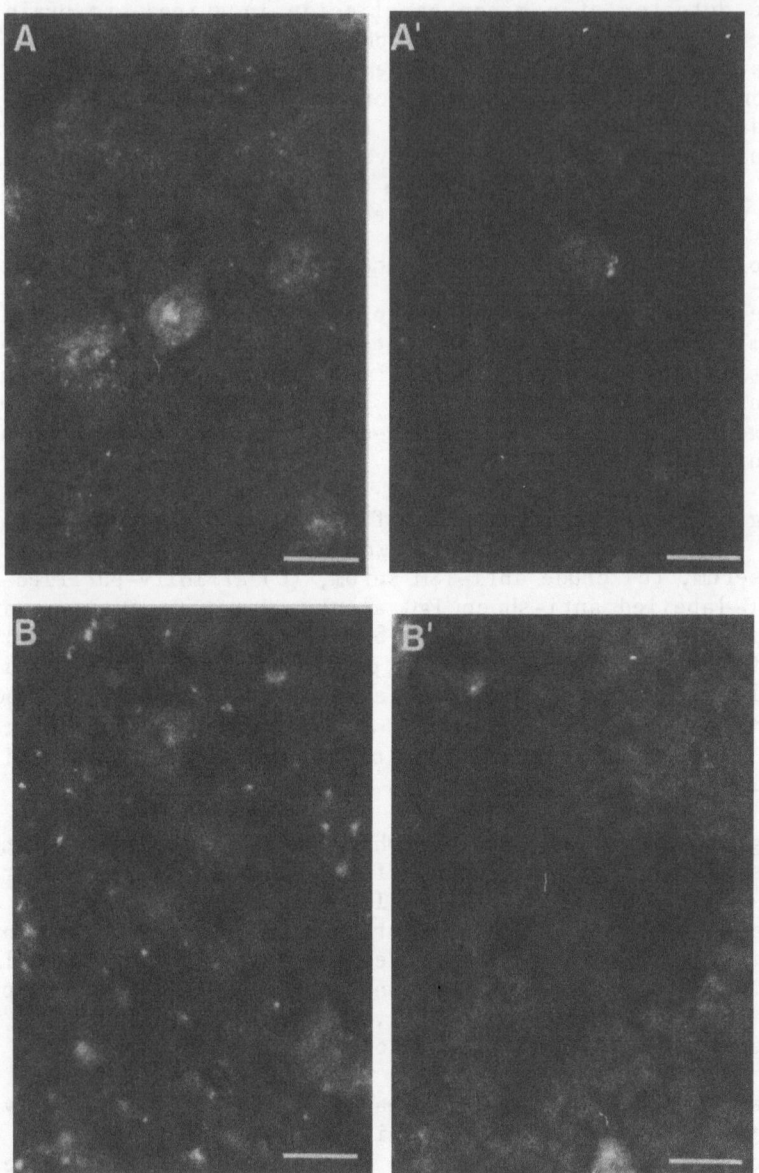

**Fig. 11.** IF staining with affinity-purified anti-TSM antiserum in cryostat sections of rat brain. **(A)**, cortex, **(B)** nucleus lateralis thalami, **(C)** nucleus medialis thalami, pars lateralis, **(D)** nucleus caudatus putamen, **(E)** tractus diagonalis, **(F)** gyrus dentatus. **A', B'**: controls for **A** & **B** using FITC-labelled anti-sheep IgG alone. For further explanation, see text. Bar is 10 μm.

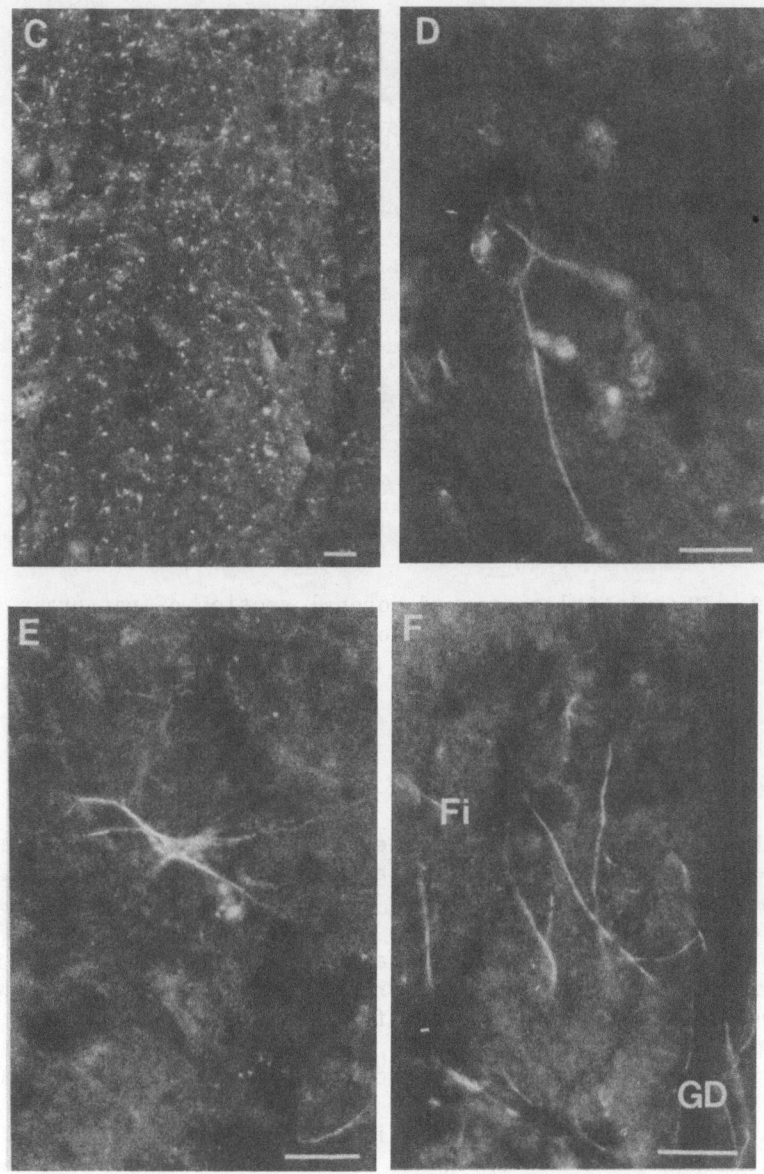

Fig. 11, *continued from opposite.*
GD, gyrus dentatus; Fi, fimbria.

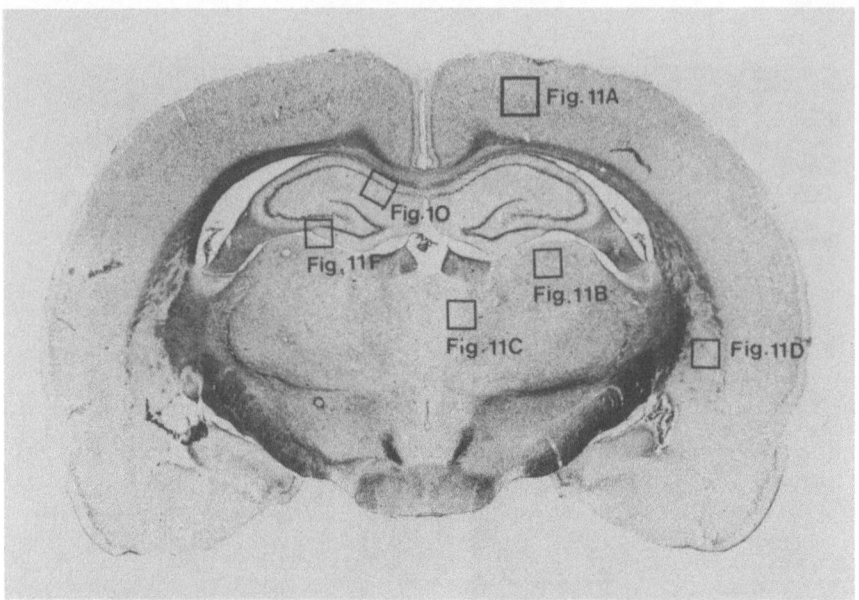

Fig. 12. Cresyl Violet-stained coronal section of rat brain with the areas shown in Figs. 10 & 11 outlined, except for Fig. 11 E which is rostral to this section.

CONCLUSIONS

Scheme 1 summarizes how antisera to transmitter-linked antigens have been obtained by means of a model cholinergic system. The various steps in the preparation of the antisera, the specificity criteria and the uses to which the antisera can be put have all been described. In principle, the same approach could be applied to other model systems, e.g. the adrenal medullary cell, a model for the adrenergic neurone. However, for many transmitters, e.g. glutamate, γ-aminobutyrate and glycine, model systems are either lacking or provide only impractically low concentrations of nerve terminals.

Is there any way round this impasse? The MAb technique should provide one. We are attempting to raise MAb's to mammalian cortical synaptosomal p.m. and testing these by their ability to induce complement lysis of selective subpopulations of synaptosomes. Hopefully other transmitter-related antigens will be discovered, and it would be interesting if these too proved to be gangliosides.

MODEL SYSTEM
*cholinergic (Torpedo electromotor terminals)*
*adrenergic (mammalian adrenal medullary cell)*

| Synaptic vesicles | Presynaptic plasma membranes |
| Storage granules | Plasma membranes |

*Antiserum*                    *Antiserum*

*Adsorption of unwanted Ab's*    *Adsorption of unwanted Ab's*

*Specific antiserum*            *Specific antiserum*

Identification of             Identification of
specific antigen            specific antigen

*Criteria of specificity*       *Criteria of specificity*
1. Copurification with       1. Copurification with membrane
   organelle                   2. Induces complement lysis of
2. Has correct location          synaptosome subpopulation
   in tissue                   3. Has correct location in tissue

*Applications*                 *Applications*
1. New anatomical method     1. New anatomical method
2. Is stable vesicle marker    2. Immunoabsorption of cholinergic
   2.1 Life-cycle of vesicle       synaptosomes
   2.2 Re-cycling of vesicle    3. Possible role in synaptogenesis
3. Synaptogenesis
[4. Immunoadsorption]

Scheme 1. The use of model systems (e.g. *Torpedo* electromotor
nerve terminals, mamalian adrenal medullary cells) to obtain antisera to transmitter-related antigens in (*left*) synaptic vesicles or
storage granules, (*right*) the plasma membranes (p.m.) through which
these storage granules discharge by exo- and endocytosis. The
method of preparing the adsorbed antisera, the criteria of specificity and investigations to which the antisera can be applied are
summarized.

*References*

1. Fox, G.C. & Richardson, G.P. (1978) *J. Comp. Neurol. 179*, 677–697.
2. Fox, G.Q. & Richardson, G.P. (1978) *J. Comp. Neurol. 185*, 293–316.
3. Stadler, H. & Tashiro, T. (1979) *J. Biochem. 101*, 171–178.
4. Whittaker, V.P., Essman, W.B. & Dowe, G.H.C. (1974) *Biochem. J. 128*, 833–845.
5. Ohsawa, K., Dowe, G.H.C., Morris, S.J. & Whittaker, V.P. (1979) *Brain Res. 161*, 447–457.
6. Walker, J.H., Obrocki, J. & Zimmermann, C.W. (1983) *J. Neurochem. 41*, 209–216.
7. Stadler, H. & Whittaker, V.P. (1978) *Brain Res. 153*, 408–413.
8. Stadler, H. & Dowe, G.H.C. (1982) *EMBO J. 1*, 1381–1384.
9. Walker, J.H., Jones, R.T., Obrocki, J., Richardson, G.P. & Stadler, H. (1982) *Cell Tiss. Res. 223*, 101–116.
10. Osborne, N.N., Beale, R.,Nicholas, D., Stadler, H., Walker, J.H., Jones, R.T. & Whittaker, V.P. (1982) *Cell Mol. Neurobiol. 2*, 151–163.
11. Jones, R.T., Walker, J.H., Stadler, H. & Whittaker, V.P. (1982) *Cell Tiss. Res. 223*, 117–126.
12. Jones, R.T., Walker, J.H., Stadler, H. & Whittaker, V.P. (1982) *Cell Tiss. Res. 224*, 685–688.
13. Zimmermann, H. & Denston, C.R. (1977) *Neuroscience 2*, 715–730.
14. Giompres, P.E., Zimmermann, H. & Whittaker, V.P. (1981) *Neuroscience 6*, 765–774.
15. Giompres, P.E. & Whittaker, V.P. (1984) *Biochim. Biophys. Acta 770*, 166–170.
16. Jones, R.T. & Fox, G.Q. (1982) *Abstracts, Eur. Symp. on Cholinergic Mechanisms, Strasbourg* (R. Massarelli, organizer), p. 59.
17. Jones, R.T., Walker, J.H., Jones, R.T. & Whittaker, V.P. (1982) *Cell Tiss. Res. 218*, 355–373.
18. Richardson, P.J., Walker, J.H., Jones, R.T. & Whittaker, V.P. (1982) *J. Neurochem. 38*, 1605–1614.
19. Richardson, P.J. (1981) *J. Neurochem. 37*, 258–260.
20. Ferretti, P. & Borroni, E. (1984) *J. Neurochem. 42*, 1085–1093.
21. Ferretti, P. & Borroni, E. (1983) *Abstracts, Symp. on Ganglioside Structure Function and Biomedical Potential* (held at Island Hall Hotel, B.C., Canada), p. 62.
22. Young, W.W., Laine, R.A. & Hakomori, S.I. (1978) *Meth. Enzymol. 50*, 137–140.
23. Tettamanti, G., Bonali, F., Marchesini, S. & Zambotti, V. (1973) *Biochim. Biophys. Acta 296*, 160–170.
24. Miettinen, T. & Takki-Lukkainen, I.T. (1958) *Acta Chem. Scand. 13*, 856–858.
25. Brockhaus, M., Masnani, J.L., Blaszczyk, M., Stephlewki, Z., Koprowski, H., Karlsson, K.A., Larson, G. & Ginsburg, V. (1981) *J. Biol. Chem. 256*, 1322–1323.

#C-5

# USE OF ANTIBODIES TO PURIFY CHOLINERGIC NERVE TERMINALS

**Peter J. Richardson**

Department of Clinical Biochemistry
University of Cambridge
Addenbrooke's Hospital, Hills Road
Cambridge CB2 2QR

*In the isolation of nerve terminals from the mammalian CNS - crucial for elucidating synaptic biochemistry [1] - a major limitation is the heterogeneity of the preparations, particularly the diversity of distinct neuronal types present. Previous attempts to purify neurotransmitter-specific nerve terminals were frustrated by similarities in density and size of most terminals [2], while the absence of known transmitter-specific surface antigens prevented the application of immunological techniques.*

*A complement-mediated cytotoxicity assay has demonstrated the presence of a cholinergic-specific surface antigen (Chol-1) on the nerve terminals of many species. Mammalian cholinergic nerve terminals sensitized with sheep anti-(Chol-1) sera have been isolated on a high-capacity immunoabsorbent [3], viz. a mouse monoclonal antibody covalently coupled to cellulose. The isolated terminals exhibited high-affinity uptake and released acetylcholine (ACh) upon depolarization in a $Ca^{2+}$-dependent manner. These immunoaffinity-purified cholinergic terminals represent the first population of neurotransmitter-specific nerve terminals to be purified from mammalian brain.*

The study of CNS biochemistry has been hampered by the difficulty of separating the constituent cells and subcellular particles into homogeneous populations. Even the separation of neurones from glia results in the isolation of very few viable neurones [4], and it has not as yet proved possible to isolate one homogeneous (e.g. transmitter-specific) population of neurones from the remainder. A similar problem is encountered with isolated nerve terminals which can be prepared in high yield from the CNS [1]. These terminals, like their parent neurones, may contain one or more of ~40 transmitters and modulators recognized in the nervous system. The absence

of such homogeneous populations has hindered the study of the biochemical basis by which nerves and their terminals integrate extracellular signals and then modulate their transmitter release.  There
is therefore a requirement for development of methods by which biochemically homogeneous cells and their constituents can be purified
from nervous tissue.

Immunoaffinity techniques have both the required  selectivity
and capacity for the purification of cells and particles from complex
mixtures [5-12].    However,  very few of the requisite cell-type-
specific surface antigens have been described in the mammalian
nervous system [13]. It was considered likely that a potential source
of such antigens would be transmitter-specific regions of nervous
systems in distantly related species, which nevertheless use the
same transmitters as mammalian nerves.  One example is the electric
organ of *Torpedo*  which has a purely cholinergic innervation [cf.
E.A. Barnard, #A-6 in Vol. 13, this series - *Ed.*]. A sheep antiserum
was therefore raised in *Torpedo* nerve-terminal membranes [14].

To determine whether the antiserum binds to cholinergic nerve
terminals and, if so, whether it was specific for these, a cytotoxicity test was used: those terminals that bear antibodies are lysed
in the presence of complement, and the lysed subpopulation identified by the release of soluble markers.

## METHODS

*Anti-(Chol-1) serum.-* The procedure of Israel et al. [15] furnished nerve terminals from *T. marmorata* electric organ.  After hypo-
tonic lysis, the membranes were suspended in 10 mM Hepes containing
0.1% Triton X-100 mixed with an equal vol. of Freund's complete adjuvant, and the sheep injected s.c. (doses 0.75 mg protein; 6 months).

*Nerve terminal labelling.-* Nerve  terminals  were  labelled  by
incubation with low concentrations (<0.5 µM) of a number of tritiated
transmitters, and with [$^3$H]deoxyglucose which served as a more general
marker for cytoplasmic space [16].  After washing, the labelled terminals were subjected to the cytotoxicity test.

*Cytotoxicity test.-* Complement-mediated lysis of nerve terminals
[16] entailed, in brief, incubation of crude mitochondrial fractions
($P_2$), from 5-20 mg rat brain [1], for 30 min at 4° with a 1:160
dilution of anti-(Chol-1) serum and centrifugation at 10,000 g for
2 min in an Eppendorf bench centrifuge.  After re-suspension in a
1:15 dilution of human serum (a source of complement), the terminals
were incubated for 20 min at 32°.  Samples (0.25 ml) were centrifuged as before, and the supernatants assayed for marker release.

*Immunoabsorbents.-* The range of immunoabsorbents tested included
Protein A and polyclonal anti-(sheep IgG) sera coupled to Sepharose,

Sephadex and cellulose.  The most useful immunoabsorbent consisted of a mouse  monoclonal anti-(sheep IgG) antibody (avidity ~$10^{11}M^{-1}$) covalently coupled to cellulose as described in [17].    Its protein content was $245 \pm 45$ (SEM, n=5) µg/mg cellulose.

*Affinity purification of cholinergic nerve terminals.-* Cholinergic nerve terminals were purified from rat caudate nucleus largely as described in [3].   Briefly, after homogenization and centrifugation at 1,000 g  for 10 min, the $S_1$ fraction [1] was incubated for 40 min with a 1:40 dilution of anti-(Chol-1) serum.  After centrifugation (16,000 g,  12 min) the  pellet was washed 3 times and resuspended in a medium of the following composition (mM): NaCl, 145; KCl, 3; glucose, 10; Hepes, 5; EDTA, 2;  pH 7.4.  Terminals derived from ~ 30 mg tissue were incubated with 1.25 mg of a cellulose immunoabsorbent bearing the mouse monoclonal anti-(sheep IgG) antibody. After 40 min, bound and free terminals were separated by centrifuging at 9,000 g  for 1 min, with 2.5 ml of 45% (v/v) Percoll made up in the above salt solution.   After a further two washes with Percoll, the immunoabsorbent was washed once with a Krebs Ringer Hepes solution, viz. (mM): NaCl, 125; KCl, 2.6; $MgCl_2$, 1; $CaCl_2$, 1.3; $NaHCO_3$, 21; glucose, 10; Hepes, 10; pH 7.4 (abbreviated KRH below).

*Choline uptake and acetylcholine release.-* Immunoabsorbent-bound terminals derived from ~20 mg tissue were pre-incubated for 5 min at 37° in Krebs Ringer Hepes (KRH), spun at $1 \times 10^4$g (30 sec) in an Eppendorf bench centrifuge and re-suspended in 0.25 ml of the same medium supplemented with 0.1 mM eserine and [$^3$H]choline (8 µCi/ml, 0.2 to 100 µM).  The incubation was stopped by dilution in 5 ml ice-cold pH 7.4 phosphate-buffered saline (PBS; 0.15 M NaCl, 10 mM $NaH_2PO_4$) containing 0.1 mM eserine, and filtration through Whatman GF/F glass fibre filters.   The filters were washed 3 times with 5 ml saline and analyzed for radioactivity.   In some experiments  the filters were extracted twice with 2 ml of chloroform/methanol/water (30:60:10 by vol.), and the extracts pooled and dried down.   The relative amounts of phosphorylated, acetylated and free choline were then determined after HPLC separation [18].  The identity of the labelled derivatives of choline was confirmed by TLC [19].

In other experiments immunoadsorbent-bound terminals were labelled by incubation for 15 min at 37° with  0.2 µM [$^3$H]choline (10 Ci/ml), washed and placed in a periperfusion chamber (0.25 ml volume). The terminals were perifused with KRH (1 ml/min), and their ability to release labelled ACh upon stimulation evaluated. When testing the influence of extracellular $Ca^{2+}$, the  zero-$Ca^{2+}$ condition consisted of KRH lacking in $Ca^{2+}$ and containing 1 mM EGTA.

**RESULTS**

*Specificity of the anti-(Chol-1) serum.-* Fig. 1 shows the ability of this serum to promote the release of two markers as a result

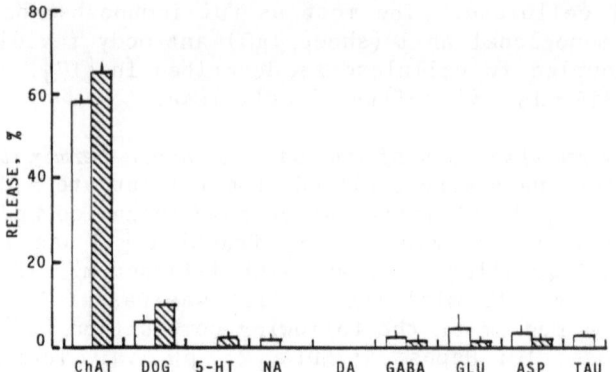

**Fig. 1.** Release by anti-(Chol-1) serum of nerve-terminal markers during complement-mediated lysis. Pre-loaded terminals were lysed as described. *White columns,* cortex; *striped columns,* hippocampus. Results are means ±S.E.M. of 3 or more experiments. ChAT, choline acetyltransferase; DOG, 2-deoxyglucose; 5-HT, serotonin; NA, noradrenaline; DA, dopamine; GABA. 4-aminobutyrate; GLU, glutamate; ASP, aspartate; TAU, taurine. *From [16], by permission.*

of complement activation. The release of ~60% of the choline acetyltransferase was accompanied by the release of only 6% of the deoxyglucose, while there was no significant release of the 7 transmitter labels studied. The antiserum therefore appears to be specific for cholinergic nerve terminals.·

*Immunoadsorbent.* - In initial experiments, with immunoadsorbents bearing polyclonal antisera or Protein A, no purification of cholinergic terminals was achieved. It was thought likely that this was due to their low IgG-binding capacity. An immunoadsorbent of high binding capacity was therefore developed using a MAb covalently coupled to cellulose. Table 1 compares the binding capacities of a number of commonly used immunoadsorbents. It is apparent that the best combination consists of a MAb with the high protein-bearing capacity of cellulose.

*Affinity purification.* - Incubation of nerve terminals, sensitized by the anti-(Chol-1) serum, with the immunoadsorbent resulted in a maximum yield of bound choline acetyltransferase of 24%. However, in order to restrict the non-specific binding of terminals (i.e. that obtained with terminals sensitized with non-immune sheep serum) to acceptably low levels (<0.5%) it was necessary to reduce the amount of immunoadsorbent present in the incubation. Table 2 shows the purification achieved from two brain regions under conditions chosen to minimize the non-specific binding. It is apparent that both choline acetyltransferase and acetylcholinesterase have been purified, and that this was accompanied by some of the general cytoplas-

**Table 1.** IgG-binding capacities of anti-IgG immunoabsorbents. Binding capacities of the cellulose and Sepharose 4B (protein A) immunoabsorbents were determined using $^{125}I$-labelled sheep IgG tracer, in the presence of excess unlabellled sheep IgG. The remaining values were deduced from the sources cited.

| Matrix | Anti-IgG | IgG binding capacity/mg |
|--------|----------|-------------------------|
| Cellulose | Monoclonal | 50-250 µg |
| Cellulose | Polyclonal | 0.5-2.5 µg |
| Sepharose 4B | Protein A | 70 µg |
| Sepharose 4B | Monoclonal | <6 µg [11] |
| Sepharose 4B | Protein A | 10-15 µg [21] |
| Polyacrylamide | Polyclonal | <1 µg [22] |

**Table 2.** Affinity purification of cholinergic nerve terminals from rat caudate nucleus and cerebral cortex. Affinity purification of nerve terminals sensitized with the anti-(Chol-1) serum was as described. Purification is expressed as the ratio of the specific activity in the purified terminals to that in the washed $P_2$ fraction.

| | Caudate nucleus | | Cortex | |
|---|---|---|---|---|
| | Yield, % | Purification | Yield, % | Purification |
| Choline acetyltransferase | 14.6 | 13.0 | 9.5 | 18.9 |
| Acetylcholinesterase | 7.6 | 6.8 | 7.1 | 14.1 |
| Glutamate decarboxylase | <0.5 | <0.5 | 0.2 | 0.5 |
| Lactate dehydrogenase | 4.8 | 4.2 | 0.8 | 1.5 |

mic marker lactate dehydrogenase. The lesser purification of acetylcholinesterase compared to that of choline acetyltransferase can be explained by the fact that although the former is present in all cholinergic synapses, it is also present elsewhere [20]. Purification of the cholinergic terminals with respect to the general cytoplasmic compartment can be assessed by inspecting the ratio of choline acetyltransferase to lactate dehydrogenase activities in the purified terminals. On this basis these have been purified 3.5-fold from the caudate nucleus and 13.5-fold from the cerebral cortex. The ratios obtained are also very similar to the ratio of the released activities upon complement-mediated lysis by anti-(Chol-1) serum [3].

*Metabolic competence of the purified terminals.-* The purified terminals can oxidize glucose to $CO_2$, and have electron-microscopic

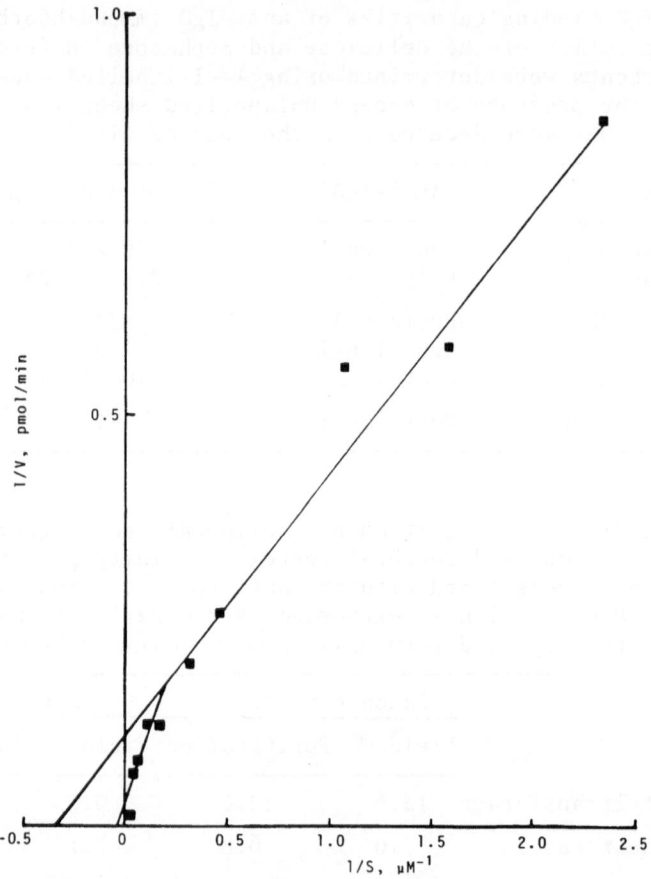

**Fig. 2.**  Uptake of choline by affinity-purified cholinergic nerve terminals.  Apparent $K_m$ of choline transport were $2.75 \pm 0.5$ (6) $\mu M$ and $38.2 \pm 7.5$ (4) $\mu M$ for the high- and low-affinity transport systems respectively.  The apparent $V_{max}$ of the high-affinity uptake system was $794 \pm 150$ (6) pmol/min/mg protein.  (The means are given with their SEM and no. of observations.)

profiles characteristic of nerve terminals [3].  Fig. 2 illustrates their ability to take up labelled choline via two kinetically different uptake systems – one with an apparent $K_m$ of 5 $\mu M$ and the other of 45 $\mu M$.  Under normal conditions most of the label (65%) taken up via the high-affinity system (i.e. in the presence of 2 $\mu M$ extracellular choline) is acetylated provided that there is a sufficient supply of substrate, e.g. glucose or pyruvate.

The ability of the purified terminals to respond to depolarizing stimuli is shown in Fig. 3.  In the presence of high $K^+$ concentrations (25 mM) and normal $Ca^{2+}$ the terminals are stimulated to

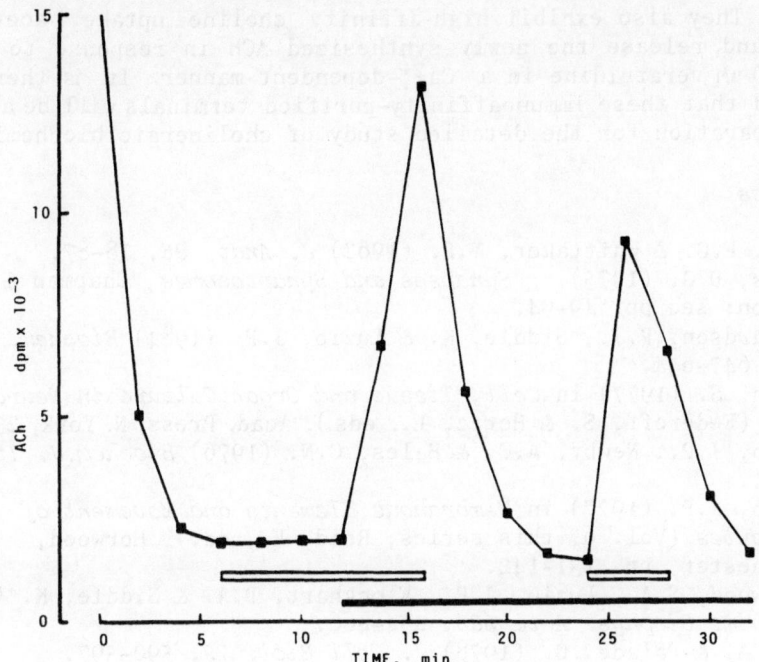

**Fig. 3.** Effect of high K$^+$ and Ca$^{2+}$ on the release of ACh from affinity-purified nerve terminals, pre-loaded with [$^3$H]choline and perifused as described. *Open bar:* 25 mM K$^+$ present in the perifusion medium; *black bar*, 1.3 mM Ca$^{2+}$ present.

release ACh. A similar Ca$^{2+}$ dependency of release is obtained with 40 µM veratridine.

## DISCUSSION

The discovery of the cholinergic-specific surface antigen Chol-1 [14] has for the first time permitted the isolation of a homogeneous population (with respect to transmitter) of nerve terminals. In the past such a development has been prevented by the absence of recognized transmitter-specific surface antigens. The other main contributing factor to this development was the use of a high-capacity immunoadsorbent which probably confers two main advantages. Firstly it decreases the influence of binding competition between free sheep IgG and the labelled terminals, and secondly it increases the chance of productive binding between the cellulose particles and the labelled terminals.

The potential usefulness of this preparation is dependent on the terminals being metabolically competent and being capable of responding appropriately to depolarizing stimuli. It has already been shown that they are osmotically sensitive and metabolize glucose [3]. They also exhibit high-affinity choline uptake, acetylate choline and release the newly synthesized ACh in response to 25 mM $K^+$ and 50 µM veratridine in a $Ca^{2+}$-dependent manner. It is therefore concluded that these immunoaffinity-purified terminals will be a suitable preparation for the detailed study of cholinergic biochemistry.

*References*

1. Gray, E.G. & Whittaker, V.P. (1962) *J. Anat. 96,* 78-87.
2. Jones, D.G. (1975) *Synapses and Synaptosomes,* Chapman & Hall, London; see pp. 49-94.
3. Richardson, P.J., Siddle, K. & Luzio, J.P. (1984) *Biochem. J. 219,* 647-654.
4. Varon, S. (1977) in *Cell, Tissue and Organ Culture in Neurobiology* (Federoff, S. & Hertz, L., eds.), Acad. Press, N. York, 236-261.
5. Luzio, J.P., Newby, A.C. & Hales, C.N. (1976) *Biochem.J. 154,* 11-21.
6. Luzio, J.P. (1977) in *Membranous Elements and Movement of Molecules* (Vol. 6, this series; Reid, E., ed.), Horwood, Chichester, pp. 131-142.
7. Westwood, S.A., Luzio, J.P., Flockhart, D.A. & Siddle, K. (1979) *Biochim. Biophys. Acta 583,* 454-466.
8. Ito, A. & Palade, G. (1978) *J. Cell Biol. 79,* 590-597.
9. Matthew, W.D., Tsavaler, L. & Reichardt, L.F. (1981) *J. Cell Biol. 91,* 257-269.
10. Luzio, J.P. & Stanley, K.K. (1983) *Biochem. J. 216,* 27-36.
11. de Kretser, T.A., Bodmer, J.G. & Bodmer, W.F. (1980) *Tissue Antigens 16,* 317-325.
12. Pontremoli, S., Melloni, E., Damiani, G., Michetti, M., Salamino, F., Sparatore, B. & Horecker, B.L. (1984) *Arch. Biochem. Biophys. 233,* 267-271.
13. Schachner, M. (1982) *J. Neurochem. 38,* 1605-1614.
14. Jones, R.T., Walker, J.H., Richardson, P.J., Fox, G.Q. & Whittaker, V.P. (1981) *Cell Tissue Res. 218,* 355-373.
15. Israel, M., Manaranche, R., Mastour-Franchon, P. & Morel, N. (1976) *Biochem. J. 160,* 113-115.
16. Richardson, P.J. (1983) *J. Neurochem. 41,* 640-648.
17. Hales, C.N. & Woodhead, J.S. (1980) *Meths. Enzymol. 70,* 334-355.
18. Potter, P.E., Meek, J.L. & Neff, N.H. (1983) *J. Neurochem. 41,* 188-194.
19. Marchbanks, R.M. & Israel, M. (1971) *J. Neurochem. 18,* 439-448.
20. Silver, A. (1974) *The Biology of Cholinesterases,* North-Holland, Amsterdam, 596 pp.
21. Ghetie, V., Mota, G. & Sjoquist, J. (1976) *J. Immunol. Meths. 21,* 133-141.
22. Ito, A. & Palade, G. (1978) *J. Cell Biol. 79,* 590-597.

*Relevant article in this book:* #C-5, by V.P. Whittaker et al., describes the characterization of the anti-(Chol-1) serum and immunofluorescence results it furnished.

#NC(C)

NOTES and COMMENTS related to the foregoing topics

Comments related to particular contributions:

#C-1, #C-2 & #C-4, p. 227
#NC(C)-1 to -3, p. 228

#NC(C)-1

*A Note on*

# AN IMMUNOLOGICAL APPROACH TO THE CELL LINEAGE PROBLEM IN THE MAMMALIAN CNS*

Martin C. Raff

Medical Research Council Neuroimmunology Project
Zoology Department, University College London
Gower Street, London WC1E 6BT

The majority of neurons and glial cells in the mammalian CNS develop from the neuroepithelial cells that form the neural tube. But what is the lineage relationship between these different cell types and what determines whether an individual neuroepithelial cell becomes a particular type of neuron or one of the several types of glial cells?

We have used conventional and monoclonal antibodies (Ab's) and tissue culture to study cell lineages in the simplest part of the CNS – the optic nerve – in which glial cells but not neurons develop. The optic nerve contains three distinct types of glial cells: type 1 astrocytes, oligodendrocytes, and type 2 astrocytes, which develop in that order from two separate cell lineages, one giving rise to type 1 astrocytes before birth and the other to oligodendrocytes and then type 2 astrocytes after birth. The latter two cell types develop from a common progenitor cell that differentiates *in vitro* into an oligodendrocyte if cultured in serum-free medium and into a type 2 astrocyte if cultured in foetal calf serum. Our immunological approach has helped elucidate factors that control the timing and direction of glial cell differentiation.

*Illustrative publications*

a. Raff, M.C., Abney, E.R., Cohen, J., Lindsay, R. & Noble, M. (1983) *Two types of astrocytes in cultures of developing rat white matter: differences in morphology, surface gangliosides and growth characteristics.* J. Neuroscience 3, 1289-1300.

---

* This Forum Abstract is supplemented, in respect of approaches, by an Editor's précis, not by author's text (it had been agreed that none would be furnished).

*b.* Raff, M.C., Miller, R. & Noble, M. (1983) *A glial progenitor cell that develops* in vitro *into an astrocyte or an oligodendrocyte depending on the culture medium. Nature 303,* 390-396.

*c.* Miller, R.H. & Raff, M.C. (1984) *Fibrous and protoplasmic astrocytes are biochemically and developmentally distinct. J. Neurosci. 4,* 585-592.

*d.* Raff, M.C., Abney, E.R. & Miller, R.H. (1984) *Two glial cell lineages diverge prenatally in rat optic nerve. Dev. Biol. 106,* 53-60.

*e.* Raff, M.C., Williams, B.P. & Miller, R.H. (1985) *The* in vitro *differentiation of a bipotential glial progenitor cell. EMBO J.,* in press.

==========

## *SOME METHODOLOGICAL POINTS (Senior Editor's excerpts)*

*Ref. a:* Cell suspensions were prepared by incubating fragments of optic nerve with trypsin and collagenase, followed by dissociation by repeated pipetting with DNAase and trypsin inhibitor present but no $Ca^{2+}$ or $Mg^{2+}$. The preparation of purified astrocytes, based on a published procedure (McCarthy & de Vellis), likewise entailed a trypsin/EDTA step, applied to adherent cells, after 9-10 day culture of the cells dissociated from the cerebral cortex of 2-day rats; re-culture gave a high proportion of cells positive for glial fibrillary acidic protein ($GFAP^+$). The range of Ab's used included rabbit anti-human GFAP serum and anti-bovine neurofilament serum, and monoclonal anti-rat neurofilament Ab. To visualize Ab's in indirect immunofluorescence assays, use was made (with appropriate choice of species) of commercial anti-Ig that had been absorbed with Ig and coupled with fluorescein or rhodamine. Foetal calf serum was present (10%) both in diluents and in stored cell suspensions. For immunofluorescence labelling, cells were incubated with ascites fluid containing the monoclonal Ab A2B5 (or with tetanus toxin followed by mouse anti-tetanus toxoid), with a heated goat serum present to saturate macrophage Fc receptors, and after fixation were labelled with rabbit anti-GFAP serum followed by a coupled anti-Ig Ab (goat anti-mouse).

*Ref. b:* In optic nerve cultures most type 2 astrocytes develop from $A2B5^+$ precursor cells, as had been shown for oligodendrocytes by strategies now again applied: "(1) if (the 7-day) cells are killed with A2B5 Ab and complement before culture, no type 2 astrocytes develop; (2) if (the) cells are prelabelled with A2B5 Ab before culture, 65-75% of the type 2 astrocytes that develop after 2 days have residual A2B5 Ab on their surface, indicating that they have developed from $A2B5^+$ cells in the original cell suspension." The concluding step in (1) is cell-counting, after fixation etc. as in *a* above, in glycerol on coverslips by phase-contrast with epi-UV illumination and selective filters for fluorescein and rhodamine.

#NC(C)-2

*A Note on*

# AN ENZYME-LINKED IMMUNOSORBENT ASSAY FOR CALMODULIN

M.E. Bardsley, P.J. Roberts and [*]I.R. Cottingham

Department of Physiology
and Pharmacology
University of Southampton
Medical & Biological Sciences
Building, Bassett Crescent E.
Southampton SO9 3TU, U.K.

[*]Department of
Biochemistry
University of Sussex
Falmer
Brighton BN1 9QG, U.K.

Calmodulin (CaM) is a $Ca^{2+}$-binding protein notably important in the translation of $Ca^{2+}$ signals inside the cell (as reviewed [1, 2]) [cf. A.R. Means & J.G. Chafouleas, Vol. 13, #A-7, this series – *Ed.*]. It is found throughout the plant and animal kingdoms and its sequence is highly conserved [3]. Measurement of CaM by its activation of 3',5'-cyclic nucleotide phosphodiesterase, and using radioimmunoassay (RIA) or enzyme-linked immunosorbent assay (ELISA), provides useful comparative measurements of the amount of protein and its activity. Methods have been described for assay by enzyme activation [4] and by RIA ([5, 6], & Vol. 13 art. cited above). Here we describe the production of an anti-CaM antibody (Ab) and its use in setting up an ELISA. Elsewhere [7] we exemplify the applicability of the ELISA.

*Purification of CaM.*    Based on a published method [8], whole fresh bovine brain was homogenized in ice-cold 4% (w/v) trichloro-acetic acid and stirred for 20 min at 4°. The homogenate was centrifuged at 6,500 g for 90 min and the pellet re-homogenized in buffer containing PMSF[*], pH 7.5, re-adjusted to pH 7.5 with NaOH and centrifuged (48,500 g, 140 min). The supernatant was loaded onto a DEAE column pre-equilibrated with B containing ME and 0.1 mM EGTA, and the column was washed with the same buffer + ammonium sulphate (to 145 mM). The CaM-containing fraction was eluted by B/0.3 M ammonium sulphate, then diluted with 1 vol. B/ME/0.5 mM $CaCl_2$ and loaded onto an epoxy-linked fluphenazine-Sepharose 4B affinity column ([8]; 5 g; equilibrated with B/ME/500 mM NaCl/0.2 mM $CaCl_2$). After washing with the equilibration buffer, $CaCl_2$ was replaced by 2 mM EGTA to elute

---

[*] Abbreviations: B = pH 7.5 20 mM Tris-HCl; ME = 5 mM 2-mercaptoethanol; PMSF (0.5 mM; in B containing 5 mM EDTA) = phenylmethylsulphonyl fluoride; PBS = phosphate-buffered saline, pH 7.4.

the CaM, with use of UV absorbance (260 & 280 nm monitoring) and SDS-PAGE to follow CaM purification. – On running adjacent PAGE tracks with 8 mM EGTA or 0.2 mM $CaCl_2$ present, and staining with Coomassie Brilliant Blue, CaM is easily identified by its characteristic sigmoid band shape (Fig. 1a). Both column steps were done at room temp.

In agreement with Kakiuchi et al. [8] we found this method of purification to to be simple, rapid and efficient (yield 70-80%). Even with our described conditions, Western blotting shows impurities that call for preparative SDS-PAGE (see below) to obviate problems in raising Ab's and (unpublished work) in immunofluorescence studies.

*Raising Ab's to CaM*. The purified CaM was oxidized with performic acid [9] and 12 mg injected s.c. into a rabbit over a 3-week period (as in [10]); the first bleeding was at 6 weeks. The serum showed measurable immunoreactivity by ELISA (cf. below) towards both native CaM purified as above, and bought (Sigma) CaM, for at least 6 months following immunization.

*Purification of Ab's*. An IgG fraction was prepared from serum by fractionation with 6,9-diamino-2-ethoxyacridine lactate and ammonium sulphate [see 11], and immunoabsorbed onto a CaM-agarose column (1 ml; purchased from Sigma) equilibrated in 125 mM borate (pH 8.4) containing 75 mM NaCl. Before loading, the IgG fraction (1.5 ml, ~39 mg protein) was dialyzed overnight against this buffer, as also used to wash the column till $A_{280}$ was zero (10 x 1 ml aliquots). The anti-CaM Ab's were eluted with pH 2.7 200 mM glycine (usually fractions 2 & 3, 2 ml), then concentrated and dialyzed against 10 mM phosphate buffer pH 7.4. With the affinity step, the anti-CaM Ab's were obtained in 55% yield with 670-fold purification as judged by ELISA.

*ELISA*. As in the 'sandwich' assay of Engvall & Perlmann [12], CaM was diluted with 50 mM pH 9.6 carbonate/bicarbonate to give 10 µg protein/ml, and 150 µl aliquots added to polystyrene cuvettes which were incubated overnight at $4°$. Subsequent steps were at room temperature; between each the cuvettes were washed 5 times with PBS containing 0.05% Tween 20 (PBST). Ab (200 µl) at the appropriate dilution was added and, after 2 h, 200 µl of a 1:500 dilution of goat anti-rabbit IgG peroxidase conjugate (Nordic Immunological Labs., Maidenhead, U.K.). After incubation for a further 2 h, peroxidase was assayed with 0.2 ml of *o*-phenylenediamine (2.2 mM)/$H_2O_2$ (3.9 mM) in 24 mM citric acid/51 mM $Na_2HPO_4$, pH 5.0. After 15 min the reaction was stopped with 200 µl of 0.3 M citrate buffer, pH 2.9. The cuvettes were read (405 nm filter) on a manual EIA reader.

Fig. 2 shows an Ab dilution curve for affinity-purified anti-CaM Ab's and the displacement by CaM of Ab binding. The basic protocol outlined was routinely extended to cross-linking CaM to BSA with glutaraldehyde prior to adsorption onto the cuvette wall and including a 2 h incubation in BSA (150 µg/ml) before adding Ab.

**Fig. 1.** *Left.*- Characteristic
$Ca^{2+}$-dependent change in
mobility of CaM (brain, purified; 4 µg/track) in SDS-
PAGE. The two pairs of adjacent tracks show respectively
runs with 10 mM EGTA *(on left)*
and with 0.8 M $Ca^{2+}$.
*Right.*- Immunoblotting to
detect CaM (brain, purified;
2 µg/track) and a high mol.
wt. impurity after SDS-PAGE
(20% polyacrylamide) and
electrophoretic transfer to
nitrocellulose sheet; blot
treated with 100 µl purified
anti-CaM Ab's (detection as
in text). One pair of tracks
*(left)* contained 0.8 mM $Ca^{2+}$,
and the other 5 mM EGTA.

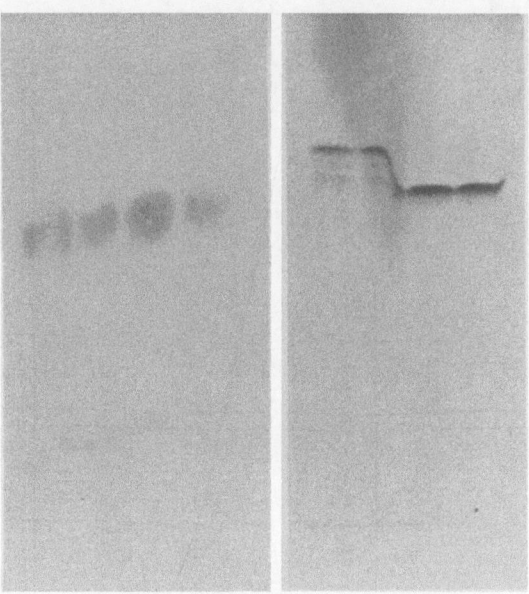

## PROBLEMS ENCOUNTERED

Owing to the low antigenicity of CAM, the presence of a small
amount of impurity in our CaM preparation gave rise to a heterogeneous population of Ab's. These impurities were manifest by staining
(not shown) only if the SDS-PAGE load were very heavy, but were easily
seen by transfer to nitrocellulose and immunodecoration (Fig. 1; left
pair of lanes in right portion). For a particular high mol. wt.
impurity that is evident, Ab's are present both in the IgG fraction
and in the affinity-purified Ab's. Curiously, this impurity also
shows $Ca^{2+}$-dependent mobility changes; this and its co-purification
with CaM suggest that it is a related protein.

The presence of impurity appeared not to interfere in our ELISA,
as checked with CaM from Sigma although not all their batches were
impurity-free. Immunoblotting gives a misleading idea of the relative amounts of the two Ab populations, as bound Ab's were detected
using the highly sensitive second-Ab biotin–straptavidin horse radish
peroxidase conjugated system (Amersham $Int^{1.}$). This detection system
gives a non-linear response to the amount of bound Ab. We therefore
suggest that in the ELISA the Ab's to the impurity are too low to
interfere significantly. Also, the impurities represent very little
of the total protein. However, to avoid such complications in future
Ab preparations, we will include preparative SDS-PAGE as a final
step in CaM purification.

*Acknowledgements.*- This work was supported by project grants
from the Wellcome Trust and the SERC. We thank D.C. Williams for gifts
of CaM and C.I. Ragan, M. Cleeter and V. Taylor for help and generosity.

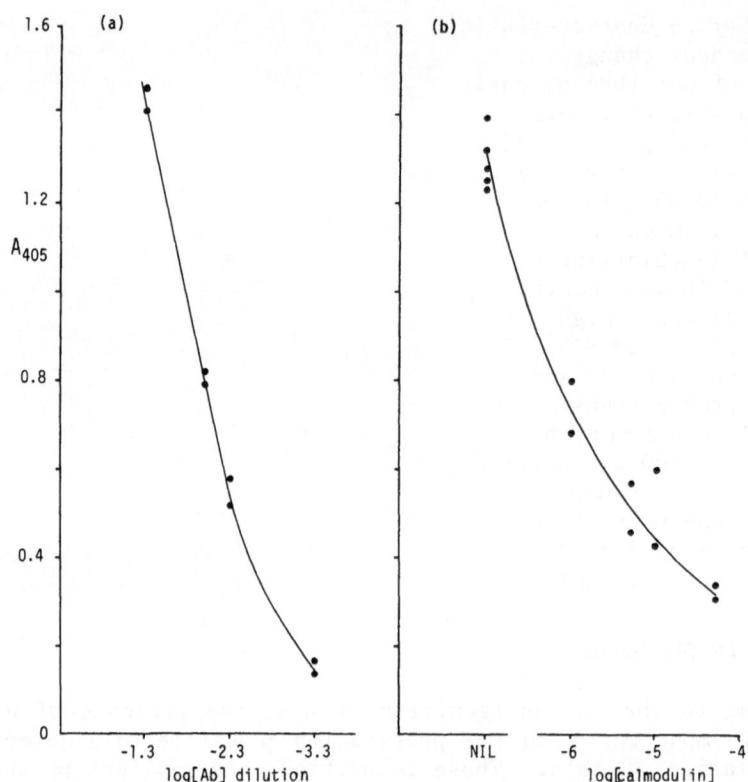

**Fig. 2.** *Left:* ELISA absorbance *vs.* Ab dilution; *right:* displacement by CaM of Ab binding in ELISA. Affinity-purified Ab was used.

*References*

1.  Klee, C.B., Crouch, T.H. & Richmann, P.G. (1980) *Ann. Rev. Biochem. 49*, 489-515.
2.  Kakiuchi, S., Hidaka, H. & Means, A.R., eds. (1982) *Calmodulin and Intracellular Calcium Receptors*, Plenum, New York, 425 pp.
3.  Vanaman, T.C. (1980) in *Calcium and Cell Function*, Vol. 1 (Cheung, W.Y., ed.), Academic Press, New York, pp. 41-58.
4.  Wallace, R.W., Tallant, E.A. & Cheung, W.Y. (1983) *Meths. Enzymol. 102*, 39-47.
5.  Chafouleas, J.G., Dedman, J.R., Munjaal, R.P. & Means, A.R. (1979) *J. Biol. Chem. 254*, 10262-10267.
6.  Wallace, R.W. & Cheung, W.Y. (1979) *J. Biol. Chem. 254*, 6564-6571.
7.  Bardsley, M.E., Roberts, P.J. & Cottingham, I.R. (1985) *Biochem. Soc. Trans.*, in press.
8.  Kakiuchi, S., Sobue, K., Yamazaki, R., Kambayashi, J., Sakon, M. & Kosaki, G. (1981) *FEBS Lett. 126*, 203-207.
9.  Hirs, C.H.W. (1967) *Meths. Enzymol. 11*, 197-199.
10. Van Eldik, L.J. & Watterson, D.M. (1981) *J. Biol. Chem. 256*, 4205-4210.
11. Hurn, B.A.L. & Chantler, S.M. (1980) *Meths. Enzymol. 70*, 104-142,
12. Engvall, E. & Perlmann, P. (1972) *J. Immunol. 109*, 129-135.

#NC(C)-3

*A Note on*

# INTERMEDIATE FILAMENT DIVERSITY AS DETECTED BY ANTIBODIES*

Werner W. Franke and Roy A. Quinlan

Institute of Cell and Tumor Biology
German Cancer Research Center
D-6900 Heidelberg, W. Germany (FRG)

Based on biochemical and immunological properties, 5 cell type-specific iF classes have been identified [1] (each having a central α-helical rod portion flanked by non-α-helical head and tail regions), according to their subunit protein(s): #Ck's, 2-10/cell (40-68 x $10^3$ apparent mol. wt. by SDS-PAGE) in ep'l and ep'l-derived cells; #Vm (57) in non-ep'l cells, esp. mesenchymal, and in some ep'l cell cultures; #desmin (53) in muscle cells; #GFAP (51) in astrocytes and gliomas; #neurofilament proteins, 3/cell (210, 160 & 68) in neurons and related tumours. Co-localization reflects co-polymerization, with molecular proximity, in two cases (Vm-desmin, Vm-GFAP [2]). PAb's and MAb's specific for each iF class are important in histology and especially in tumour diagnosis [3, 4]. There are as many as ~20 Ck's [5], the set depending on the ep'l type [6] - as exploited in tumour and cell-line typing [7]. By various criteria including antigenic determinants [8], Ck's fall into 2 groups which each contribute to an iF-assembly pair: type I, acidic; type II, more basic with i.e.p. in 9.5 M urea up to 8 (*vs.* ~5.5 for all other iF types). Purification of iF's relies on non-extractability by (e.g.) Triton-X-100. Solubilization of iF proteins needs an agent such as citric or acetic acid, SDS, or buffer rich in urea or guanidine [9,10]. As the polymer-monomer equilibrium is very biased towards the polymer under physiological buffer conditions the study of iF assembly *in vivo* has been severely inhibited; but below we indicate progress in identifying possible differences among iF's in assembly requirements *in situ* and in exploring how iF's either segregate or co-polymerize *in vivo*. First we consider changes in the iF network *in situ* with respect to the cell cycle, illustrating the dynamic nature of iF's although the *in vitro* stability of iF's and especially Ck iF's is suggestive of unchanging structure *in vivo*.

*Senior Editor's condensation (not cleared with the authors) of a late MS., with abbreviations introduced:* Ck, cytokeratin; ep'l, epithelial/epithelium; GFAP, glial filament acidic protein; iF *(not* IF), intermediate-sized filaments; IF, immunofluorescence; MAb/PAb, mono-/polyclonal antibody; Vm, vimentin. *Many refs. had to be omitted.*

*Cell cycle-dependent changes in iF's.-* In interphase ep'l cells
the iF's or iF bundles are spread throughout the cytoplasm;  but in
many (cultures, varying in origin) mitosis entails collapse of this
network and formation of microscopically visible (0.2-2 μm) dense
bodies ([11]; Fig. 1a) with, as shown by e.m., accompanying iF fragments and protofilament-like (Ck iF) fibres.  We have studied MDBK
cells especially, with IF microscopy using Ab's specific for Vm or
Ck iF's [11].  The concept of an inter-relationship between iF's and
iF protein aggregates is supported by a very close association often
seen during early mitosis.  Moreover, certain Ck Ab's preferentially
stain the aggregates and early prophase iF bundles rather than interphase iF's (Fig. 1b), whereas others can detect the mitotic aggregates too [11, 12]. Thus, for RVF-SM (non-ep'l) cells double-label IF has
shown that a Vm MAb recognizes only the mitotic aggregates and not
the iF network in interphase cells, whereas Ab's from a Vm-immunized
guinea pig (and several commercially available Vm MAb's) do not differentiate between the two states.  Other examples of rearrangements
during mitosis include intestinal mucosa cells *in situ* [13].  Seemingly
there is a physiological process by which iF integrity drastically
alters within minutes and correlated with specific cell-cycle events.
The apparent intact-iF ⇄ mitotic-aggregates change may also account
for other antigenic-determinant unmasking phenomena seen with iF's.

Even more surprising are IF results    exemplified in Fig. 2 for
cells whose iF network is evident during mitosis too and is only then
stainable by an MAb. Though interphase and mitotic iF's appear identical morphologically, protein-region exposure on the iF-cytosol inter-
face seems to change subtly.  The idea of iF changes on entering and
leaving mitosis accords with other evidence [14-16].   Epitopes masked
in interphase cells may yield to modified fixation etc. (Fig. 2 legend).
Evidently negative immunocytochemical results should be distrusted. We
are elucidating [cf. 17] the possible nature of iF structural changes.

---

*Opposite:* **Fig. 2.** Double IF microscopy of PtK$_2$ cells (whose iF network does not change dramatically during mitosis as in other cell
lines) using Ck-specific Ab's - (a) guinea pig PAb's, (b) K$_G$ 8.13
(IgG$_2$), a murine MAb.  Whereas both interphase and mitotic cells
show filament staining in (a), the MAb epitope is exposed (b) only
during mitosis.  This differential exposure of an epitope is not a
function of shape changes alone, since (not illustrated) both Ab's
reveal Ck iF's in parts of mitotic cells which have remained flatte-
ned.  See [18] for unmasking in interphase cells (e.g. by Triton or
proteases) of the epitope recognized by K$_G$ 8.13. Bars denote 10 μm.

**Fig. 3.**  Co-assembly of muzzle Ck's into common Ck iF, investigated
by immuno-e.m. using Ab's specific for muzzle Ck labelled with 5 nm
gold particles and PtK$_2$ [rat kangaroo-derived ep'l cell line, kidney]
Ck-specific Ab's labelled with 20 nm gold particles. Both sizes of
gold labels are distributed within the iF network and actually, in
ideal cases *(inset)* appear on the same iF.  Bars denote 50 μm.  The
e.m. studies [21] confirmed co-polymerization (IF merely established
co-localization); muzzle Ck poly(A)$^+$ mRNA had been micro-injected.

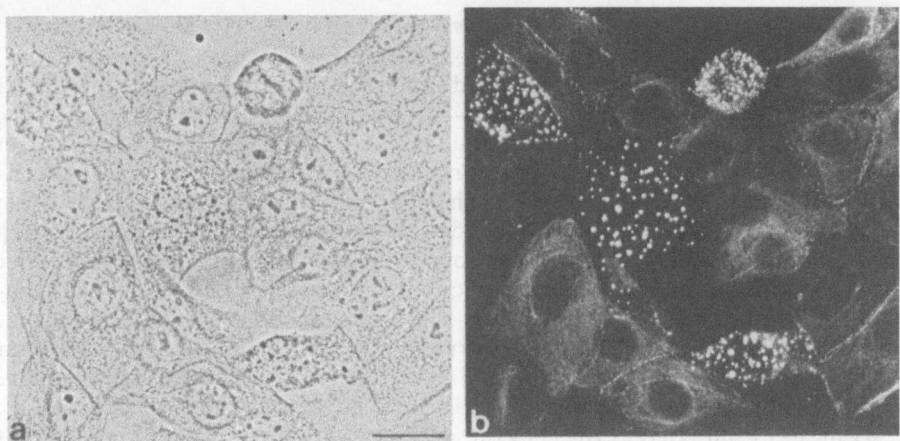

**Fig. 1.** Cell cycle-dependent changes in the iF network: MDBK cells, visualized by phase-contrast (a) after preparation for IF microscopy (b) using guinea pig Ab's against electrophoretically purified Ck's of tonofilaments from bovine muzzle epidermis. The Ab's preferentially stain aggregates in mitotic cells (b) which appear in (a) as cytoplasmic dense bodies, and only weakly stain the Ck iF network of interphase cells. Various Ck PAb's and MAb's have given similar results [11]. Bars represent 20 µm.

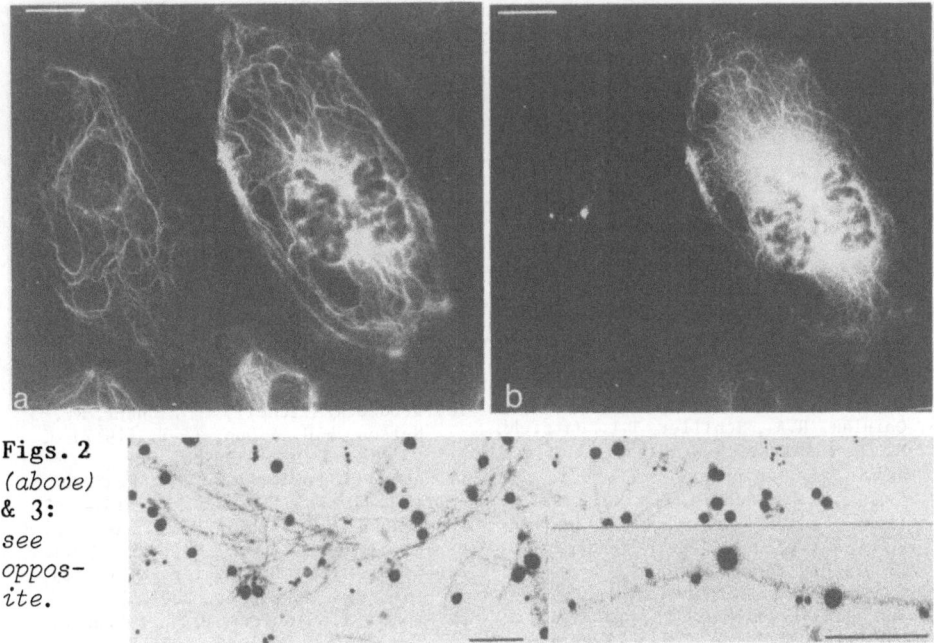

**Figs. 2**
*(above)*
**& 3:**
*see*
*oppos-*
*ite.*

*De novo synthesis and iF assembly in vivo.*- Micro-injection of fluorescently labelled proteins for studying *in vivo* assembly (cf. tubulin, actin [19]) is precluded due to the low solubility of iF proteins, circumventable by micro-injecting their mRNA and detecting the intracellular translation products with specific Ab's. Thus, when bovine lens-forming cells, which normally express only Vm-type (mesenchymal) iF, were micro-injected with a Ck mRNA, within 1 h positively labelled small 'whiskers' were evident throughout the cytoplasm [20]. A comprehensive fibril array as found in ep'l cells was established after 12 h which, from double IF microscopy using Vm, actin and tubulin Ab's, appeared different from the endogenous filament networks. Introducing an alien iF protein system had not affected cell shape or prevented normal cellular events, e.g. cell division. Similar results have been obtained with Vm-expressing cells from other species [20], and also [21] on injecting bovine epidermal mRNA into ep'l cells of the same (MDBK) or a different species (PtK$_2$): the translated Ck's were incorporated, at least largely, into the endogenous Ck network, actually co-polymerized (immuno-e.m.; Fig. 3). As Ck's do not co-polymerize with Vm *in vivo* there seems to be a mechanism segregating these two iF proteins into different iF structures. In general we also conclude, concerning iF protein synthesis *in vivo*, that cell-type specific synthesis of iF proteins is not controlled translationally, that alien iF proteins in a cell need not interfere with its growth and proliferation, and that filament assembly begins at multiple sites; there is no evidence for an intermediate filament-nucleating centre.

*Acknowledgements*

    We thank Drs. Thomas Kreis and Benjamin Geiger (Weizmann Inst., Rehovot), Jose L. Jorcano (this Institute) as well as Erika Schmid, Sybille Mittnacht and Christine Grund for cooperation and discussions. The work has been supported in part by the Deutsche Forschungsgemeinschaft (to W.W.F.).

*References*

1. Lazarides, E. (1982) *Ann. Rev. Biochem. 51*, 219-250.
2. Quinlan, R.A. & Franke, W.W. (1983) *Eur. J. Biochem. 132*, 477-484.
3. Ramaekers, F.C.S., Puts, J.J.G.,...... & Vooijs, G.P. (1982) *Cold Spring Harb. Symp. Quant. Biol. 46*, 331-339.
4. Osborn, M. & Weber, K. (1983) *Lab. Invest. 48*, 372-394.
5. Moll, R., Franke, W.W., Schiller, D.L., Geiger, B. & Krepler, R. (1982) *Cell 31*, 11-24.
6. Quinlan, R.A., Schiller, D.L., Hatzfeld, M.,...... & Franke, W.W. (1984) *Ann. N.Y. Acad. Sci.*, in press.
7. Franke, W.W. , Schmid, E. & Moll, R. (1982) in *Human Carcinogenesis* (Harris, C.C. & Autrup, H.N., eds.), Academic Press, New York, pp. 3-34.
8. Tseng, S.C.G., Jarvinen, M.J.,...... & Sun, T.T. (1982) *Cell 30*, 361-372.
9. Steinert, P., Idler, W.W. & Zimmerman, S.B. (1976) *J. Mol. Biol. 108*, 547-567.

10. Quinlan, R.A., Cohlberg, J.A.....& Franke, W.W. (1984) *J. Mol. Biol. 178*, 365-388.
11. Franke, W.W., Schmid, E., Grund, C. & Geiger, B. (1982) *Cell 30*, 103-113.
12. Franke, W.W., Grund, C.,......& Virtanen, I. (1984) *Exp. Cell Res. 154*, 567-580.
13. Brown, D.T., Anderton, B.H. & Wylie, C. C. (1983) *Int. J. Cancer 32*, 163-169.
14. Dulbecco, R., Allen, R., Okada, S. & Bowman, M. (1983) *Proc. Nat. Acad. Sci. 80*, 1915-1919.
15. Lazarides, E., Granger, B.L., Gard, D.L. ......& Danto, S.I., as for 3., 351-378.
16. Woodcock-Mitchell, J., Eichner, R., Nelson, W.G. & Sun, T.-T. (1982) *J. Cell Biol. 95*, 580-588.
17. Aebi, U., Fowler, W.E., Rew, P. & Sun, T.-T. (1983) *J. Cell Biol. 97*, 1131-1143.
18. Franke, W.W., Schmid, E.......& Geiger, B. (1983) *J. Cell Biol. 97*, 1255-1260.
19. Kreis, T.E. & Birchmeier, W. (1982) *Int. Rev. Cytol. 75*, 209-227.
20. Kreis, T.E, Geiger, B...... & Franke, W.W. (1983) *Cell 32*, 1125-1137.
21. Franke, W.W., Schmid, E.......& Jorcano, J.L. (1984) *Cell 36*, 813-825.

## Comments on material in #C

*Comments & annotation on*    #C-1, C.R. Birkett & K. Gull - MICROTUBULES

K. Gull, answering D.J. Morré.- The post-translational modification giving $\alpha_3$ seems to be a late event, just before or after entry
of the unmodified subunit into the exoneme.  The acyltransferase has
not been localized.  Recent evidence on role of microtubules *(noted
by Editor; see also p. 150).-* They serve in amoeba for 'purposeful'
intracellular movement of mitochondria and vesiculated materials [1],
and (filaments likewise) for organelle movement in squid axoplasm [2].

1. Koonce, M.P. & Schliwa, M. (1985) *J. Cell Biol.* 100, 322-326.
2. Scheetz, M.P., et al., & Vale, R.D., et al. (1985) *Cell* 40, 449-454/455-462.

*Comments on* #C-2, R.F. Foster & S.J. Kaufman - MYOGENESIS

S.F. Kaufman, answering D.R. Headon.- The proteins of mol. wt.
95,000 & 110,000 detected in antigen immunopurification are both glycoproteins, resistant while in the membrane to the diverse proteolytic enzymes tried.  **Reply to A. Fotedar.-** Muscle cultures showed
aberrant fusion if virus- or PEG-induced; we haven't tested  our  anti-
muscle MAb's coupled to Sepharose for this capability (cf. immune-
cell activation). **Remark by C.M. Lewis,** on fusogenic-negative variants: one wonders whether on co-culturing them there is any complementation and the emergence of fusion, and whether treatment of the
variants with the MAb's induces maturation into myotubes.

*Comments on* #C-4, V.P. Whittaker et al. - CHOLINERGIC NERVE TERMINALS

E.F. Hounsell asked: what is the evidence that the anti-Chol-1
antiserum is recognizing a series of gangliosides with different
degrees of sialylation, and what are the sizes of the carbohydrate
chains?  **Reply.-** We used the blotting technique on TLC plates in
which the gangliosides were separated.  Gangliosides co-migrating
with GD, GT and GQ stained.  This is in the *Torpedo.*  In the guinea-
pig, only one ganglioside stained when adsorbed serum was used. This
runs just behind the GQ group.  So far nothing is known about the
chemical structure  of Chol-1.

Question by G.E. Isom.- Concerning the life cycle of the ACh
vesicle in the mature neurone, can this technique be  utilized to
study the synthesis in the cell body and the dynamics of movement
down the axon to the nerve terminal?  **Whittaker's reply.-** Yes.  I
showed micrographs in which vesicles are accumulating in the axon
hillock and are accumulating above a ligature in the axon: this is
the first evidence that cholinergic vesicles are formed in the cell

body and transported to the synapse. We believe this is a 'topping-up' process to replace worn-out vesicles after they have recycled an unknown number of times. Using $^{35}$S-labelled vesicular proteoglycan, my colleague Dr. Stadler has shown that the vesicles are transported at the fast rate.

*Comments on* #NC(C)-1, M.C. Raff - CNS CELL LINEAGE

   **A. Fotedar asked:** can positional information (like Top antigen) be retained when cells are grown in culture? **Reply.-** Yes, for Top. **D.J. Morré asked:** do the glial progenitor cells respond to a puri-fied mitogen to get away from using 'astrocyte soup' to sustain division and retard differentiation? **Reply.-** The astrocyte mitogen is the only one known at the moment. **Remark by B.F. Erlanger.-** In the pulse-chase experiments it is conceivable that the Ab is influ-encing the differentiation process.

*Comment on* #NC(C)-2, M.E. Bardsley et al. - CALMODULIN ASSAY

   **To D.J. Morré:** the binding of CaM-inhibiting drugs requires Ca$^{2+}$ (thus any speculation about drug occupancy of the Ca$^{2+}$ pocket is in-applicable). **To Editor:** a study of CaM stability [Guerini, D. & Krebs, J. (1983) *FEBS Lett. 164*, 105-110] indicates the undesirability of heat treatment in CaM purification, as in the Vol. 13 art. cited in #NC(C)-2.

*Comments on* #NC(C)-3, R.A. Quinlan - INTERMEDIATE FILAMENTS

   **Question by D.J. Morré.-** How general is the specificity of the cytokeratin MAb to the dividing cells? **Answer.-** It may not be very specific; some tumour cells also react. **Remark by V.P. Whittaker.-** An illustration of the reliability of intermediate filament type as diagnostic for cell type is the fact that the electrocytes, derived embryologically from myoblasts, contain desmin as their intermediate filament protein. **M.A. Raff asked:** can you suggest any function for intermediate filaments besides being integrators of intracellular space? **S.J. Kaufman asked:** are the dynamics of intermediate filament reassembly consistent with the dynamics of cell movement? **Reply.-** Really not known, but the tools are now available to address this. **Answer to Raff.-** The mRNA in our micro-injection experiment was ~80% pure. **Remark by J. Brundell.-** Cytokeratin (fragments) can be iden-tified in patients carrying tumours of epithelial origin. This fragmentation seems to be responsible for the solubility of the usually very insoluble intermediate filament.

Section #D

ANTIBODY APPROACHES TO MEMBRANE COMPONENTS
INCLUDING THE T-CELL RECEPTOR

#D-1

# EXPLOITATION OF ANTIBODIES IN THE STUDY OF CELL MEMBRANES

O.J. Bjerrum, *J.C. Selmer, *F.S. Larsen
and S. Naaby-Hansen

The Protein Laboratory          *NOVO Industri
University of Copenhagen         Novo alle
34 Sigurdsgade                   DK-2880 Bagsvaerd
DK-2200 Copenhagen N             Denmark
Denmark

*Some methodological possibilities for the analysis of membrane proteins based on different types of antibody are described. Detergent immunoelectrophoresis (D-IE) is surveyed as a technique which works with polyspecific antibodies (Ab's), and immunoblotting as one which functions with oligo- and mono-specific Ab's. Finally, the application of monoclonal Ab's in immunosorbent chromatography and ELISA techniques is shown.*

The analysis of membrane proteins by means of Ab's is promoted by their natural binding to the insoluble carrier: the lipid bilayer. For many years this capacity has been utilized in agglutination and complement fixation tests  and in immunohistochemistry (techniques which will not be dealt with in this article).  On the other hand, the study of individual membrane proteins with Ab's in solution has been highly restricted by the difficulty of solubilizing  the proteins under non-denaturing conditions.  Approaches such as detergent immunoelectrophoresis (D-IE), immunoblotting and immunoassays such as ELISA and RIA have coped with this difficulty.

## 1. EXPLOITATION OF POLYSPECIFIC ANTI-MEMBRANE ANTIBODIES

### 1.1  Strategy

If a preparation of Ab against a membrane protein is desirable for a given research project, it is seldom necessary to wait until a pure protein is available before raising the Ab.  Instead, a polyspecific Ab against a crude membrane fraction should be used at an early stage to provide information about the protein in question with respect to its antigenicity, solubility and biomolecular characteristics; such information may facilitate the purification of the

Table 1.  Determination of some molecular and supramolecular parameters of electro-immunoprecipitated membrane antigens.

| Parameter | Approach | Ref. |
|---|---|---|
| Lipid | Staining (Sudan Black) | 8 |
| | Chemical/enzymatic modification | 9 |
| Carbohydrate | Staining, labelling; chemical/enzymatic modification; lectin binding | 1, 5 |
| Protein | Staining, labelling | 1 |
| *Charge (pI)* | Electrofocusing | 1, 4 |
| *Heterogeneity* | Precipitate shape (in both dimensions) | 1, 3 |
| *Stability* | Sign of degradation | 5 |
| *Amphiphilicity* | Aggregation upon removal of detergent | 3, 5 |
| | Mixed micelles containing Sudan Black | 4, 8 |
| | Radiodetergent and autoradiography | 3, 10 |
| | Crossed hydrophobic IE with phenyl-agarose | 1, 5, 11 |
| | Charge-shift crossed IE | 1, 3, 12 |
| *Enzymic activity* | Zymogram staining | 1, 13 |
| *Polypeptide content* | Excision → SDS-PAGE | 1, 5 |
| *Ultrastructure* | Excision, agarase treatment→electron microscopy | 1, 5 |
| Histological specificity | Absorption with solubilized tissue of the compartment | 13, 14 |
| Inside/outside localization | External labelling; Ab absorption | 10, 13 |
| Protein-protein interaction | Application of dissociative media | 3, 7 |
| Ligand binding | Affinity electrophoresis | 1, 5 |

protein.  It is also possible to produce monospecific Ab's against
precipitates excised from a gel precipitate with such an Ab [1].
This approach has recently been employed to characterize membrane
receptors [2].  An alternative approach is to produce a monoclonal
antibody (MAb) with the crude membrane fraction as immunogen (see 3.).

A polyspecific Ab is easily prepared.  The membrane fraction is
mixed with 1 vol. of Freund's incomplete adjuvant, and ~500 µg of
protein is injected intracutaneously into rabbits at 2-week intervals; bleedings (30 ml) are started 7 days after the 3rd injection [1-4].

The exploitation of the resulting polyspecific Ab (leading to
identification procedures, to be considered on the basis of Table 1)
depends on the availability of high-resolution techniques such as
immunoblotting (2. below) or crossed IE, which is one of the few

techniques that permit analysis of membrane proteins solubilized under non-denaturing conditions with a non-ionic detergent [4-7]. The individual protein antigens of the precipitation pattern are designated on the basis of their position, shape, size, staining intensity and of the morphology, e.g. sharp/blurred. A number of techniques, some of which are shown in Table 1, have been developed to identify the individual proteins in the precipitation pattern. Since two recent manuals [1, 3] describe these techniques, only a few important principles will be illustrated with new experiments below.

## 1.2 Chacterization of acetylcholinesterase (AChE) of human erythrocyte membranes by crossed IE

The presence of AChE on erythrocyte membranes, ~6,000 copies per cell, was discovered as early as 1973 [15]. In 1975 Bjerrum et al. [16] described the enzyme as a distinct molecular entity not interacting with other membrane proteins. Crossed IE of erythrocyte membrane material solubilized with a non-ionic detergent using a polyspecific rabbit Ab (cf. below, Fig. 2A) gave a symmetrical bell-shaped immunoprecipitate which was discovered after staining for enzyme activity [16].

Whilst still in the milieu of the solubilized membrane material the esterase was demonstrated to be a sialoglycoprotein because the precipitation pattern changed upon mixing of the solubilized membrane material with lectins such as wheat-germ agglutinin, concanavalin A and phytohaemagglutinin P [13]. The observations made with wheat-germ agglutinin were later confirmed with crossed affinity IE [2]. The sialic acid content was demonstrated as a reduction in electrophoretic migration after treatment of the membranes with neuraminidase [13].

The amphiphilic nature of erythrocyte AChE was demonstrated for the first time by charge-shift IE [7, 12]. When the hydrophilic molecule haemoglobin migrated 20 mm, AChE showed significant bidirectional migrational shifts of 12 mm and 8 mm in the presence of 0.5% (v/v) Triton X-100 along with 0.2% (w/v) deoxycholate and of 0.5% Triton-X-100 along with 0.0125% (w/v) cetyltrimethylammonium bromide respectively [12]. In crossed hydrophobic interaction IE with phenyl-Sepharose, AChE was bound to the hydrophobic matrix like other amphiphilic proteins [11]. The amphiphilicity of AChE was later verified with the purified enzyme [17, 18].

Concerning the localization of AChE in the erythrocyte membrane, it was shown in 1975 by crossed IE that it was not possible to precipitate any esterase-active antigen with an Ab preparation which had been absorbed with intact erythrocytes [16]. This suggested that AChE was located solely on the erythrocyte's outer surface[*],

_____
[*] *Co-Editor's comment.* - A possible location on the inner surface would hardly be excluded by this approach.

and prompted the question of whether the anchor penetrated the membrane or was merely inserted in the outer lipid leaflet.    The 1975
approach was therefore repeated in a refined form, using a purified
enzyme preparation (3.1, below) and the 40 times more sensitive technique of crossed IE with an intermediate gel containing the absorbed
Ab (Fig. 1) [1]. The evident extension of the anodic leg down into
the intermediate gel containing the AChE Ab preparation absorbed
only once with intact erythrocytes indicated residual Ab activity
(Fig. 1A, *arrow*).    However, this activity disappeared after a second
absorption  (Fig. 2B), indicating that AChE carries no epitopes on
the inner side of the membrane.    This suggestion accords with the
finding, from experiments involving proteolytic cleavage, of a relatively small hydrophobic anchor for AChE (mol. wt. ~2,000–4,000) [18].

## 1.3  Demonstration of protein–protein interaction and ligand binding of erythrocyte membrane proteins in crossed IE

Electroimmunoprecipitation normally takes place at an unphysi-
ological pH around 8.7, because the average migration of the Ab's
has to be zero.    Therefore the first published crossed IE  reference
pattern of Berol EMU-043-solubilized human erythrocyte membrane proteins,   on which the numbering of the precipitates is  based, was
performed at pH 8.7 (cf. Fig. 2B) [13].  However, when the analysis was
performed at pH 7.7 [20] the precipitation pattern contained fewer
precipitates (Fig. 2A).    Two new precipitates, nos. 2.1 and 18,
appeared in  the transition from pH 7.7 to pH 8.7 or higher values.
Furthermore, a reaction of partial identity  between precipitates
2.1 and 16 (Fig. 2A) was observed, since the cathodic leg of precipitate 2.1 never crossed the anodic leg of precipitate 16 (Fig. 2B,
*arrow*).  This can be explained by either (a) the existence of common
epitopes on the two molecules (true partial identity) or (b) the
presence of the same antigen in free and complexed form [1, 7].    The
latter explanation was supported by the fact that a further increase
in pH (to pH 11)  completely converted precipitate 16 to 18 and 2.1
(cf. Fig. 2D).

For further study of this phenomenon, the polypeptide content
of the excised precipitates 2.1, 16 and 18 was analyzed by SDS–PAGE
[1, 5, 7].    The presence of ankyrin (band 2.1) in precipitate 2.1,
of band-3 protein in 18, and of band-3 plus ankyrin in precipitate
16, could be demonstrated.    The electroimmunochemical analysis of
erythrocyte membrane material  solubilized in non-ionic detergent
at two different pH values thus confirmed that the binding of the
cytoskeleton protein ankyrin to the lipid-anchored band-3 molecule,
as seen at pH 8.7, was also present after solubilization at pH 7.7
[21] but becomes dissociated at higher pH values.

Simultaneously with the pH-induced dissociation of ankyrin from
band 3, precipitate 16 is converted to 18 (cf. Fig. 2A & 2B).  The
question was whether this was  due  to  dissociation  of  ankyrin  or

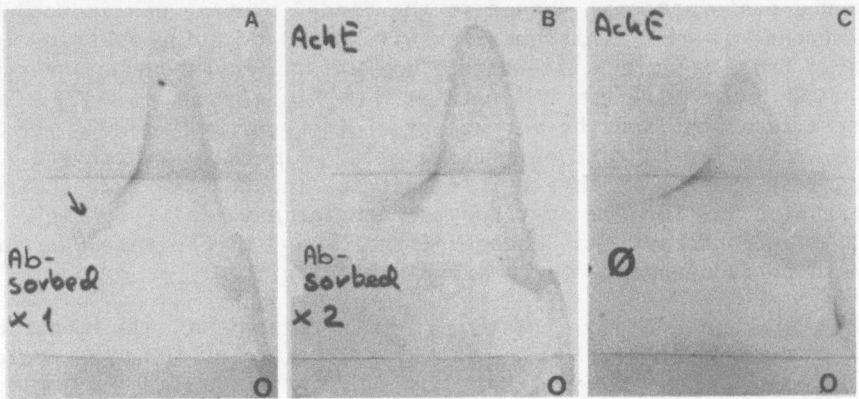

Fig. 1.  Immunoabsorption experiment for demonstration of membrane penetration of an amphiphilic protein: crossed IE of purified AChE (0.8 µg) with anti–AChE Ab's (3 µl/cm$^2$) in the upper gel.  The intermediate gel contains anti–ChE Ab's absorbed once (A) or twice (B) with intact erythrocytes [16] (5 µl/cm$^2$), or is blank (C; ∅).  Plates stained for esterase activity with 1-naphthyl acetate as substrate and Fast Red as coupling salt [16].  The anode for the first-dimension electrophoresis is to the right.  Due to residual Ab activity in the absorbed Ab preparation [1], the anode leg is deflected *(arrow)*, i.e. inward feet reaction [1].

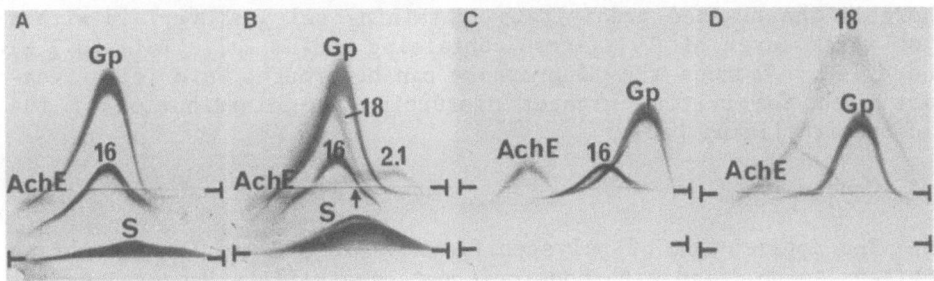

Fig. 2.  Demonstration of protein-protein interaction in crossed IE: human erythrocyte membranes solubilized with 1% (w/v) Berol EMU-043 at pH 7.7 (A) or 10.5 (B) [7], or erythrocyte membrane-derived dimyristoylphosphatidyl choline vesicles [18] solubilized with 1% (v/v) Triton X-100 at pH 7.7 (C) or 11.0 (D).  After titration to pH 7.7, electrophoresis in agarose gel with reduced electroendosmosis (mol. wt. = -0.10) in 17 mM Tris/80 mM phosphate buffer pH 7.7 [20]; intermediate gels *(between marks)* contained anti–spectrin Ab's (7 µl/cm$^2$).  The 2nd-dimension gel contained anti–erythrocyte membrane Ab's (6 µl/cm$^2$).  Plates stained for protein (Coomassie Brilliant Blue R) and AChE (as in Fig. 1).  Gp, glycophorin; S, spectrin; 2.1, ankyrin; 16 & 18, band-3 protein in different conformations.  *Arrow* points to reaction of partial identity between the precipitates 16 and 2.1.

whether a conformational change in the band-3 protein was induced by the increase in pH? This question was solved by using ankyrin- and spectrin-free vesicles liberated from intact erythrocytes treated with dimyristoylphosphatidyl choline [18]. As shown in Fig. 2C, spectrin is absent and the morphology of the vesicles' band-3 precipitate is similar to that of precipitate 16 of the erythrocyte. An increase in the pH of the vesicle material induces the same changes in morphology as for the erythrocyte membrane material (cf. Figs. 2C & 2D). The conformational change of the band-3 precipitate was thus shown to be independent of the ankyrin binding.

Measurement of the relative concentrations of the membrane proteins present in the vesicles - by planimetry of the areas below precipitates [1] - showed that the AChE concentration increased 3-fold compared to the glycophorin concentration (compare Figs. 2A & 2C). This means that AChE differs from transmembrane proteins by having a larger lateral mobility in the membrane [18].

The precipitation pattern after crossed IE is a kind of immobilized replica of the membrane proteins. This enables analysis of the binding of various ligands to individual proteins provided it is possible to trace the ligand, e.g. by fluorography (Fig. 3). Evidently tritiated Triton X-100 binds to the amphiphilic proteins glycophorin and band 3 (Fig. 3A) but not to the hydrophilic membrane proteins spectrin and ankyrin [14]. The experiment further illustrates how it is possible to detect the weak β-emitter ($^3$H) in IE. After electrophoresis the pressed precipitate-containing gel was overlaid with a 2 mm thick layer of 2% agarose containing sodium salicylate (0.7 M) and dried. Thereby the fluorophore can be brought into close contact with the tritium without disturbing the distribution of the radioactive ligand [10].

## 1.4  Conclusion

The application of polyspecific Ab's in crossed IE of membrane antigens solubilized with non-ionic detergent allows precise identification, molecular characterization, and quantitation of major membrane antigens as well as functional protein complexes. Protein-protein interactions and ligand binding can be studied under non-denaturing conditions. The technique is inexpensive and does not call for any special equipment.

## 2. EXPLOITATION OF MONO- AND OLIGO-SPECIFIC ANTIBODIES

In IE these Ab's function as for their polyspecific counterparts. However, the recently introduced Western immunoblotting technique [24-26] offers an even wider range of applications.

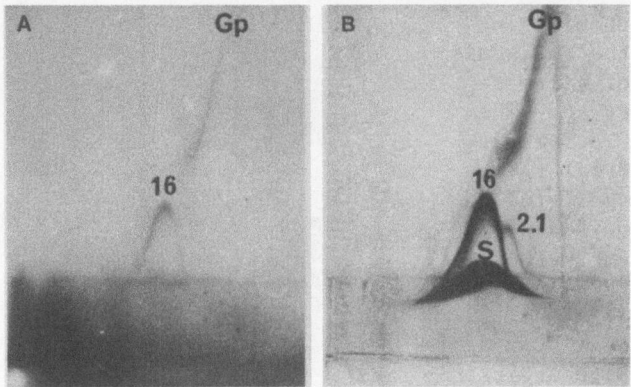

**Fig. 3.** Demonstration of reversible ligand binding to immunoprecipitated proteins. Crossed IE of 15 μg human erythrocyte membrane
material solubilized in 1% Triton X-100 and analyzed in a gel containing, per ml, 2 mg Triton X-100 plus $6 \times 10$ cpm of $^3$H-Triton X-100
(1.6 mCi/mg; NEN Inc.). A, fluorograph of the plate after overlaying with agarose (2% w/v) containing 0.7 M sodium salicylate. B, the
same plate after washing for 24 h and staining with Coomassie Brilliant blue. The 2nd dimension gel contained anti-erythrocyte membrane Ab's (7 μl/cm$^2$). Designations are as for Fig. 2. Exposure
time: 3 weeks.

## 2.1 Immunoblotting technique

The direct combination of, e.g., SDS-PAGE analysis with the
binding specificity of Ab's endows the immunoblotting technique
with a very high resolution. As for IE, it works with polyspecific
Ab's, and requires less antiserum (1-200 μl per analysis). Very
dilute as well as non-precipitating Ab's can also be applied [24].
The antigens become exposed to the Ab's through electrophoretic
transfer from the separation gel to a nitrocellulose membrane. After
blocking excess binding sites, the immobilized antigens can be
selected immunochemically by Ab and identified by using secondary
labelled Ab's or protein A. Under these conditions the SDS-induced
denaturation of the antigens only seems to affect the binding of
polyclonal Ab's to a minor degree [24, 25]. For practical descriptions of the technique see, for example, refs. [23, 24, 26 & 27].

As for the IE techniques described earlier, immunoblotting can
be used to characterize the membrane antigens present in complex
mixtures. The lipid, carbohydrate and protein contents of the antigens can be determined according to the principles given in Table 1.
The mol. wt. and pI may be estimated from the chosen separation gel,
and isoelectric focusing gels can also be immunoblotted. For SDS-
resistant receptors, ligand binding can be performed, e.g. with lectins. Histological specificity and membrane topography can be determined in Ab absorption experiments [24].

Fig. 4. Immunoblotting
analysis of the antigenic
composition of a cross-
linked polymer. A.- The
high mol. wt. polymeric
membrane material (X)
generated in $Ca^{2+}$-loaded
human erythrocytes was
separated from the normal
membrane polypeptides on
composite SDS gels of 2%
polyacrylamide & 0.5% agarose. Lane CBB: proteins
stained as for Fig. 2. After
electroblotting to nitrocellulose the proteins were
stained as indicated with
monospecific rabbit Ab's
against individual polypeptides of the erythrocyte
including haemoglobin (Hb).
B.- For these Ab's mono-
specificity is proved with
normal erythrocyte membrane
proteins separated by SDS-
PAGE [4]. Peroxidase-conjugated swine anti-rabbit Ab's
served as secondary Ab [24].
Epitopes of all the examined
polypeptides except glycoprotein are present on the
polymer (cf. X in Fig. 4A).

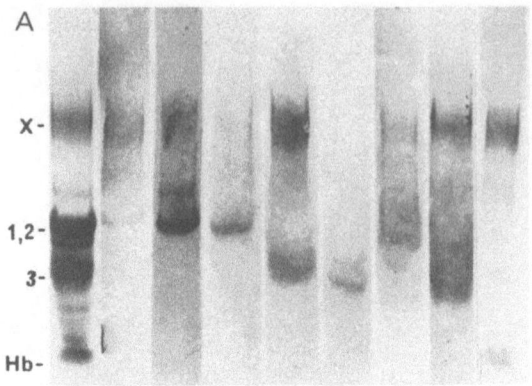

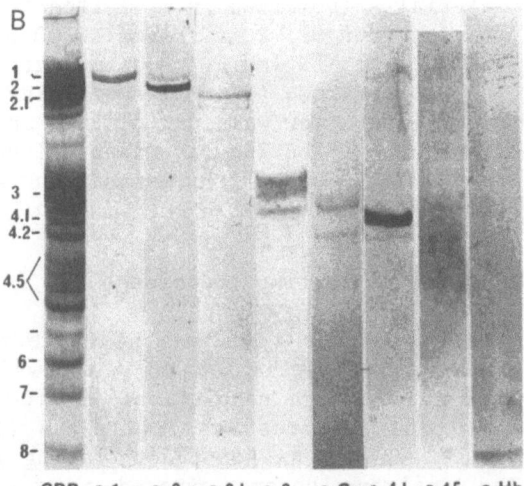

Another important application of the immunoblotting technique
is the analysis of antigens showing partial identity. Thus, it is
possible to identify by means of a polyclonal monospecific Ab the
degradation products of a given protein. The degradation may be due
either to the presence of proteolytic enzymes naturally occurring in
the membrane preparation or to deliberately added enzymes [24]. In
a similar way, pre- and pro-forms can be detected in genetic engineering experiments. Due to the high sensitivity, even very weak
cross-reacting Ab's can give a positive identification, e.g. for sort-
ing out proteins in other species. For applications, see refs. [24-27].

## 2.2 Immunoblotting with monospecific Ab's for identifying protein constituents of cross-linked membrane polymers

As exemplified in Fig. 4, protein complexes can also be analyzed
by immunoblotting if monospecific Ab's are available. When intact

erythrocytes are loaded with $Ca^{2+}$ by means of the ionophore A 23187, a transglutaminase is activated which generates γ-glutamyl-ε-lysine cross-linked polymers in the membrane [28], involving proteins which cannot be identified directly from staining-profile changes in SDS-PAGE because $Ca^{2+}$-dependent proteases become activated as well. Disappearance of a band in the SDS-PAGE profile can thus be due either to incorporation into polymers or to degradation. Strictly monospecific rabbit Ab's against individual erythrocyte membrane proteins were raised by immunizing rabbits with isolated bands from preparative SDS-PAGE [13] and with immunoprecipitates excised from crossed IE [4]. After immunoabsorption with whole membranes, the specificity was tested by immunoblotting (Fig. 4B). The polymers generated in $Ca^{2+}$-loaded cells were clearly separated from membrane proteins of lower mol. wt. (Fig. 4A) by electrophoresis in an open-pored SDS-containing gel. Incubation of the electroblots of such gels with monospecific Ab's showed the presence on the polymer of epitopes of bands 1, 2, 2.1, 3, 4.1, 4.5 (glucose transporter) and Hb, but not glycophorin.

Besides illustrating the immunoblotting technique, this study exemplifies nearest-neighbour analysis of membrane proteins via the functioning of the intrinsic enzyme transglutaminase, because demonstrated cross-linking indicates a close contact between the involved proteins. With the same Ab's, immunoblotting has also been used to study the $Ca^{2+}$-induced degradation of the erythrocyte membrane proteins [24].

## 2.3 Analysis of auto-Ab's against human spermatozoal membrane proteins

The immunoblotting technique is also a valuable tool for the study of Ab's. The reactivity of xeno-Ab's can be tested against relevant antigen preparations irrespective of whether the Ab's were obtained in laboratory animals or were generated in man by infection or vaccination. The study of allo- and auto-Ab's represents another rewarding field for immunoblotting in medicine [27]. The topographical or histological specificity of an antiserum can be tested by analysis of various extracts, e.g. tissue, organ or membrane fractions. Furthermore, it is possible to determine the class of the reacting Ab by using secondary Ab's of different Ig-class specificity. Due to the high sensitivity of the technique, it is at present the best way to test the monospecificity of a given polyclonal Ab preparation (cf. Fig. 4B).

For many years circulating auto-Ab's have been known to contribute to some cases of human infertility, but the antigens involved have not been characterized. Recent immunoblotting studies succeeded in describing the reactivity of the Ab's of sera from such patients against spermatozoal membrane proteins (Fig. 5) [27]. Binding patterns of IgG from 2 immunoinfertile and 4 fertile individuals show that all the sera reacted with the spermatozoal proteins. However,

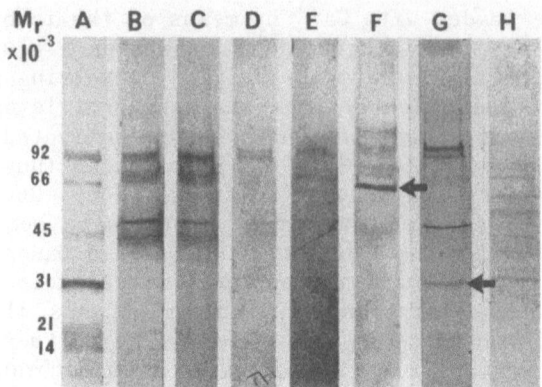

**Fig. 5.** Analysis of auto-Ab's by immunoblotting: binding of human IgG auto-Ab's to human spermatozoal protein antigens separated by SDS-PAGE. Lane A contains mol. wt. markers. Lanes B-F show the immunoblots obtained with sera from fertile males (B, E), females (C, D) and infertile males (F, G). Electroblotted spermatozoal protein in Amido Black staining is shown in lane H. Peroxidase-conjugated rabbit anti-human IgG served as secondary Ab [27]. Antigens specifically related to infertility are *arrowed*.

a larger series of analyses of well-characterized sera from 28 fertile and 28 immunoinfertile persons made it possible to correlate the cases of infertility with auto-Ab's against one or more auto-antigens with mol. wts. of 120,000, 78,000, 64,000, 41,000 or 32,000 (Fig. 5, *arrows*) [27].

## 2.4  Conclusion

The use of oligo- and mono-specific Ab's in conjunction with immunoblotting makes it possible to study the supramolecular organization and bimolecular characteristics of membrane antigens without prior purification. In particular, it is easy to establish antigenic relationships between antigens of various mol. wts. Also for descriptions of an Ab response, the high sensitivity makes immunoblotting very suitable.

## 3.  EXPLOITATION OF MONOCLONAL ANTIBODIES (MAb's)

### 3.1  General applications

The hybridoma technique for producing MAb's [29] was a major break-through in immunochemistry in general. All at once it became possible to monitor, for example, lymphocyte surface markers with the same Ab everywhere and to isolate homogeneous lymphocyte preparations by cell sorters. Such features have largely expanded our knowledge of the immune system, as indicated earlier in this book;

labelled MAb's have also helped to study cell differentiation and to monitor changes in subcellular components during the cell cycle.

However, analysis of solubilized membrane antigens with MAb's in the mode described for the afore-mentioned Ab-based approaches is encumbered by two major limitations. Firstly, the MAb's will precipitate only if the antigens contain repetitive epitopes. However, under certain conditions it is possible to use MAb's in conjunction with crossed IE. Thus, if the MAb is mixed with the antigen or incorporated in the first-dimension gel, characteristic changes might occur, e.g. reduction in size, skew shapes with tailing and retarded position of the precipitate. The MAb can also be co-precipitated with rabbit Ab's or bound post-electrophoretically to already formed precipitates whereupon its presence can be detected with secondary labelled anti-mouse IgG Ab's [5, 30]. However, in all probability Ab's of high affinity can be detected in this way.

The second limitation becomes evident in the use of MAb's for immunoblotting. It is our experience that with some antigens less than one-fifth of the IgG MAb's reacting with the native protein also react with the SDS-treated protein, e.g. human erythrocyte membrane proteins and β-galactosidase from *E. coli*. On the other hand, with other human proteins such as α-foetoprotein (21 clones investigated), albumin (50), IgG (15) and thyroglobulin (20) more than four-fifths of the IgG MAb's reacted with both the native and the SDS-treated protein. The article by D.P. Lane et al. (#B-1) is relevant.

## 3.2 Isolation of human erythrocyte membrane AChE by immunoabsorbent chromatography

Besides the analytical advantages described above, the introduction of MAb's also has greatly improved the possibilities of performing preparative isolation of membrane proteins [31, 32]. Used for immunoabsorbent chromatography, they confer the following advantages over polyclonal Ab's.- (1) The purity of the eluate from the column is improved due to the monospecificity of the attached Ab. (2) The elution is sharp and the recovery is high since multiple attachment is avoided with only one epitope on the protein involved in the binding. (3) The capacity of columns coupled with MAb's is high. Thus, with 5 mg/ml gel, it has been possible to obtain a molar ratio of 1:1 between coupled Ab and eluted antigen. (4) The MAb may also be used as catching Ab in an enzyme immunoassay with a polyclonal or monoclonal detecting Ab, whereby it is possible to measure accurately the relative antigen concentration in the applied extract, the run-through and the eluate from the column. Such an immunoassay is easy to perform (3.3, below), and is valuable for determining the conditions that allow a gentle and efficient elution of the antigen [33].

The purification of membrane-bound erythrocyte AChE illustrates a practical application of the immunosorbent technique. In this

experiment compacted human erythrocytes, washed free from serum proteins with 0.154 M NaCl, were lysed by addition of 2 vol. of 45 mM
EDTA, pH 7.4, containing 3% (v/v) Triton X-100, 2 mM benzamidine,
1 mM sodium azide and 0.5 mM iodoacetamide. After 1 h of gentle
mixing, insoluble material was removed by centrifugation (36,000 $g_{av}$,
30 min; Sorvall RC 5B). The pooled supernatant (3.9 l) was then
percolated at a flowrate of 36 ml/h through an immunoabsorbent column
equipped with an adaptor of 1.5 cm diameter. The immunosorbent
matrix consisted of 5 ml CNBr-activated Sepharose 4B coupled in a
concentration of 4 mg/ml gel with a monoclonal anti-human AChE Ab
(F18), isolated by protein A chromatography from cell cultures [34].

Under these conditions, 25% of the applied esterase activity,
as measured by activity towards acetylthiocholine (3.3, below) passed
through the column. The column was washed with 100 ml 1% (v/v) Triton
X-100 in 5 mM EDTA, pH 7.4, then with 100 ml 50 mM phosphate buffer,
pH 7.4, containing 1% Triton X-100, and finally with 100 ml 0.1%
Triton X-100 in 5 mM EDTA. The bound AChE was eluted with 0.1 M
phosphate, pH 11.0, containing 0.15% Triton X-100. To the tube for
each 2 ml fraction, 0.3 ml 0.5 M sodium phosphate, pH 6.5, had been
added to restore the pH to 8.0. The temperature was 5° throughout.

The major contaminant was found to be Hb (Fig. 6, *arrow*) on the
basis of Coomassie Brilliant Blue staining (lane b) and immunoblotting (lane a) with a polyspecific rabbit anti-AChE Ab (cf. Fig. 1).
Based on enzyme activity the recovery of the column-bound material
was 70%. The specific activity of the peak fraction was 1000 IU/mg
[35] as compared to 0.02 IU/mg protein in the lysate.

Since erythrocyte- and brain-derived AChE have common epitopes
[35] it has been possible to isolate both AChE's on the same column
(O.J. Bjerrum et al., to be published). The purification efficien-
cies of NaCl and Triton X-100 extracts of nucleus caudatus from
human brain [35] were equally high.

## 3.3  AChE quantitation by an enzyme immunoassay with MAb's

In enzyme immunoassays (ELISA) the MAb is commonly used as the
catching Ab [34]. Thereby the high specificity of the Ab ensures
that only the appropriate antigen is bound. The detecting step emp-
loys an enzyme-labelled polyclonal or monoclonal Ab; the former gives
better sensitivity because several labelled Ab's can be bound per
antigen molecule [36]. In selecting MAb's for ELISA, normal screening of hybridomas by binding to immobilized antigen gives Ab's with
affinity constants in the range $10^5$-$10^7$ 1/mol corresponding to a
sensitivity of 0.5-5 ng antigen per ml if used as catching Ab. More
sensitive systems (0.01-0.1 ng/ml) require catching Ab's with affinity constants >$10^{10}$ 1/mol, which call for special screening systems
or affinity determination.

Fig. 6.  Immunoabsorbent
chromatography with an MAb:
elution profile of the AChE
activity desorbed with 0.1 M
phosphate buffer, pH 11.0,
from a 5 ml anti-AChE column
containing 4 mg MAb/ml and
percolated with 3.9 l Triton
X-100 containing human erythrocyte lysate. *Insert:*
lane b, SDS-PAGE of the pair
of peak fractions (Coomassie
Brilliant Blue); lane **a**, an
immunoblot of the same preparation with a monospecific Ab
to AChE (cf. Fig. 1). Most of
the protein bands react with
the Ab, indicating them to be
AChE degradation products.
*Arrow:* Hb, the main contaminant.

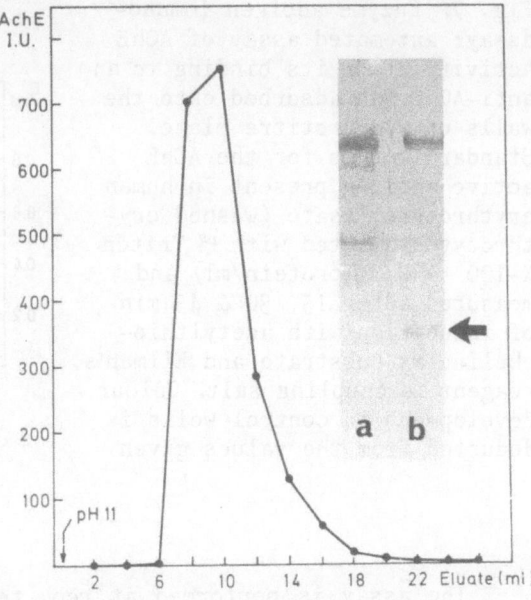

As the catching MAb reacts with only one epitope on the antigen, antigen quantitation depends on factors that affect the integrity of this epitope.  Proteolytic degradation of the epitope, conformational changes or steric hindrance due to complexes between
enzyme and inhibitor may affect the antigen-Ab reaction.  Thus, a
capturing Ab recognizing a polymorphic determinant cannot be used
to assay the antigen generally in different individuals.  Yet such
Ab's may be advantageous, e.g. for measuring the antigen subpopulation, functionally active/inactive antigens and degradation products.

MAb physico-chemical properties vary considerably.  Non-specific binding to plastic or other proteins, and MAb denaturation
after adsorption onto plastic, may limit MAb use in ELISA.  The
problems are avoidable by appropriate screening procedures.

Fig. 7 shows such an application of the anti-AChE Ab.  In pre-
natal diagnostics, AChE is a marker for detecting open neural tube
defects, because shed enzyme can be detected in the amniotic fluid
of these malformed foetuses [37].  However, AChE assay based directly
on measurements of its own catalytic activity is impeded by the lack
of specific substrates that do not react with other esterases normally present in the body fluids.  Specific inhibitors exist but are
difficult to work with [37].  In the present single-layer immunoassay
the specific AChE enzyme is caught by means of the monospecific Ab
and other  esterases are washed away, after which the retained AChE
is measured by its own catalytic activity.

Fig. 7. Enzyme antigen immunoassay: automated assay of AChE
activity after its binding to an
anti-AChE MAb adsorbed onto the
walls of a microtitre plate.
Standard curves for the AChE
active antigen present in human
erythrocyte lysate (washed erythrocytes diluted with 1% Triton
X-100 to 2 mg protein/ml) and
measured after 15, 30 & 45 min
of incubation with acetylthiocholine as substrate and Ellman's
reagent as coupling salt. Colour
development in control wells is
deducted from the values given.

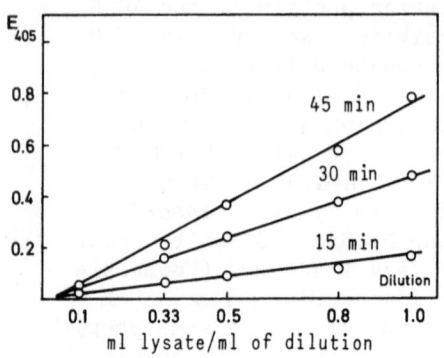

The assay is performed at room temperature as follows: 200 µl
of the anti-AChE MAb (3 µg/ml) in 50 mM sodium carbonate buffer, pH
9.6, is added to each well of a 96-well microtitre plate (Nunc) and
incubated (room temp.) overnight. The wells are then washed 3 times
with 0.05% (v/v) Tween 20 in phosphate-buffered saline, pH 7.2, and
200 µl of the esterase-containing solution is added and allowed to
react with the Ab for 2 h. Then the wells are washed in the above-
mentioned Tween solution. With 200 µl of acetylthiocholine (0.3 mg/
ml) as substrate and Ellman's reagent (0.2 mg/ml) as coupling salt
in 50 mM phosphate buffer, pH 7.0, the colour development at 405 nm
is measured (Minireader, Dynatech) after 15-60 min.

Standard curves showed linearity with respect to incubation
time and, inversely, to the dilution of the lysate (Fig. 7). Other
preparations, e.g. NaCl and Triton X-100 extracts of human brain,
likewise showed linearity. Binding specificity was ensured by app-
lying the AChE specific inhibitor BW 284C51 [37] which completely
inhibited the bound enzyme activity. The detection limit of the
assay was ~1 ng AChE/ml of extract. However, this sensitivity was
impaired by the presence of AChE in the MAb preparation. Evaluated
on the basis of the activity measurements, one-tenth of the binding
sites were occupied by bovine AChE picked up from the cell-culture
media, necessitating a correction (Fig. 7 legend).

Exemplifying the individual properties of MAb's with the same
specificity, the anti-AChE Ab F18 which worked efficiently on the
immunoabsorbent column (3.2, above) did not bind AChE in immunoassay
when spiked into plasma, whereas another MAb worked well under such
conditions.

## 3.3 Conclusion

With MAb's it is possible to identify very precisely individual membrane antigens and to describe their dynamic behaviour during differentiation and cellular cycles. In immunoabsorbent chromatography they are very powerful tools for purification of membrane components, and applied in ELISA techniques they allow automated determination of minute amounts of antigen.

*Acknowledgements*

The authors thank Miss V. Kruse Pagh, Mrs. I. Vaarst Andersen and Mr. K. Pii Larsen for their skilful technical assistance. Financial support provided by the Danish Medical Research Council (Grant nos. 12-4305, 12-4401), Lundbeckfonden, Harboefonden, and Gerda og Aage Hench Fond is gratefully acknowledged.

*References*

1. Axelsen, N.H., ed. (1983) *Handbook of Immunoprecipitation in Gel Techniques* (Blackwell, Oxford)/*Scand. J. Immunol. 17, Suppl. 10*, 383 pp.
2. Bjerrum, O.J., Ramlau, J., Bock, E. & Bøg-Hansen, T.C. (1980) in *Membrane Receptors: Techniques for Purification and Characterization* (Jacobs, S.J. & Cuatrecasas, P., eds.), Chapman & Hall, London, pp. 113-156.
3. Bjerrum, O.J., ed. (1983) *Electroimmunochemical Analysis of Membrane Proteins*, Elsevier, Amsterdam, 176 pp.
4. Bjerrum, O.J. (1983) *as for* 3., 3-43.
5. Bjerrum, O.J. & Hagen, I. (1983) *as for* 3., 77-115.
6. Bjerrum, O.J. & Gianazza, E. (1983) *as for* 3., 127-153.
7. Bjerrum, O.J., Bjerrum, P.J., Larsen, K.P., Norrild, B. & Bhakdi, S. (1983) *as for* 3., 173-200.
8. Bjerrum, O.J., Gerlach, J., Bøg-Hansen, T.C. & Hertz, J.B. (1982) *Electrophoresis 3*, 89-98.
9. Blomberg, F. & Raftell, M. (1974) *Eur. J. Biochem. 49*, 21-30.
10. Heegaard, N.H.H., Hebsgaard, K. & Bjerrum, O.J. (1984) *Electrophoresis 5*, 263-269.
11. Bjerrum, O.J. (1978) *Anal. Biochem. 90*, 331-348.
12. Bhakdi, S., Bhakdi-Lehnen, B. & Bjerrum, O.J. (1977) *Biochim. Biophys. Acta 470*, 35-44.
13. Bjerrum, O.J. & Bøg-Hansen, T.C. (1976) *Biochim. Biophys. Acta 455*, 66-89.
14. Owen, P. (1983) in *Electroimmunochemical Analysis of Membrane Proteins* (Bjerrum, O.J., ed.), Elsevier, Amsterdam, pp. 56-76.
15. Juliano, R.L. (1973) *Biochim. Biophys. Acta 300*, 341-378.
16. Bjerrum, O.J., Lundahl, P., Brogren, C.-H. & Hjertén, S. (1975) *Biochim. Biophys. Acta 394*, 173-181.
17. Ott, P. & Brodbeck, U. (1978) *Eur. J. Biochem. 88*, 119-125.

18. Weitz, M., Bjerrum, O.J., Ott, P. & Brodbeck, U. (1982) *J. Cell Biochem. 19*, 179-191.

19. Weitz, M., Bjerrum, O.J. & Brodbeck, U. (1984) *Biochim. Biophys. Acta 776*, 64-74.

20. Pluzek, K.J. & Bjerrum, O.J. (1982) *J. Biochem. Biophys. Meth. 6*, 261-265.

21. Bennett, V., & Stenbuck, P.J. (1980) *J. Biol. Chem. 255*, 6424-6432.

22. Burnette, W.N. (1981) *Anal. Biochem. 112*, 195-203.

23. Towbin, H., Staehelin, T. & Gordon, J. (1979) *Proc. Nat. Acad. Sci. 76*, 4350-4354.

24. Bjerrum, O.J., Larsen, K.P. & Wilken, M. (1983) in *Modern Methods in Protein Chemistry* (Tschesche, H., ed.), W. de Gruyter, Berlin, pp. 79-124.

25. Gerson, J.M. & Palade, G.E. (1982) *Anal. Biochem. 131*, 1-15.

26. Haid, A. & Suissa, M. (1983) *Meth. Enzymol. 96*, 192-205.

27. Naaby-Hansen, S. & Bjerrum, O.J. (1985) *J. Reprod. Immunol. 7*, 41-57.

28. Lorand, L., Weissmann, L.B., Epel, D.L. & Bruner-Lorand, J. (1976) *Proc. Nat. Acad. Sci. 73*, 4479-4481.

29. Köhler, L. & Milstein, C. (1975) *Nature 256*, 495-497.

30. Skjødt, C., Schone, C. & Kock, C. (1984) *J. Immunol. Meth. 72*, 243-249.

31. Schecher, D.S. & Burke, D.C. (1980) *Nature 285*, 446-450.

32. Vockley, J. & Harris, H. (1984) *Biochem. J. 217*, 535-541.

33. Selmer, J.C., Larsen, F.S., Hertz, J. & Parton, R. (1984) *Acta Path. Microbiol. Immunol. Scand. Sec. C, 92*, 279-284.

34. Larsen, F.S., Selmer, J.C. & Hertz, J. (1984) *Acta Path. Microbiol. Immunol. Scand., Sec. C, 92*, 271-277.

35. Sørensen, K., Gentinetta, R. & Brodbeck, U. (1982) *J. Neurochem. 39*, 1050-1060.

36. Anderson, L.J., Godfrey, E., McIntosh, K. & Hierholzer, J.C. (1983) *J. Clin. Microbiol. 18*, 463-468.

37. Nørgaard-Pedersen, B., Hangaard, J. & Bjerrum, O.J. (1983) *Clin. Chem. 29*, 1061-1064.

#D-2

# USE OF POLYCLONAL AND MONOCLONAL ANTIBODIES IN STUDYING COMPLEMENT COMPONENT C9 AS A TRANS-MEMBRANE PROTEIN

J. Paul Luzio, [1]Keith K. Stanley, Peter Jackson,
Kenneth Siddle, [2]B. Paul Morgan and [2]Anthony K. Campbell

Department of Clinical Biochemistry
University of Cambridge
Addenbrooke's Hospital
Hills Road, Cambridge CB2 2QR, U.K.

[1]European Molecular
Biology Laboratory
Meyerhofstrasse 1, 10.2209
D-6900 Heidelberg
W. Germany

[2]Department of Medical
Biochemistry, University
of Wales College of
Medicine, Heath Park
Cardiff CF4 4XN, U.K.

*C9 is a serum protein that acts as the final component of the complement membrane attack complex (MAC). For its purification, assay, and immunofluorescent localization in biopsy specimens, use has been made of 5 mouse monoclonal antibodies (MAb's) that recognize different antigenic sites. The Ab's have also been used to show that C9 becomes a transmembrane protein on insertion into the MAC, and one Ab has been used to identify a cloned cDNA coding for C9 in a bacterial expression vector. With proteolytic cleavage and immunoblotting, along with functional inhibition experiments, the folding of C9 through the target membrane is being mapped.*

Immune attack on cells and tissues may involve antibody (Ab), complement or cell-mediated responses and can result in either cell lysis or impaired metabolic function. For the complement cascade considerable information is available concerning the proteins involved and their interaction with cell-bound Ab to cause formation of the complex MAC. This complex is made up of the complement components C5b, C6, C7, C8 and C9, though in the last few years there has been considerable controversy about its exact nature and the lesion caused in biological membranes [1-3]. C9 is a protein of mol. wt. 71,000, present in normal human plasma at 60 µg/ml, that acts as the final component of the MAC and is necessary for maximal rates of cell lysis. It is thought that components C5b-C8 catalyze polymerization of C9 to form a hollow protein cylinder inserted into the plasma membrane of the target cells [2]. Ion movement through [4] or round [5] this protein cylinder leads to an alteration of colloid osmotic pressure [6] and eventual cell lysis.

Although the most studied consequence of MAC formation has been cell lysis, particularly of erythrocytes, non-lytic effects may also occur. Previously we have shown that a rise in intracellular free $Ca^{2+}$ concentration is a very early event following C9 insertion into the MAC, is specific to C9 incorporation, and precedes the release of other ions and macromolecules [7-9]. We have proposed that the initial rise in intracellular free $Ca^{2+}$ mediates non-lytic effects of complement and also contributes to the specificity of membrane damage [10-13].

C9 is clinically important since it is involved in the prevention of infection [14, 15] and in auto-immune disease [10, 16, 17]. It is also a fascinating protein biochemically since it undergoes a transformation from being a hydrophilic serum protein to an amphiphilic integral membrane protein in the MAC. MAb's and polyclonal Ab's (PAb's) were prepared to human C9 so as to investigate its structure, fate on binding to the target membrane, and clinical function. It is hoped that a greater understanding of C9 insertion into and damage to the target membrane will also clarify cell-mediated cytotoxicity, which has similarities to complement-mediated cell damage [1, 6].

## PREPARATION AND PROPERTIES OF ANTIBODIES TO C9

Five mouse MAb's to partially purified human C9 [18] were prepared by standard methods [19, 20] and coded as follows (with affinities; all values $\times 10^9$ $M^{-1}$): C9-8 (0.1); C9-34 (0.3); C9-36 (3.1); C9-42 (1.6); C9-47 (2.6). Epitope analysis (Table 1) showed that at least 4 distinct antigenic sites on C9 were detected by these MAb's with C9-36 and C9-47 binding to the same or closely related sites. Different MAb's were used to purify C9 (C9-8) [20], for immunoradiometric assay (C9-34 and C9-47) [16, 21], for immunofluorescence [17] and to show internalization of cell surface-bound C9 (C9-47) [22]. Rabbit polyclonal antiserum to C9 immunoaffinity-purified on an MAb immunoadsorbent [20] was prepared by standard methods [11].

## MEMBRANE TOPOLOGY OF C9 INSERTED INTO THE MAC

Using indirect labelling with $^{125}I$-labelled sheep anti-mouse IgG, 4 of the MAb's have been shown to bind to the cell surface when added extracellularly to pigeon erythrocytes containing the MAC [17]. Three of these (C9-36, C9-42 and C9-47) will inhibit release of marker $^{14}C$-sucrose when added extracellularly to pigeon erythrocyte 'ghosts' containing MAC [22]. One Ab (C9-34) which will not bind to the MAC when added extracellularly inhibits complement-stimulated marker release when re-sealed inside pigeon erythrocyte 'ghosts' [22]. This has provided definitive proof that C9 becomes a

**Table 1.** Epitope analysis of monoclonal antibodies. Binding of [125]I-C9 to MAb immunoadsorbents was measured after pre-incubation with soluble MAb's [20]. The values represent % inhibition of binding to each adsorbent, relative to binding in the absence of soluble Ab; each value is the mean of duplicate determinations. The total radioactivity was 42800 cpm, and control (no soluble Ab) binding to each adsorbent was: C9-8, 6400; C9-34, 9670; C9-36, 20900; C9-42, 12400; C9-47, 20600.

| Soluble Ab | Solid phase: C9-8 | C9-34 | C9-36 | C9-42 | C9-47 |
|---|---|---|---|---|---|
| C9-8 | 74 | 19 | 8 | 4 | 1 |
| C9-34 | 0 | 90 | 7 | 11 | 2 |
| C9-36 | 0 | 7 | 90 | 17 | 96 |
| C9-42 | 0 | 7 | 5 | 91 | 5 |
| C9-47 | 0 | 0 | 60 | 0 | 73 |

transmembrane protein in the MAC and shows that one of the antigenic sites is expressed on the inner membrane face after C9 insertion whereas the others are expressed at the extracellular membrane face.

## IMMUNOBLOTTING OF C9 AND OF C9 FRAGMENTS

Chemical and enzymatic cleavage of C9 has allowed characterization of those parts of the molecule that react with different MAb's and hence should lead to a better definition of the folding of C9 through the target cell membrane. Western blotting [23] of C9 after reduction and alkylation (with SDS-PAGE and electrophoretic transfer onto nitrocellulose) allowed recognition of C9 only by MAb's C9-42 and C9-47 (data not shown); C9-36 did not react in these blots despite its binding to C9 being indistinguishable from C9-47 by epitope analysis. C9-34 reacted in a Western blot only if C9 was not reduced and alkylated before electrophoresis. Further analysis of the C9-47 binding site was carried out after chemical cleavage of C9 by 2-(2-nitrophenylsulphenyl)-3-methyl-3'-bromoindolenine ('BNPS-skatole') which acts at tryptophan [24]. This produced several smaller polypeptides detectable by Coomassie Blue staining after SDS-PAGE. Electrophoretic transfer of the stained peptides to nitrocellulose [25] and treatment with MAb C9-47 followed by peroxidase-labelled second Ab stained 3 fragments of mol. wt. 60,000, 49,000 and 38,000 (Fig. 1).

## MOLECULAR CLONING OF C9

MAb C9-47 has also been used to detect a human liver cDNA coding for C9 cloned in a bacterial expression vector [26]. The optimized expression vector was one of a family, pEX1-3, derived from a cro-lac Z gene fusion plasmid (Fig. 2), which expressed large quantities of fusion protein that is insoluble and can be detected using an

Fig. 1.  Western blotting of
intact C9 and C9 fragments
with MAb C9-47.  C9 was
cleaved at Trp residues by
digestion with 'BNPS-
skatole'.  The tracks show:
a, Coomassie blue stained
gel;
b, Coomassie blue stained
nitrocellulose blot;
c, Western blot, C9-47 plus
peroxidase-labelled second
Ab viewed through a Wratten
47B gelatin filter to remove
interference from the
Coomassie blue stain [25].

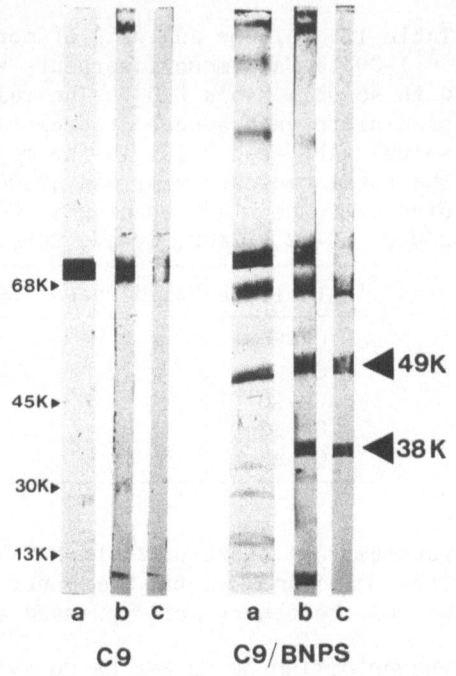

Fig. 2.  Bacterial expression vector pEX.  The pEX vectors [26]
express a β-galactosidase hybrid protein which can account for up
to 25% of total *E. coli* protein.  a: A linear representation of the
hybrid gene and its corresponding expressed polypeptide.  b: The
same vector containing a short open reading frame cDNA (solid segment).  P = $P_R$ promoter; Z = cro-lacI-lacZ gene fusion; L = oligonucleotide linker containing cloning sites; S = oligonucleotide
containing translation stop codons in all reading frames; T = transcription terminator fragments from  phage fd.

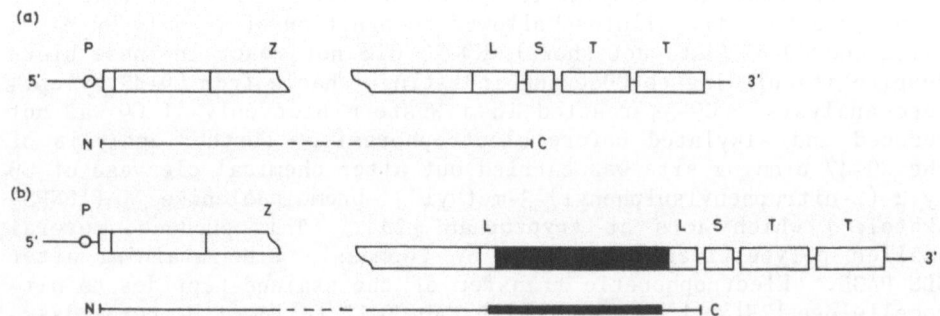

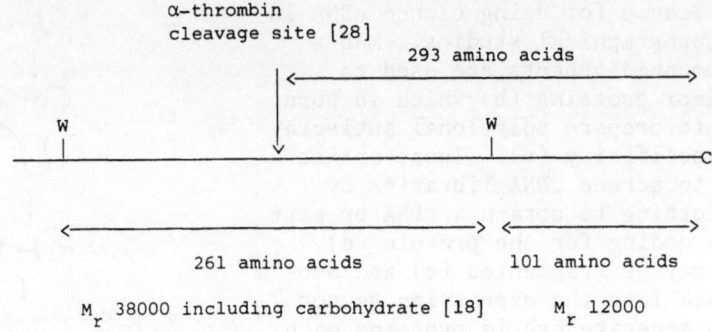

Fig. 3. Localization on the C9 primary sequence of the binding site for MAb C9-47. The diagram summarizes information from cleavage, cloning, sequencing and MAb binding studies, and shows the C-terminal end of the C9 molecule.

immune 'colony blot' procedure [19]. Several cDNA clones coding for C9 were detected with anti-C9 PAb's and one, clone 7, was also detected with MAb C9-47. Clone 7 contains a cDNA insert of 1350 base pairs, which has been sequenced and found to contain the C-terminal end of C9. The sequence was confirmed by comparison with the known amino acid composition and partial sequence of an α-thrombin cleavage product of C9 [28]. MAb C9-47 reacts with an epitope at the C-terminal end of C9. However, examination of the primary sequence in combination with the immunoblotting of cleavage fragments suggests this may not include the C-terminus itself (Fig. 3).

Additional benefits will accrue as a result of the expression cloning of C9. The complete cDNA can be cloned downstream of a strong promoter so that the synthesis and assembly of the enzyme may be studied in an *in vitro* transcription and translation system. Moreover, domains of the protein may be expressed *in vitro* and used to raise Ab's of pre-determined specificity. These can then be used to map protein topography with respect to the DNA sequence. Fig. 4 shows the overall approach of using Ab's to purify a protein and identify cDNA clones which in turn are used to generate Ab's to different parts of the protein molecule for topographical studies.

CONCLUSIONS

Polyclonal antisera and the 5 MAb's characterized have proved invaluable in investigating the clinical significance, structure and properties of complement component C9. The different properties and

**Fig. 4.** Scheme for using cloned cDNA in protein topographical studies. MAb's (**a**) on immunoadsorbents are used to purify minor proteins (**b**) which in turn are used to prepare polyclonal antisera of mono-specificity (**c**). These antisera are used to screen cDNA libraries by colony blotting to obtain a cDNA or part of a cDNA coding for the protein (**d**). This DNA may be fragmented (**e**) and sub-cloned back into the expression vector system to generate hybrid proteins each containing discrete domains of the original protein (**f**). These hybrid proteins may be used for raising Ab's of pre-determined specificity (**g**) or used for affinity-purifying a particular specificity of Ab from the original polyclonal serum. These mono-specific Ab's can then be used in structural and functional studies on the original protein (**h**).

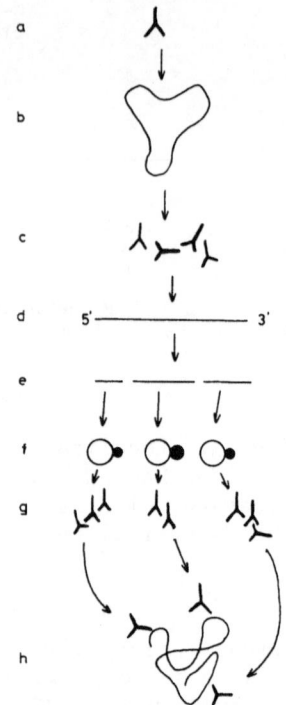

**Table 2.** Summary of the properties and uses of 5 MAB's to human C9.

|                                  | C9-8 | C9-34 | C9-36 | C9-42 | C9-47 |
|----------------------------------|------|-------|-------|-------|-------|
| Affinity × $10^9$ $M^{-1}$       | 0.1  | 0.3   | 3.1   | 1.6   | 2.6   |
| Purification of C9               | ✓    |       |       |       |       |
| Immunoradiometric assay          |      |       | ✓     |       | ✓     |
| Immunofluorescence               |      |       | ✓     | ✓     | ✓     |
| Binding to outer membrane domain | ✓    |       | ✓     | ✓     | ✓     |
| Binding to inner membrane domain |      | ✓     |       |       |       |
| Western blotting                 |      | ✓     |       | ✓     | ✓     |
| Expressed cDNA                   |      |       |       |       | ✓     |

uses of the MAb's are summarized in Table 2.  The prospect of using bacterial expression vectors to generate Ab's of pre-defined specificity is exciting if such an approach can be combined with fragmentation and functional studies to allow us to map the folding of C9 in the MAC.

*Acknowledgements*

The Medical Research Council, the European Molecular Biology Organization and the Arthritis & Rheumatism Council provided support.

*References*

1.  Lachmann, P.J. (1983) *Nature 305*, 473-474.
2.  Podack, E.R. & Tschopp, J. (1984) *Mol. Immunol. 21*, 589-603.
3.  Bhakdi, S. & Tranum-Jensen, J. (1984) *J. Immunol. 133*, 1453-1463.
4.  Mayer, M.M. (1972) *Proc. Nat. Acad. Sci. 69*, 2954-2958.
5.  Esser, A.F., Kolb, W.P., Podack, E.R. & Muller-Eberhard, H.J. (1979) *Proc. Nat. Acad. Sci. 76*, 1410-1414.
6.  Mayer, M.M. (1977) *J. Immunol. 119*, 1195-1203.
7.  Luzio, J.P., Daw, R.A., Hallett, M.B., Richardson, P.J. & Campbell, A.K. (1979) *Biochem. Soc. Trans. 7*, 1066-1068.
8.  Campbell, A.K., Daw, R.A. & Luzio, J.P. (1979) *FEBS Lett. 107*, 55-60.
9.  Campbell, A.K., Daw, R.A., Hallett, M.B. & Luzio, J.P. (1981) *Biochem. J. 194*, 551-560.
10. Campbell, A.K. & Luzio, J.P. (1981) *Experentia 37*, 1110-1112.
11. Hallett, M.B., Luzio, J.P. & Campbell, A.K. (1981) *Immunology 44*, 569-576.
12. Hallett, M.B. & Campbell, A.K. (1982) *Nature 295*, 155-158.
13. Richardson, P.J. & Luzio, J.P. (1980) *Biochem. J. 186*, 897-906.
14. Harriman, G.R., Esser, A.F., Podack, E.R., Wunderlich, A.C., Braude, A.I., Lint, T.F. & Curd, J.G. (1981) *J. Immunol. 127*, 2386-2390.
15. Joiner, K.A., Brown, E.J. & Frank, M.M. (1984) *Ann. Rev. Immunol. 2*, 461-491.
16. Morgan, B.P., Compston, A. & Campbell, A.K. (1984) *Lancet ii*, 251-255.
17. Morgan, B.P., Sewry, C.A., Siddle, K., Luzio, J.P. & Campbell, A.K. (1984) *Immunology 52*, 181-188.
18. Biesecker, G. & Muller-Eberhard, H.J. (1980) *J. Immunol. 124*, 1291-1296.
19. Galfre, G. & Milstein, C. (1981) *Meths. Enzymol. 73*, 3-46.
20. Morgan, B.P., Daw, R.A., Siddle, K., Luzio, J.P. & Campbell, A.K. (1983) *J. Immunol. Meths. 64*, 269-281.
21. Morgan, B.P., Campbell, A.K., Luzio, J.P. & Siddle, K. (1983) *Clin. Chim. Acta 134*, 85-94.
22. Morgan, B.P., Luzio, J.P. & Campbell, A.K. (1984) *Biochem. Biophys. Res. Comm. 118*, 616-622.
23. Burnette, W. (1981) *Anal. Biochem. 112*, 195-203.
24. Fontana, A. (1972) *Meths. Enzymol. 25*, 419-423.
25. Jackson, P. & Tommpson, R.J. (1984) *Electrophoresis 5*, 35-42.
26. Stanley, K.K. & Luzio, J.P. (1984) *EMBO J. 3*, 1429-1433.
27. Stanley, K.K. (1983) *Nucleic Acids Res. 11*, 4077-4092.
28. Biesecker, G., Gerard, C. & Hugli, T.E. (1982) *J. Biol. Chem. 257*, 2584-2590.

#D-3

# Ca$^{2+}$-BINDING PROTEINS LOCATED ON THE CYTOPLASMIC FACE OF THE LYMPHOCYTE PLASMA MEMBRANE

Raymond J. Owens[*], Adelina A. Davies,
Christopher J. Gallagher[†], Mark Hexham
and Michael J. Crumpton

Imperial Cancer Research Fund
Lincoln's Inn Fields, London WC2A 3PX, U.K.

*Three polypeptides of mol. wts. ~68,000, 33,000 and 28,000 have been identified in Ca$^{2+}$-chelator extracts of purified preparations of lymphocyte plasma membrane (p.m.). They are preferentially associated with the actin-rich, Nonidet P40-insoluble fraction of the membrane and are not labelled by lactoperoxidase-catalyzed iodination of whole lymphocytes, suggesting that they are located on the cytoplasmic face of the p.m. The 68,000 mol. wt. protein, the most abundant of the three in the EGTA extract, has been purified from pig and human lymphocyte p.m. and partially characterized. It is a monomeric acidic protein with a single high-affinity Ca$^{2+}$-binding site ($K_D = 1.2 \mu M$). From their solubility with Ca$^{2+}$-chelators present, the 33,000 and 28,000 components seem to be Ca$^{2+}$-binding proteins too.*

*An affinity-purified polyclonal antibody (Ab) has been prepared against the 68,000 mol. wt. protein. Immunofluorescence (IF) and immunoprecipitation studies have confirmed its intracellular localization in lymphocytes, and shown its presence also in various non-lymphoid cells. The protein, from immunological cross-reactivity, seems identical to Ca$^{2+}$-binding proteins of similar mol. wt. that have been isolated from bovine liver and adrenal medulla and that promote the Ca$^{2+}$-dependent aggregation of isolated chromaffin granules. Immunological results indicate that the 33,000 mol. wt. lymphocyte protein is related to a phosphotyrosine-containing protein, 'p36', which can be phosphorylated by the pp60$^{src}$ kinase of Rous sarcoma virus. Thus, the lymphocyte proteins are representative of a a group of membrane-associated Ca$^{2+}$-binding proteins which are widely distributed in mammalian tissues.*

Present addresses: [*] Sir William Dunn School of Pathology, South Parks Road, Oxford OX1 3RE; [†] Royal Marsden Hospital, Fulham Road, London SW3 6JJ.

Changes in the cytoplasmic concentration of $Ca^{2+}$ have been implicated in the regulation of a number of cellular processes that involve the cell surface membrane, e.g. secretion by exocytosis, which relies on the fusion of intracellular membrane vesicles with the p.m. [1], and the internalization of cell-surface receptor-ligand complexes by endocytosis ([2, 3], & arts. in Vol. 13, this series). The effects of calcium appear to be generally mediated via its interaction with specific binding proteins. Thus, $Ca^{2+}$-mediated events at the p.m. probably involve membrane-associated $Ca^{2+}$-binding proteins. It seems likely that such proteins are located on the cytoplasmic face of the p.m.

Except for erythrocytes, very little is known about either the components comprising the cytoplasmic face of the p.m. or their arrangement. In erythrocytes, a macromolecular complex of proteins comprising mainly spectrin, ankyrin and actin underlies the surface membrane and is defined as the erythrocyte cytoskeleton [4]. By using the extraction conditions employed by Steck and his colleagues to define the erythrocyte cytoskeleton [4], a non-ionic detergent-insoluble complex of proteins has recently been isolated from purified preparations of human and pig lymphocyte p.m. [5]; a similar complex has been separated from mouse lymphocytes [6]. By analogy with the erythrocyte, the detergent-insoluble complex is most probably associated with the cytoplasmic side of the lymphocyte p.m. A number of $Ca^{2+}$-binding proteins have been identified as major components of the lymphocyte detergent-insoluble complex. These proteins appear to be good candidates for mediating $Ca^{2+}$-dependent processes at the cytoplasmic face of the p.m.

In this article we describe the experimental approach we have used to identify the membrane-associated $Ca^{2+}$-binding proteins of lymphocytes. The procedure, which involves extraction of the non-ionic detergent-insoluble fraction of the purified p.m. with $Ca^{2+}$ chelators, is generally applicable to other cell types. In addition, we have used polyclonal Ab's to investigate the cellular and species distributions of these proteins.

## PREPARATION AND NONIDET P40 EXTRACTION OF LYMPHOCYTE PLASMA MEMBRANE

Human B lymphoblastoid cells* (cell lines BRI 8, Maja and RPMI 1788) in Dulbecco's phosphate-buffered saline (PBS; 10 mM Na phosphate buffer, pH 7.2, 0.17 M NaCl, 3 mM KCl, 1 mM $CaCl_2$, 1 mM $MgCl_2$) were homogenized using a cell-disrupting pump [7]. Proteolysis was inhibited by the addition of phenylmethanesulphonylfluoride and iodoacetamide to final concentrations of 1 mM and 10 mM respectively. The p.m. was separated from the homogenate by a combination of differential rate and sucrose-density centrifugation [7]. As judged morphologically and biochemically, the purified pm. fraction was not

---

* Art. #D-4 in Vol. 8, this series (Owen & Crumpton) describes lymphocyte isolation.- *Ed.*

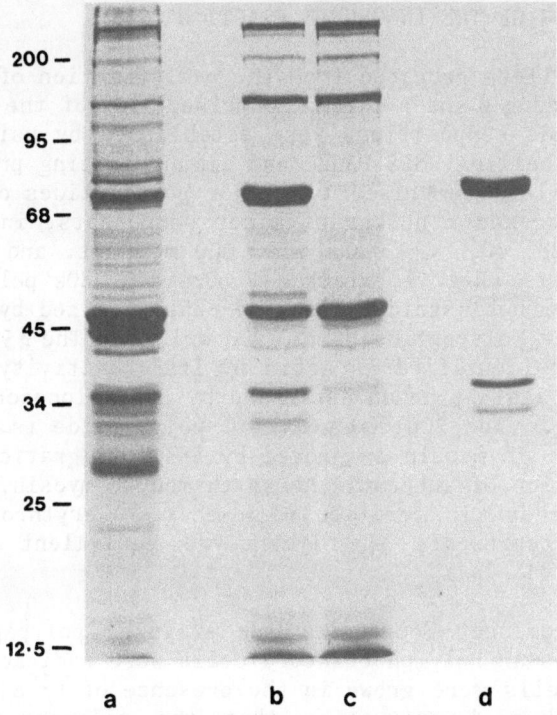

**Fig. 1.** Polypeptide compositions of sub-fractions of lymphocyte p.m., as revealed by SDS-PAGE under reducing conditions, with Coomassie Blue staining. Track a, a purified preparation of the p.m. of human B lymphoblastoid BRI 8 cells. Tracks b and c show the Nonidet P40-insoluble residue (the '20k pellet') or BRI 8 p.m. prepared in the presence of 1 mM $Ca^{2+}$ or 1 mM EGTA respectively. Track d: the supernatant obtained by re-extracting the 20k pellet shown in track b with 1 mM EGTA in 1% Nonidet P40.

contaminated by significant amounts of other subcellular membranes [7, 8]. The purified p.m. (Fig. 1, track a) was washed twice with PBS and then extracted (0.5 mg of membrane protein/ml) with 0.5% (v/v) Nonidet P40 in PBS for 30 min at 0°. The extract was centrifuged at 4° for $10^6$ g-min ($20 \times 10^3$ rpm for 30 min; Beckman 50.2 Ti rotor). The resulting detergent-insoluble fraction is referred to as the '20k pellet' [5]. The foregoing procedure has also been used to prepare the Nonidet P40*-insoluble fraction from the p.m. of human peripheral blood lymphocytes as well as the following lymphoid tissues: human tonsil and spleen, and pig mesenteric lymph node. In the case of the lymphoid tissues the cells were disrupted by homogenizing the tissue in a tissue press [8].

---

* non-ionic detergent (an octyl phenol ethylene oxide condensate)

## CHARACTERIZATION OF THE INSOLUBLE FRACTION

The 20k pellets prepared from the p.m. fraction of various lymphocytes, including B and T cells, comprised ~10% of the p.m. protein. Their polypeptide compositions were established by using a combination of radiolabelling, SDS-PAGE, and immunoblotting procedures. In general the pellets comprised two major polypeptides of ~45,000 and ~68,000 mol. wt. and a number of minor components, including polypeptides of ~28,000, ~33,000, ~200,000 mol. wt. and a doublet of ~240,000 mol. wt. (Fig. 1, track b). Besides, 20k pellets prepared from human B lymphoblastoid cells were characterized by the presence of a third major polypeptide of 120,000 mol. wt. The 45,000 mol. wt. polypeptide was identified as actin by its reactivity with a monoclonal Ab (MAb) against human actin and by Staphylococcus V8 protease peptide-mapping. The 200,000 mol. wt. polypeptide is most probably the heavy chain of myosin as judged by its co-migration on SDS-PAGE with a preparation of authentic human thymocyte myosin. The 240,000 doublet coincides in position with that of erythrocyte spectrin and probably represents the lymphocyte equivalent of spectrin, namely fodrin [9].

All the prominent Coomassie Blue staining polypeptides of the 20k pellet of human B lymphoblastoid cells were radioactively labelled when the cells were grown in the presence of [$^{35}$S]methionine or [$^{3}$H]leucine, thus demonstrating that the polypeptides were not derived from extraneous sources. In contrast, these polypeptides did not incorporate detectable amounts of radioactivity when the cells were biosynthetically labelled with [$^{3}$H]mannose or labelled at the surface by lactoperoxidase-catalyzed iodination. Since apparently all protein molecules exposed on the cell surface are glycosylated [10], these results suggest that the 20k pellet's polypeptides are not located on the lymphocyte surface and thereby argue strongly for their location on the cytoplasmic face of the p.m.

## ISOLATION/IDENTIFICATION OF CALCIUM-BINDING PROTEINS

The polypeptide composition of the 20k pellet depended directly upon the concentration of $Ca^{2+}$ in the buffer used to extract the p.m. fraction. As shown in Fig. 1 (track c), the 20k pellet prepared in the presence of a $Ca^{2+}$ chelator (EGTA) lacked the 68,000 mol. wt. polypeptide together with the polypeptides of mol. wt. 33,000 and 28,000 (cf. track b in Fig. 1, showing the composition of the 20k pellet prepared in the presence of 1 mM $CaCl_2$). The concentration of $Ca^{2+}$ required to maintain the association of the 68,000 mol. wt. polypeptide with the 20k pellet was estimated as >10 μM.

The polypeptides of mol. wt. 68,000, 33,000 and 28,000 were separated from the 20k pellet by making use of their $Ca^{2+}$-dependent solubilities. Thus, a 20k pellet which had been prepared in the presence of 1 mM $Ca^{2+}$ was re-extracted with 1% (v/v) Nonidet P40,

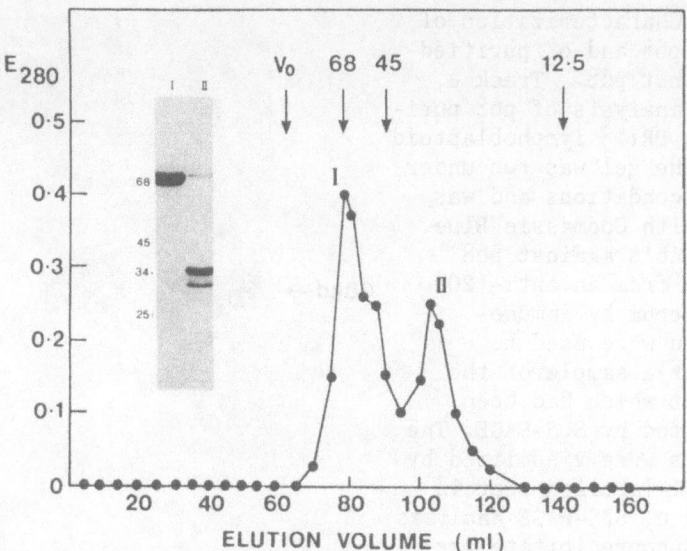

Fig. 2.　Fractionation on a Sephacryl S-200 column [11] of the EGTA extract prepared from the 20k pellet of BRI 8 p.m.　The column (160 ml) was eluted with 20 mM Tris-HCl buffer, pH 7.4. *Arrows* indicate the elution positions of mol. wt. markers ($V_0$, blue dextran; 68, bovine serum albumin; 45, ovalbumin; 12.5, cytochrome C). *Insert* shows the polypeptide compositions of the two major peaks (I and II) as revealed by SDS-PAGE.

1 mM EGTA, 1 mM $Mg^{2+}$ in PBS containing no added $Ca^{2+}$, for 30 min at $0°$ and then centrifuged at 100,000 $g_{av}$.　Fig. 1 (track d) shows that, under these extraction conditions, the supernatant consisted mainly of the 68,000 mol. wt. polypeptide together with smaller amounts of the 33,000 and 28,000 mol. wt. polypeptides.　Analysis of the EGTA extract by 2-D isoelectric focusing (IEF) SDS-PAGE indicated that the 68,000, 33,000 and 28,000 mol. wt. polypeptides had isoelectric points of ~5.7, ~7.2 and ~5.2 respectively.　The 68,000 and 33,000 mol. wt. polypeptides also showed three closely aligned, regularly spaced isoelectric variants which focused at ±0.1 pH units of their respective pI's.

The IEF data indicated that the three polypeptides of the EGTA extract could potentially be separated by ion-exchange chromatography. More simply, it proved possible to separate the 68,000 mol. wt. polypeptide by gel filtration as in the legend to Fig. 2, which shows that the protein was eluted at the position expected for a globular protein of mol. wt. 68,000.　No association of this protein with any of the other polypeptides in the extract was apparent under the conditions used.

Fig. 3. Characterization of
purified p68 and of purified
Ab's against p68.  Track a,
SDS-PAGE analysis of p68 purified from BRI 8 lymphoblastoid
cells.  The gel was run under
reducing conditions and was
stained with Coomassie Blue.
Track b, Ab's against p68
separated from an anti-(20k
pellet) serum by immuno-
adsorption were used to
immunoblot a sample of the
20k pellet which had been
fractionated by SDS-PAGE. The
bound Ab's were visualized by
using $^{125}$I-labelled protein
A.  Track c, SDS-PAGE analysis
of an immunoprecipitate prepared from a Nonidet P40-1 mM
EGTA lysate of [$^{35}$S]methionine-
BRI 8 cells (biosynthetically
labelled) by using the purified Ab's against p68.

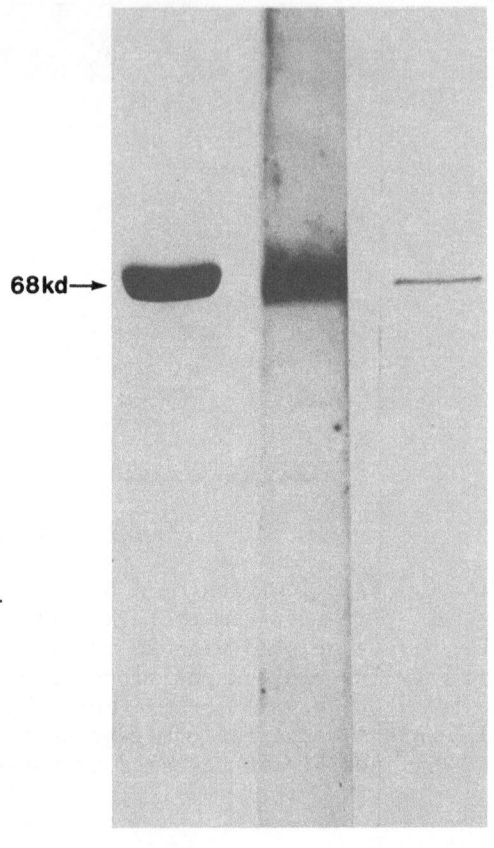

68kd→

a          b          c

The 68,000 mol. wt. protein, termed p68, has been purified to
apparent homogeneity, as judged by 1-D and 2-D SDS-PAGE, from the
p.m. fractions of both human and pig lymphocytes.  Thus, as shown in
Fig. 3 (track a), the purified protein from BRI 8 lymphocytes gave
only one band when analyzed by SDS-PAGE.  By a number of criteria,
including amino acid composition and peptide mapping, the protein is
highly conserved between humans and pigs.  Using equilibrium gel
filtration, it was shown that each molecule of p68 has a single high-
affinity $Ca^{2+}$-binding site ($K_D$ = 1.2 μM).  It seems likely on the
basis of their similar solubilities in the presence of $Ca^{2+}$ chelators
that the 33,000 and 28,000 mol. wt. polypeptides of the 20k pellet
are also $Ca^{2+}$-binding proteins.

## CHARACTERIZATION OF p68

In order to investigate the cellular distribution of p68 an Ab
was raised against the protein.  As p68 could be obtained only in
small amounts (~10 μg/10$^{10}$ cells), an indirect method was used to

produce the Ab. Thus, rabbits were immunized with the 20k pellet
prepared from human B lymphoblastoid cells (BRI 8), and sera from
successive bleeds were monitored by immunoblotting against 20k pellets prepared in either the presence or the absence of $Ca^{2+}$ (i.e.
with or without p68) for reactivity against p68. Ab's against p68
were then separated from the positive sera by using purified p68
attached to Sepharose 4B as an immunoadsorbent. The results of
immunoblotting and immunoprecipitation experiments (Fig. 3, tracks
b & c respectively) indicated that the affinity-purified Ab was specific for p68 and hence could be used to investigate the cellular
distribution of the protein [12].

The affinity-purified Ab stained fixed and permeabilized  human
B lymphoblastoid cells, peripheral blood lymphocytes and sections of
tonsil. Whole cells, however, were not stained, indicating that the
protein was not represented at the cell surface. This assignment is
consistent with the detection of p68 in immunoprecipitates from biosynthetically labelled, but not surface-labelled, cells. These
results confirmed those obtained from labelling studies on the whole
20k pellet, and collectively argue strongly in support of p68, along
with the other major proteins of the 20k pellet, being located on
the cytoplasmic face of the p.m. in lymphocytes.

An important question is whether p68 occurs exclusively in lymphoid cells. This was explored by analyzing immunoprecipitates prepared from Nonidet P40 lysates of various biosynthetically labelled
lymphoid and non-lymphoid cell lines by using the affinity-purified
Ab's against p68. The results demonstrated that the protein is
synthesized in a variety of human B and T cell lines as well as fibroblasts and epithelial cells. Most interestingly, Burkitt lymphoma
cell lines invariably contained much less p68 than mature B lymphocytes. The protein was also detected in an erythroblastoid cell
line (K562) although p68 is present neither in erythrocytes, as
judged by immunofluorescence (IF), nor in platelets, as judged by
immunoblotting. The distribution of p68 in fibroblasts was studied
further by IF staining. Anti-(p68) Ab's stained diffusely the cytoplasm of mouse 3T3 fibroblasts which had been fixed and permeabilized
whereas a reticulate pattern of staining was observed using fibroblasts which had been extracted with Nonidet P40 prior to fixation
to give a cytoskeletal preparation [13]. Similar IF staining patterns have been observed  for spectrin-related proteins and a phosphotyrosine-containing protein, p36 [14], that is a major substrate
for phosphorylation by $pp60^{src}$ kinase [15-17]. The latter  proteins
are considered to be associated with the cytoplasmic side of the
p.m.   Likewise, it appears that  p68 has a similar localization in
fibroblasts as well as other non-lymphoid cells.

Apparently, then, p68 is a p.m.-associated $Ca^{2+}$-binding protein
which is widely distributed amongst different mammalian cell-types.
Further  supporting  evidence  has  come  from  an  independent  study.

A polypeptide of mol. wt. ~67,000 has been purified from an EGTA extract of bovine liver membranes [18]. Its reported amino acid composition and pI are very similar to those of lymphocyte p68. The liver protein also binds $Ca^{2+}$ with a similar affinity and moreover appears to promote the aggregation of chromaffin granules isolated from bovine adrenal medulla in a calcium-dependent manner [18]. Confirmation that the liver protein is related to p68 has come from immunoprecipitation and immunoblotting experiments. Thus, Fig. 4 shows that an antiserum raised against the bovine liver protein [18] immunoprecipitated p68 from human BRI 8 lymphoblastoid cells (track b) whereas in the reciprocal experiment anti-(human p68) immunoprecipitated a 67,000 mol. wt. polypeptide plus a lower band of mol. wt. ~32,000 from bovine liver microsomes (track c).

## IDENTIFICATION OF THE 33,000 MOL. WT. POLYPEPTIDE AS p36

The question remains as to the nature of the lower mol. wt. putative $Ca^{2+}$-binding proteins. Neither the 33,000 nor the 28,000 mol. wt. polypeptide reacted by immunoprecipitation (Fig. 4, track d) with the purified Ab's against p68, indicating that they are probably not related to p68. Although we have no further data concerning the nature of the 28,000 mol. wt. polypeptide, a variety of evidence points to the identity of the 33,000 polypeptide.

By a procedure similar to that used to isolate lymphocyte p68, a protein of ~80,000 mol. wt. has been separated from the detergent-insoluble cytoskeleton of the brush-border membrane of pig intestinal epithelial cells [19]. This protein comprised two subunits of ~34,000 and one of 10,000 mol. wt. and bound $Ca^{2+}$ with high affinity ($K_D$ ~1 µM). Interestingly, as shown initially by Gerke & Weber [19], the larger subunit of the protein is antigenically related to p36, the major intracellular substrate for tyrosine phosphorylation by $pp60^{src}$ kinase. As shown in Fig. 5 (tracks a, b), $Ca^{2+}$-chelator extracts of the Nonidet P40-insoluble residue of pig intestinal epithelium brush-border membranes comprised two polypeptides which comigrated on SDS-PAGE with the 33,000 and 28,000 mol. wt. polypeptides of a comparable extract of human tonsil lymphocyte p.m. In contrast, p68 was barely detected in the pig intestinal epithelial extract.

The above result suggests that the 34,000 mol. wt. subunit of the intestinal epithelial protein and the 33,000 mol. wt. polypeptide from lymphocyte p.m. are related. Confirmation that the 33,000 lymphocyte polypeptide is indeed related to p36 has been obtained by immunoblotting. Thus, an antiserum produced in rabbits by immunization with p36 purified from chick embryo fibroblasts [20] recognized a polypeptide of ~33,000 mol. wt. in EGTA extracts prepared from Nonidet P40-insoluble residues of both human and pig intestinal epithelial cell p.m. (Fig. 5, tracks c, d).

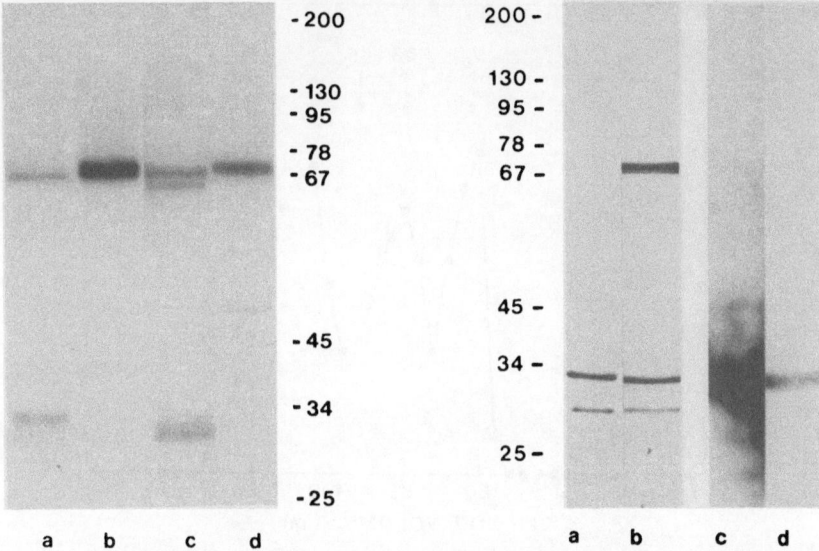

**Fig. 4.** Compositions of immuno-precipitates prepared from [125]I-labelled bovine liver microsomes (tracks a, c) and BRI 8 lympho-cyte p.m. (b, d) using an anti-(bovine p67) serum [18] (tracks a, b) and purified Ab's against human p68 (c, d). SDS-PAGE analysis under reducing condi-tions, and autoradiographic visualization.

**Fig. 5.** Polypeptide compositions of EGTA extracts of Nonidet P40-insoluble residues of pig intesti-nal brush-border membrane (tracks a, c) and human tonsil (b, d). SDS-PAGE under reducing conditions. Tracks a & b stained for protein by Coomassie Blue; c & d immuno-blotted with an anti-(chicken fibroblast p36) serum.

The detection of a protein related to p36 in lymphocytes is contrary to some recent reports [21, 22]. The latter interpreta-tions were based mainly on immunohistochemical procedures, and it appears likely that the discrepancy between the studies merely reflects the relative sensitivities of the different detection methods.

In spite of the apparent identity by SDS-PAGE of the p36 poly-peptides of lymphocytes and intestinal epithelial cells, there appears to be a major difference between their respective p36-related proteins. As shown in Fig. 6, the p36 protein of pig epithelial cells is eluted from Sepharose S-200 in a position coincident with a mol. wt. of ~70,000. This behaviour is consistent with the previous report [19] that this protein comprises two subunits of 34,000 and one of 10,000 mol. wt. In contrast, the lymphocyte p36 protein behaves as a monomer on gel-filtration (Fig. 2). Further work is necessary to determine whether this apparent difference in behaviour

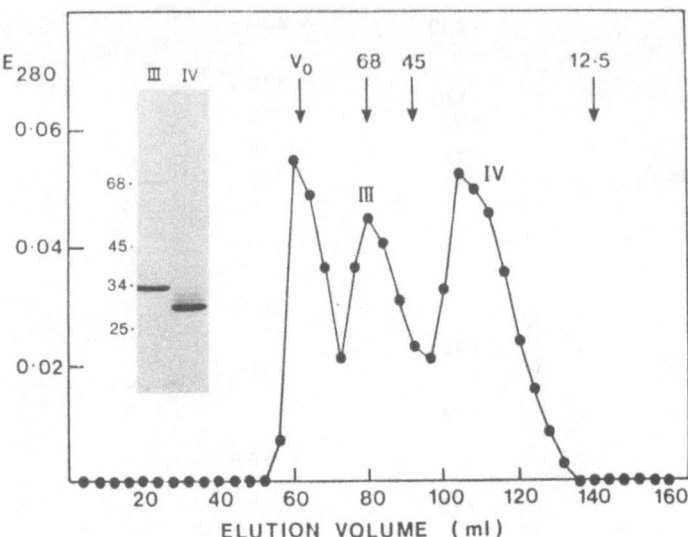

**Fig. 6.** Fractionation on a Sephacryl S-200 column of the EGTA extract of the Nonidet P40-insoluble residue of pig intestinal epithelium brush-border membrane. The experimental conditions were identical with those employed in Fig. 2. The *insert* shows the polypeptide compositions of peaks III and IV as revealed by SDS-PAGE.

has a trivial explanation, such as concentration-dependent dissociation, or a more profound and interesting explanation.


## CONCLUSION

Insolubility in non-ionic detergents has been used to identify a group of proteins which are located on the cytoplasmic face of the lymphocyte p.m. Apart from actin, the group includes three novel $Ca^{2+}$-binding proteins. On the basis of immunological cross-reactivity, these proteins are present in a spectrum of mammalian types, including fibroblasts and epithelial cells. Their wide cell distribution suggests that they may have a common function in different cells. One of the membrane-associated $Ca^{2+}$-binding proteins is related to p36, the principal intracellular substrate of the tyrosine-specific protein kinase, pp60[src]. This raises the possibility that this $Ca^{2+}$-binding protein may be involved in lymphocyte transformation. One of the other $Ca^{2+}$-binding proteins, namely p68, is differentially expressed during B-lymphocyte differentiation and may be involved in membrane-membrane fusion during secretion.

*References*

1.  Metzger, H. (1978) *Immunol. Rev. 41*,  186–199.
2.  Salisbury, J.L., Condeelis, J.S. & Satir, P. (1980) *J. Cell Biol. 87*, 132–141.
3.  Kretsinger, R.H. (1979) *Adv. Cyclic Nucleotide Res. 11*,  1–26.
4.  Yu, J., Fishman, D.A. & Steck, T.L. (1973) *J. Supramol. Struct. 1*,  233–248.
5.  Davies, A.A., Wigglesworth, N.M., Allan, D., Owens, R.J. & Crumpton, M.J. (1984) *Biochem. J. 219*, 301–308.
6.  Mescher, M.F., Jose, M.J. & Balk, S.P. (1981) *Nature 289*, 139–141.
7.  Crumpton, M.J. & Snary, D. (1974) *Contemp. Top. Mol. Immunol. 3*, 27–56.
8.  Snary, D., Woods, F.R. & Crumpton, M.J. (1976) *Anal. Biochem. 74*,  457–465.
9.  Glenney, J.R. & Glenney, P. (1983) *Cell 34*, 503–517.
10. Bretscher, M.S. & Raff, M.C. (1975) *Nature 258*, 43–49.
11. Owens, R.J. & Crumpton, M.J. (1984) *Biochem. J. 219*, 309–316.
12. Owens, R.J., Gallagher, C.J. & Crumpton, M.J. (1984) *EMBO J. 3*, 945–952.
13. Lenk, R. & Penman, S. (1979) *Cell 16*, 289–301.
14. Greenberg, M.E. & Edelman, G.M. (1983) *Cell 33*, 767–779.
15. Lehto, V.-P., Virtanen, I., Passivuo, R., Ralston, R. & Alitalo, K. (1983) *EMBO J. 2*, 1701–1705.
16. Nigg, E.A., Cooper, J.A. & Hunter, T. (1983) *J. Cell Biol. 96*, 1601–1609.
17. Erickson, E. & Erickson, R.L. (1980) *Cell 21*, 829–836.
18. Sudhof, T.C., Ebbecke, M., Walker, J.R., Fritzsche, U. & Boustead, C. (1984) *Biochemistry 23*, 1103–1109.
19. Gerke, V. & Weber, K. (1984) *EMBO J. 3*, 227–233.
20. Courtneidge, S., Ralston, R., Alitalo, K. & Bishop, J.M. (1983) *Mol. Cell. Biol. 3*,  340–350.
21. Gould, K.L., Cooper, J.A. & Hunter, T. (1984) *J. Cell Biol. 98*, 487–497.
22. Greenberg, M.E., Brackenbury, R. & Edelman, G.M. (1984) *J. Cell Biol. 98*, 473–485.

#D-4

# THE T-CELL RECEPTOR

Phillipa Marrack, Kathryn Haskins, Neal Roehm,
Janice White, Willi Born, Jordi Yagüe, Edward Palmer
and John W. Kappler

Department of Medicine
National Jewish Hospital & Research Center
3800 East Colfax Avenue, Denver, CO 80206, U.S.A.

*Use has been made of monoclonal antibodies (MAb's) that appear to recognize all or part of the T-cell receptor, features of which are discussed. We have devised a new approach to receptor specificity investigation, and encountered problems that are discussed. We consider the contribution of α and β chains to T-cell specificity, and the expression of T-cell receptors on thymocytes.*

The molecule on T cells responsible for antigen in association with products of the major histocompatibility complex (MHC) has been isolated by us using MAb's [1-3]. Every property of these Ab's and the molecules they recognize supports the idea that it is indeed the antigen/MHC receptor that has been identified. Thus, we have prepared a MAb, KJ1-26, which recognizes the receptor on a mouse T-cell hybridoma, DO-11-10. KJ1-26 was originally picked as interesting because Ab therefrom inhibited the recognition of chicken ovalbumin (cOVA) and I-A$^d$ by DO-11-10. The T-cell hybridoma normally responds to cOVA in the presence of I-A$^d$-bearing antigen-presenting cells by secreting the lymphokine interleukin-2 (IL-2), but this secretion does not occur in the presence of KJ1-26 Ab. Strikingly, the Ab inhibits antigen/MHC responses by DO-11.10 only. Other T-cell hybridomas with similar but not identical specificities are unaffected, for example 3DO-54.8, a T-cell hybridoma derived from the same types of source as DO-11.10 - i.e. BALB/c, cOVA-immune T-cells fused to the AKR thymoma, BW5147 - is unaffected by KJ1-26 and, indeed, the MAb does not bind to the latter T-cell hybrid. The specificity of our MAb, which had been noted previously using antisera by Infante et al. [4], strongly suggested that the Ab was recognizing all or part of the T-cell receptor since no other surface molecule would be so clone-specific in its distribution.

Other properties of KJ1-26 serve only to confirm our conclusion. In polyvalent form, e.g. coupled to Sephadex beads, the Ab stimulates IL-2 production by DO-11.10 in the absence of any other stimulus, i.e. the Ab substitutes for both antigen and MHC, suggesting that it binds a single receptor responsible for recognizing both entities. The Ab has also been used to find a second T-cell hybridoma to which it binds. The latter, 7DO-286.2, has the same specificity for antigen and MHC as DO-11.10, again proving that the Ab recognizes the molecule on these T-cell surfaces responsible for recognition of both cOVA and I-A$^d$ [5, 6].

## FEATURES OF THE RECEPTOR

The receptor has a number of interesting properties. Immuno-precipitation of $^{125}$I-surface-labelled material showed that it is a disulphide-linked heterodimer, made up in mouse of an acidic α chain and a basic β chain each of mol. wt. ~43,000. In man, differential glycosylation leads to an α chain of mol. wt. 48,000-50,000, and a β chain of mol. wt. ~39,000. Non-equilibrium pH gradient electrophor-esis and peptide finger-printing of tryptic digests reveals that both α and β chains vary in amino acid sequence when the receptors from different T-cell hybridomas are compared. Finger-prints do reveal, however, that α chains from different T-cell clones share peptides, as do β chains [2, 3, 5, 7].

Overall, then, the structure of the receptor is reminiscent of, although by no means identical with, that of the immunoglobulins (Ig's), since the molecule is a disulphide-linked heterodimer, and each chain appears to have both constant and variable region sequen-ces.

The final proof that this molecule is indeed responsible for binding both antigen and MHC would involve the demonstration that isolated receptor protein does indeed engage these entities. So far we and others have been unable to demonstrate such binding. Thus the isolated $^{125}$I-labelled receptors of a T-cell hybridoma, 3DT-52.5, which is specific for D$^d$ in the absence of antigen, fail to show significant binding to D$^d$-bearing cells. Likewise $^{125}$I-labelled receptors from DO-11.10 do not bind any better to I-A$^d$ cells which have been pulsed with cOVA than they do to non-pulsed I-A$^d$-bearing cells. There are a number of explanations for these results. The binding of monoclonal isolated receptors may be of such low affinity as to be undetectable above the noise of binding of this hydrophobic membrane-associated molecule to any cell. The antigen/MHC binding site of the receptor may have been denatured by our isolation pro-cedures (receptor isolation involves immunoprecipitation with anti-receptor Ab-coated beads and dissociation from this complex with diethylamine at high pH). Finally, the receptor may not be exactly what we think it is, and some essential clonally variant component may have been removed by the isolation methods.

**Table 1.** Responses of T-cell hybridomas fused with receptor-bearing liposomes. The values are units/ml IL-2 secreted in response to the agents indicated.

| T-cell hybridoma | Specificity | Source of receptor in liposomes | OVA/I-A$^d$ | KLH/I-A$^k$ | KJ1-26 beads |
|---|---|---|---|---|---|
| SKK-45.10 | KLH/I-A$^k$ | none | <10 | 640 | <10 |
| SKK-45.10 | KLH-I-A$^k$ | DO-11.10 | <10 | 640 | <10 |
| DO-11.10 | OVA/I-A$^d$ | none | 640 | <10 | 160 |

## RECEPTOR SPECIFICITY: A NEW APPROACH, AND INVESTIGATIVE PROBLEMS

To circumvent some of these problems we have devised another method of analyzing the specificity of these receptors. Although this method has likewise not so far revealed the specificity we hope for, it does illustrate some of the problems of studying these molecules, and thus warrants consideration here.

The receptor from DO-11.10 is isolated by binding NP-40 lysates of the T-cell hybridoma to beads coupled to KJ1 Ab. Receptor protein is eluted from these beads with pH 10.4 diethylamine and is then dialyzed with a mixture of Sendai haemagglutinin and neuraminidase proteins and various lipids against phosphate-buffered saline (PBS). Liposomes are thus formed* which include DO-11.10 receptor and Sendai fusion proteins. These liposomes are then fused to recipient T-cell hybridomas using a combination of polyethylene glycol (PEG) and the Sendai fusion proteins.

Analysis on a cytofluorograph with fluorescein-coupled KJ1-26 shows that under the right conditions the recipient T-cell hybridoma cells equal or surpass DO-11.10 in the number of DO-11.10-specific receptor molecules per cell that they bear. The recipient cells are then tested for their ability to respond to the antigen/MHC combination for which the T-cell fusion partner was specific, for response to cOVA/I-D$^d$ and for responses to KJ1-26-bearing Sephadex beads, these last two specificities being those expected of the DO-11.10 receptors. Our results are exemplified in Table 1.

The recipient cell in this experiment, SKK-45.10, responded very well to keyhole limpet haemocyanin (KLH) plus I-A$^k$ both before and after fusion to the receptor-bearing liposomes. After fusion, SKK-45.10 cells stained brightly with KJ1-26. This staining was stable for at least 24 h (data not shown), suggesting that the DO-11.10 receptor was firmly anchored to the surface of the SKK-45.10 cells. Disappointingly, SKK-45.10 bearing DO.11.10-derived receptors

* See Vol. 13, this series, for liposome/receptor methodology.-*Ed.*

failed to respond to cOVA/I-A$^d$, whereas control DO-11.10 cells responded well. This result could imply that the entire specific portion of the DO-11.10 receptor had not been isolated with KJ1-26. Alternatively it could imply that the receptor protein had been isolated but that fusion to the membrane of SKK-45.10 was not sufficient to satisfy all the intracellular activation requirements of the cell, i.e. the fused receptor was not properly connected to the signal-transducing machinery of SKK-45.10. This second conclusion was supported by the observation, shown in Table 1, that SKK-45.10 bearing DO-11.10 receptors also failed to respond to KJ1-26-coupled Sepharose beads. In this case it is clear from the cytofluorographic data that the Ab binds to liposome-fused SKK-45.10 cells and yet fails to induce the cell.

What can be the explanation for these results? The most likely suggestion is the idea that the transferred receptor is not properly associated with T3 proteins, known to be in intimate contact with the receptor under normal circumstances. Since there is evidence that T3 is concerned with signal transduction after the receptor has been engaged, this may represent the fatal flaw in our experiments.

## CONTRIBUTION OF α AND β CHAINS TO T-CELL SPECIFICITY

Recently Mark Davis and his colleagues [9, 10; cf. #NC(D)-3, this vol.] and Yanagi et al. [11] have reported the discovery of T-cell specific-cDNA clones which code for Ig-like varying membrane-bound proteins. The work of Acuto et al. [12] has shown that these cDNA's encode the β chain of the T-cell receptor. Using methods similar to those used by Hedrick et al. [9, 10] and Yanagi et al. [11], Saito et al. [13] have described a second type of cDNA clone in T cells which also has some of the characteristics expected of a nucleotide sequence encoding a receptor mRNA. Specifically the mRNA is expressed only in T cells and is derived from genes which rearrange only in T cells.

Several features of the predicted protein sequence encoded by this cDNA, however, suggest that it does not code for the α chain. First, the sequence does not include a site for N-linked glycosylation although it is clear that in order to achieve, from the protein backbone predicted to be ~30,000, final mol. wts. in man and mouse respectively of 50,000 and 43,000, N-linked glycosylation must almost certainly occur; indeed, there is some evidence from other sources that the α chain does bear N-linked glycosylations. Secondly, the isolated clone has been reported to have relatively little variability although it is clear from peptide maps that α chains must be extensively variable. Finally we have studied effects of cyanogen bromide cleavage on isolated receptor chains, and our results indicate that the methionines in the molecule are not positioned as predicted by the sequence of Saito et al.

**Table 2.**  Properties of DO-11.10 and related clones.

| T-cell hybridoma | Presence of DO-11.10-specific 17Kb C band | Response to: | |
| --- | --- | --- | --- |
| | | cOVA/I-A | KJ1-26 beads |
| DO-11.10 | + | + | + |
| DO.11.10.3 | − | − | − |
| DO.11.10.7 | − | − | − |
| DO-11.10-15 | − | − | − |
| DO-11.10-24 | + | + | + |
| 7DO-286.2 | + | + | + |

From these results  we  conclude that Saito et al. have not cloned cDNA encoding the  α chains but rather cDNA encoding some other, very interesting, T-cell-specific molecule.

This being so, we have set out on a series of experiments  that are intended to examine the contribution of α and β chains to T-cell specificity.     To do this we grew the cloned T-cell hybridoma, DO-11.10, for ~12 weeks and then subcloned the progeny cells.  Subclones were screened for their ability  to respond to cOVA/I-A$^d$, for the presence of KJ1-reactive material on their surfaces, and, by Southern blots, for the presence of rearranged  β chain genes in their DNA.  Some of our results are shown in Table 2.

As shown, DO-11.10, of course, responds to cOVA/I-A$^d$ and bears KJ1-26-reactive receptors.  Hind III digestion of the DNA of this hybridoma reveals a C$\beta$2  band in the expected germ-line position (Hind III cleaves between the J$\beta$2 cluster and C$\beta$2 and therefore cannot be used to demonstrate rearrangements  to the J$\beta$2 cluster if a C$\beta$ probe is used, as was the case in our experiments).    The Southern blots of Hind III-digested DO-11.10 DNA also showed 2 rearranged bands identical to the rearrangements present in BW5147, the tumour cell parent of this cell line.  Only one additional band was present at ~17 Kb,  and characteristic of DO-11.10.  This presumably represented the functionally rearranged DO-11.10 V$\beta$J$\beta_1$C$\beta_1$ DNA.  A band of identical size was formed after Hind III digestion of a cOVA/I-A$^d$-recognizing, KJ-26-reactive  subclone of DO-11.10, DO-11.10.24, and was  also found in 7DO-286.2.  By contrast, 3 subclones of DO-11.10 which had lost the ability to react with KJ1-26 had lost the 17 Kb β chain, Hind III band characteristic of DO-11.10.       These  results strongly  suggested  that antigen/MHC reactivity,  reaction with a clone-specific Ab and presence of  β chain DNA were interdependent.

## EXPRESSION OF T-CELL RECEPTORS ON THYMOCYTES

The discovery of T-cell receptor protein and at least one of the genes now allows a proper examination of the expression of T-cell receptors on thymocytes, a problem of considerable theoretical importance in view of the discovery by Zinkernagel and his colleagues [14] and Fink & Bevan [15] that the T-cell repertoire is profoundly affected by what these cells encounter as they mature in this organ. We have taken two different approaches to the examination of this problem. In the first we have prepared a collection of hybridomas by fusing BW5147 to thymocytes from foetal mice of different ages. These cells can then be examined using Southern blots for the presence of various types of DNA rearrangement at the β locus. Our results so far are preliminary, but several conclusions can already be drawn. First, we have not found β chain rearrangements in the 20-30 foetal liver-cell hybrids we have examined. Secondly, there are very infrequent rearrangements in hybrids made from 14- and 15-day foetal thymocytes, but rearrangements gradually increase in frequency thereafter, such that in adult thymuses nearly all the hybrids we have examined contain rearrangements on both chromosomes inherited from the normal-cell partner. Rearrangements to Cβ1 seem to precede those to Cβ2.

To complement these experiments we have begun to examine the expression of receptors as proteins on the surfaces of thymocytes. Clearly our clone-specific Ab's are not of much use to us in these studies, since they bind to undetectably few cells in normal uncloned populations of peripheral T cells and thymocytes.

To perform these experiments we have prepared a rat MAb, KJ16.133, from the lymph node cells of a rat hyperimmunized with the isolated receptor of DO-11.10, bound by its Ab, KJ1-26 and precipitated with *S. aureus*. KJ16-133 has a number of very interesting properties, including the fact that it binds only 20% of BALB/c peripheral T cells and, in fact, ~20% of our antigen-specific, MHC-restricted T-cell hybridomas. The Ab thus seems to be recognizing only a subset of all antigen/MHC receptors in BALB/c mice. Recent evidence suggests that the determinant recognized by this Ab is on the β chain, though we have no idea what part of the β chain is involved. Surprisingly, the Ab does not bind any T cells in several strains of mice including SJL, SJA, SWR, C57BR and C57L. Although the Ab is not ideal for thymus experiments, since it does not recognize 100% of receptors in BALB/c mice, this is the only MAb with anything like the required properties to have been identified so far. Accordingly we thought experiments with it were worth pursuing.

Thymus cells from BALB/c and SJL mice were incubated with KJ16-133 Ab and, after washing, with fluorescein-conjugated mouse anti-rat κ (FL-MAR18), the kind gift of Drs. L. Arnold and L. Lanier. No significant staining was seen with thymocytes from the KJ16-133-negative strain, SJL. BALB/c thymocytes stained with Ab, but the pattern was different from that of peripheral T cells in two respects. Only ~10% of the cells stained, and the stained cells reacted with the Ab less well than peripheral T cells.

When immature and mature thymocytes were studied, either onto-logically or after separation with peanut agglutinin, or after treating mice with cortisone, the following points were apparent. The very first Thy1-positive cells in the developing thymus bear no receptors. As the thymus matures, receptors gradually appear, star-ting at day 17 of foetal life. At no time do immature thymocytes bear more receptors than mature thymocytes and peripheral T cells, and usually they bear a good deal less. By extrapolation from the KJ16-133 staining data we estimate that ~50% of immature thymocytes bear no receptors at all. Within the limits of estimation all mature cortisone-resistant thymocytes bear receptors, and at levels/cell equivalent to peripheral T cells [16].

## CONCLUDING COMMENT

With these findings in hand we hope that we can go on to apply them to an understanding of how T-cell reactivity can be selected in the thymus, and how the receptor functions in the recognition of antigen plus self MHC in the periphery.

## *Acknowledgements*

The authors would like to thank Ella Kushnir, James Leibson, Lee Niswander and Virginia Barr for excellent technical assistance, and Kelly Bakke and Donna Thompson for secretarial help. This work was supported by U.S. Public Health Service Grant AI-18785 and HD-17717 and ACS research grant IM-49. It was done during the tenure by J.K. of the American Cancer Society Faculty Research Award, and whilst J.Y. was on leave from the Servei d'Immunologia, Hospital Clinic, Barcelona, Spain, partially supported by a fellowship from CIRIT, Generalitat de Catalunya, Spain.

## *References*

1.  Allison, J., McIntyre, B. & Bloch, D. (1982) *J. Immunol. 129*, 2293-2300.
2.  Meuer, S., Fitzgerald, K., Hussey, R., Hodgdon, J., Schlossman, S. & Reinherz, E. (1983) *J. Exp. Med. 157*, 705-719.
3.  Haskins, K., Kubo, R., White, J., Pigeon, M., Kappler, J. & Marrack, P. (1983) *J. Exp. Med. 157*, 1149-1169.
4.  Infante, A., Infante, P., Gillis, S. & Fathman, G. (1982) *J. Exp. Med. 155*, 1100-1107.
5.  Kappler, J., Kubo, R., Haskins, K., White, J. & Marrack, P. (1983) *Cell 34*, 727-737.
6.  Marrack, P., Shimonkevitz, R., Hannum, C., Haskins, K. & Kappler, J. (1983) *J. Exp. Med. 158*, 1635-1646.
7.  Kappler, J., Kubo, R., Haskins, K., Hannum, C., Marrack, P., Pigeon, M., McIntyre, B., Allison, J. & Trowbridge, I. (1983) *Cell 35*, 295-302.

8. Weiss, A., Imboden, J.,Shoback, D. & Stobo, J. (1984) *Proc. Nat. Acad. Sci. 81*, 4169-4173.
9. Hedrick, S., Cohen, D., Nielson, E. & Davis, M. (1984) *Nature 308*, 149-153.
10. Hedrick, S., Nielson, E., Kavaler, J., Cohen, D. & Davis, M. M. (1984) *Nature 308*, 153-158.
11. Yanagi, Y., Yoshikai, Y., Leggett, K., Clark, S., Aleksander, I. & Mak, T. (1984) *Nature 30*, 145-149.
12. Acuto, O., Fabbi, M., Smart, J., Poole, C., Protentis, J., Royer, H., Schlossman, S. & Reinhers, E. (1984) *Proc. Nat. Acad. Sci. 81*, 3851-3855.
13. Saito, H., Kranz, D., Takagaki, Y., Hayday, A., Eisen, H. & Tonegawa, S. (1984) *Nature 309*, 757-762.
14. Zinkernagel, R., Callahan, G., Althage, A., Cooper, S., Klein, P. & Klein, J. (1978) *J. Exp. Med. 147*, 882-896.
15. Fink, P. & Bevan, M.J. (1978) *J. Exp. Med. 148*, 766-775.
16. Roehm, N., Herron, L., Cambier, J., Digiusto, D., Haskins, K., Kappler, J. & Marrack, P. (1984) *Cell 38*, 577-584.

#NC(D)

## NOTES and COMMENTS related to the foregoing topics

Comments related to particular contributions:

#NC(D)-1

*A Note on*

NEW LIGHT ON THE T-CELL RECEPTOR IN
RELATION TO ANTIBODY BINDING SITES

E. S. Golub

Department of Biological Sciences
Lilly Hall of Life Sciences, Purdue University
W. Lafayette, IN 47907, U.S.A.

For the reader who is not a cellular immunologist, I now seek to put the following contributions into perspective. N.A. Mitchison remarked at the start of the Forum (cf. Foreword) that this is a very good time to be studying immunology because things have become so simple and answers are arriving fast. I have much respect for his shrewdness. T-cell receptor knowledge is indeed advancing fast. The question is, are the results making things simpler? We can look at the past decade of immunology as a series of plateaus. When we first reach the plateau, there is a discernible sigh of relief that at last we have got it right. But we soon realize that out in front of us is not the horizon but yet another climb to yet another plateau. The T-cell receptor has had more plateaus than any other aspect of immunology; but Mitchison and I would agree that while there is more of a climb ahead before the horizon is reached with the T-cell receptor, the rate at which the picture is emerging gives us hope that we will soon be seeing the horizon. Hope beats eternal.

Why should there be such problems with the T-cell receptor? After all, the B-cell receptor has not presented a problem to immunologists, and B-cells and T-cells shouldn't be *all* that different. They both look alike in the microscope and they come from the same cell lineage. They both recognize and react with antigen in a specific manner. Granted, they carry out different functions; but the road to becoming reactive with an antigen so that the function can be carried out doesn't seem to be all that different. Yet there are some very fundamental differences. The one which has been known for a very long time is that when we look at the surface of these cells with fluorescent anti-Ig probes we find that the B-cell is loaded with surface immunoglobulin but the T-cell has none. If the accepted wisdom is correct that the surface Ig on the B-cell is used to recognize antigen, then we have to ask, how does a T-cell recognize anti-

gen without these molecules?  Is there a separate set of recognition
molecules which the T-cell uses?  This would not be a parsimonious
explanation but it is certainly a possibility.  The break-through in
understanding the difference comes about when we realize that while
both cells react with antigen they do so in different ways.  The B-
cell can react with free antigen through its surface Ig receptor
while the T-cell can react with antigen only in association with an
MHC [major histocompatibility complex] molecule.  Indeed, it can
react with antigen only when it is in association with *self* MHC.

This then casts the question  of the nature of the T-cell
receptor into a new light. Is there a receptor which sees a combined
determinant composed of self MHC and antigen, or are there separate
receptors for self MHC and for foreign antigens?  What is the nature
of the beast – Ig-like or something new to us?  How is this diversity
generated?   Within the last few years we have begun to be able to
make a direct approach to some of these questions.  The advent of
the monoclonal antibodies and the techniques of molecular biology
have allowed the chemical nature and the genetic organization of the
T-cell receptor to start to come into view.  Some aspects are dealt
with in the following two Notes,  from the laboratories of M.J.
Crumpton and M.M. Davis.  Especially relevant is the preceding contribution, #D-4, in which Phillipa Marrack presents what she and J.
W. Kappler said at the Forum.

#NC(D)-2

*A Note on*

# MOLECULAR NATURE OF THE T3 ANTIGEN OF HUMAN T LYMPHOCYTES

M.J. Crumpton, O.P. Chilson and J.M. Kanellopoulos

Imperial Cancer Research Fund
Lincoln's Inn Fields
London WC2A 3PX, U.K.

T-lymphocytes are normally quiescent but are stimulated by contact with antigen to grow, divide and differentiate into immuno-competent cells. A particular antigen, however, stimulates only the small proportion of antigen-specific T cells (usually <0.1% of the total population). Various polyclonal mitogens stimulate the major-ity of human T-lymphocytes, irrespective of their antigen specificity, to respond in an apparently identical manner with that induced by specific antigen. Such mitogens include the lectins Concanavalin A (Con A) and phytohaemagglutinin (PHA), and monoclonal antibodies (MAb's) of the OKT3 series (viz. OKT3, Leu-4 and UCH-T1). One explana-tion for their mitogenicity is that they react with the T-cell anti-gen receptor, thereby promoting biochemical responses identical with those initiated by the interaction of antigen with its specific receptor.

Studies of the T-lymphocyte surface antigen that is recognized by the OKT3 series of MAb's (viz. T3) indicate that this explanation is in principle correct, although not in detail. Thus, it is appa-rent from the reactivities of these MAb's, as well as the results of co-capping and immunoprecipitation experiments, that the T3 molecule is not the T-cell antigen receptor, but that it is associated (i.e. interacts) with the antigen receptor on the T-cell surface. T3 binds Con A but not PHA, whereas the antigen receptor binds both Con A and PHA. As a result, stimulation of T-cell growth by different lectins is most probably mediated via their interaction with the same macro-molecular cell-surface complex.

Immunochemical and biosynthetic studies show that the T3 mole-cule comprises 3 distinct polypeptides of mol. wt. ~19,000, 21,000 and 26,000. The 19,000 mol. wt. polypeptide is apparently non-glycosylated, whereas both the others possess $N$-linked carbohydrate moieties; all 3 appear to have a trans-membrane orientation. It seems likely that the cell-surface T3 antigen comprises the 19,000 mol. wt. polypeptide associated non-covalently with either the 21,000 or the 26,000 glycosylated polypeptide, although other associations are also possible [1].

These studies emphasize that valuable information on the molecular nature of cell-surface antigens can be derived from the analysis of immunoprecipitates prepared from radiolabelled cells and from characterizing the effects of glycosidase digestion of immuno-precipitates.

*Reference*

1. Kanellopoulos, J.M., Wigglesworth, N.M., Owen, M.J. & Crumpton, M.J. (1983) *EMBO J. 2*, 1807-1814.

#NC(D)-3

*A Note on*

# ORGANIZATION AND EXPRESSION OF THE MURINE T-CELL RECEPTOR β-CHAIN GENE COMPLEX

M.M. Davis, Y. Chien, N.R.J. Gascoigne and *S.M. Hedrick

Department of Medical
Microbiology
Stanford University
School of Medicine
Stanford, CA 94305, U.S.A.

*Department of Biology
University of California
San Diego
La Jolla, CA 92093, ·U.S.A.

Our groups are investigating the structure and genetics of the murine T-cell receptor for antigen. Using subtractive cDNA hybridization and taking into account that B and T lymphocytes differ by only a small fraction of their gene expression, we have been able to isolate one chain (β) of the 2-chain T-cell receptor molecule [1]. cDNA clones encoding this chain exhibit variable (V), constant (C), joining (J) and diversity (D)-like elements, precisely analogous to those of immunoglobulins [2, 3]. The constant region locus is of the form $J_T1-7 - C_T -J_T1-7' - C_T'$ with a total of at least 6 and possibly as many as 12 functional $J_T$ elements [4]. The $C_T$ elements are nearly identical, differing by only 4 amino acids, and do not appear to represent functional isotypes [4]. This gene complex is rearranged in most cytotoxic T-cell lines and hybrids that we have examined, and in all helper T-cell lines and hybrids, but in very few T suppressor hybridomas [5]. Chromosomal mapping using somatic cell hybrids has localized this gene to chromosome 6 of the mouse [6].

*Acknowledgements*

This work was supported by the National Institutes of Health, the American Cancer Society, and the UCSD Cancer Center.

*References*

1. Hedrick, S.M., Cohen, D.I., Nielsen, E.A. & Davis, M.M. (1984) *Nature 308*, 149-153.
2. Hedrick, S.M., Nielsen, E.A., Kavaler, J., Cohen, D.I. & Davis, M.M. (1984) *Nature 308*, 153-158.
3. Chien, Y., Gascoigne, N.R.J., Kavaler, J., Lee, N.E. & Davis, M.M. (1984) *Nature 309*, 322-326.

4. Gascoigne, N.R.J., Chien, Y., Becker, D.M., Kavaler, J. & Davis, M.M. (1985), submitted.
5. Hedrick, S.M., Germain, R.N., Bevan, M.J., Dorf, M., Fink, P., Gascoigne, N.R.J., Green, M., Kapp, J., Kauffman, Y., Melchers, F., Pierce, C, Sorensen, C., Taniguichi, M. & Davis, M.M. (1985) submitted.
6. Lee, N.E., D'Eustachio, P, Pravtcheva, D., Ruddle, F.H., Hedrick, S.M. & Davis, M.M. (1985) *J. Exp. Med.*, in press.

#NC(D)-4

*A Note on*

ANTIGEN-SPECIFIC T-CELL ACTIVATION

Arun Fotedar[*], Michel Boyer,
Wallace Smart and Bhagirath Singh

Department of Immunology
MRC Group on Immunoregulation
845E MSB University of Alberta
Edmonton, Alberta, Canada T6G 2H7

Extensive investigations into the phenomenon of activation of antigen-specific 'IA'-restricted T cells have led to the emergence of two main research areas. One is antigen processing by antigen-presenting cells (APC), prior to actual presentation to T cells in the context of IA on the APC surface. The second involves the study of the ternary interaction between antigen, IA and the T cell receptor. This Note concerns some of our results which impinge on these two overlapping areas of interest. We have generated T cell hybridomas by fusing antigen-primed T-cell blasts with BW5147, using procedures essentially described earlier [1]. T-cell hybridomas specific for beef insulin [2] selected for their ability to release IL2 in the presence of appropriate APC and antigen [3] were used in these studies.

One of the hybridomas studied by us, A20.2.15, was found to be restricted to IA$d$, as MKD6 [3] blocked activation. It was reactive with beef and sheep insulin but not pork insulin (Table 1), suggesting a role for A8 and A10 amino acid residues in the 'A chain loop' in activation of this T hybridoma. Since neither oxidized A chain nor a synthetic A-chain loop determinant [4] could activate A20.2.15, this suggested the involvement of additional determinants. There was no question of a B-chain sequential determinant, since neither oxidized A and B chains nor synthetic A-chain determinant cross-linked to oxidized B chain could activate A20.2.15. The implicit suggestion of a conformational determinant (probably an interactional determinant involving both A and B chains) was confirmed because beef insulin was unable to activate A20.2.16 after denaturation by 3 cycles of pH 2 ↔ pH 11 treatment in saline (Table 1).

---

[*] addressee for any correspondence

**Table 1.**  Activation of A20.2.15.  T hybridoma cells ($10^5$) were incubated with irradiated spleen cells ($10^6$; 4000 rads) and antigen (40 µg) for 48 h.  The IL2 in the supernatant was assayed by thymidine incorporation by an IL2-dependent cell line or thymocytes as described earlier [10], and IL2 units calculated [11].

| Antigen / Antibody *if applicable* | IL2 units |
|---|---|
| *None* | 10 |
| Beef insulin | 1640 |
| Sheep insulin | 2810 |
| Pork insulin | <10 |
| Ox A chain & Ox B chain (beef insulin) | <10 |
| Conformational determinant of insulin (CDI) | <10 |
| CDI-Lys | <10 |
| Denatured beef insulin (pH | <10 |
| Beef insulin / Anti-IA*d* (MKD6) | <10 |
| Beef insulin / Anti-beef insulin (AD-2.2) | 2000 |

Using glutaraldehyde-treated APC's as cells which can present but not process antigen [5], we could demonstrate that beef insulin had to be processed to be able to be presented to A20.2.15.  Evidently the conformational determinant seen by A20.2.15 on beef insulin was resistant to the processing event.  This conclusion was important since it adds a note of caution in our current simplistic description of antigen processing as merely protein fragmentation, leading to primarily sequential determinant recognition by T cells.  It has not escaped our attention that these findings can also be explained by suggesting that the conformation is not seen by T cells but could be a requirement for the antigen-processing event.  Currently we do not have evidence to distinguish between these two possibilities.

We have attempted to analyze the effector T-cell activation by the ability of a panel of antagonists (e.g. pork insulin, A or B chain of beef insulin, A-chain loop determinant, etc.) to compete with the agonist (beef insulin) in activating A20.2.15.  Since pork insulin has the same conformation as beef insulin [1, 7], it was important to test whether pork insulin could compete in the antigen-processing event [6].  Similarly the ability of oxidized A and B chains of beef insulin to compete with beef insulin would reveal whether a competitive step in T-cell activation could be defined where the sequential determinants were involved.  The results were uniformly negative.  None of these antagonists could compete with beef insulin at any dose level tested.  This was in contrast to the earlier results of Rock & Benacerraf [6] who clearly showed competition between structurally related antigens (GT & GAT) at the APC level in the activation of a GAT-reactive T hybridoma.

Our negative results, though difficult to interpret, would seem to suggest that both the sequential A-chain loop determinant and the conformational determinant are being seen at the same time. Since antigen processing is a non-specific event, it is conceivable that these determinants are involved in the presentation step and not processing. Thus we could conclude that the conformational determinant on beef insulin was 'resistant' to processing.

We have also been investigating the effect on the T-cell activstion of using monoclonal antibodies (MAb's) against the components of the ternary interactional complex of antigen-IA-T cell receptor. To this end we have used a panel of anti-insulin MAb's [7] to screen for a blocking effect on beef insulin/ IA $d$ -induced IL2 release by A20.215. All of them including AD.2.2 which recognized the A-chain loop of beef insulin [7] caused no inhibition of T-cell activation (Table 1).

This was in contrast to anti-IA$d$ MAb (MKD6) which inhibited IL2 release from this hybridoma. We also raised MAb's to the T-cell receptor of this hybridoma in rats neonatally rendered tolerant to an unrelated T hybridoma [8]. These anti-idiotypic antibodies were screened for their ability to inhibit IL2 release induced by antigen/ Ia$d$ inactivation of A20.2.15.' With successive freeze-thawings, one of these MAb's repeatedly lost the ability to inhibit but could now activate A20.2.15. This ability of anti-idiotypic Ab's, that are specific for T-cell receptors, to activate T-cells, could be speculated to have a role analogous to the idiotypic network suggested for the regulation of B cells [9].

*Acknowledgements*

We thank the Medical Research Council of Canada and the Alberta Heritage Foundation for Medical Research for support, T.G. Wegmann for helpful discussions and Dr. J. Schroer for anti-insulin MAb's.

*References*

1. Glimcher, L.H., Schroer, J.A., Chan, C. & Shevach, E.M. (1983) *J. Immunol. 131*, 2868-2874.
2. Fotedar, A., Smart, W., Boyer, M., Fraga, E., Shevach, E.M. & Singh, B. (1985), submitted for publication.
3. Kappler, J.B., Skidmore, B., White, J. & Marrack, P. (1981) *J. Exp. Med. 153*, 1198-1214.
4. Singh, B. & Fraga, E. (1981) in *Basic and Clinical Aspects of Immunity to Insulin* (Keck, K. & Erb, P., eds.), Walter de Gruyter, Berlin, pp. 45-57.
5. Shimonkevitz, R., Kappler, J., Marrack, P. & Grey, H. (1983) *J. Exp. Med. 158*, 303-316.
6. Rock, K.L. & Benacerraf, B. (1983) *J. Exp. Med. 157*, 1618-1634.
7. Schroer, J.A., Bender, T., Feldmann, R.J. & Kim, K.J. (1983) *Eur. J. Immunol. 13*, 693-700.

8.  Fotedar, A., Smart, W., Holowachuk, E.W., Wegmann, T.G. &
     Singh, B. (1984) *J. Cell Biochem. 8A,* 201.
9.  Rajewsky, K. & Takemoi, T. (1983) *Ann. Rev. Immunol. 1,* 596-607.
10. Heber-Katz, E., Schwartz, R.H., Matis, L.A., Hannum, C.,
     Fairwell, T., Appella, E. & Hansburg, D. (1982) *J. Exp. Med.
     155,* 1086-1099.
11. Hedrick, S.M., Matis, L.A., Hecht, T.T., Samelson, L.E., Longo,
     D.L., Herber-Katz, E. & Schwartz, R.H. (1982) *Cell 30,* 141-152.

## Comments on material in #D

*Comments on* #D-1, O.J. Bjerrum et al. - Ab's IN MEMBRANE STUDIES

**Remarks by C.M. Lewis.**- Western blotting enables the interaction of Ab with a specific protein to be examined, but I wonder whether anyone has tried modifying the technique to look specifically for interaction with other proteins, e.g. the T-cell receptor-T3 complex. If a particular protein is isolated on a gel, conceivably interactions between this protein and others might be picked up in a similar way. **Comment by M.E. Bardsley.**- Such adaptation of the Western blot to look at the binding of other proteins to the protein of interest has been tried successfully with calmodulin-binding proteins using $^{125}$I-labelled calmodulin.

*Comments on* #D-2, J.P. Luzio et al. - A TRANS-MEMBRANE PROTEIN
         #D-3, R.J. Owens et al. - INWARDS-FACING PROTEINS

**N.A. Mitchison, to J.P. Luzio.**- Can you tell us anything about genetic variation in C9? Do any of your Ab's react with C9 from species other than man? Cross-reactions with mouse would be particularly interesting.-Cf. Borel's work on C5 (which occurs in serum at about the same concentration as C9), indicating that mice should have no difficulty in making such self-cross-reacting Ab's. **Reply.**- No information is available at present; the MAb's have all been selected for reactivity with the human protein. **Query by R.J. Owens to J.P. Luzio:** how efficient is the bacterial expression/Ab detection system in terms of the molecular cloning of C9? - From a library of ~200,000 colonies at least as many C9-positive clones were detected with the polyclonal Ab as expected from the estimated abundance of the C9 mRNA in the mRNA preparation; this implies that the method is very efficient.

**V.P. Whittaker, commenting to R.J. Owens.** - My colleague John Walker has isolated from electric organ a protein of mol. wt. 34,000 with similar properties to the one you have described; we have named it 'calelectrin'. An Ab to this protein cross-reacts with a similar polypeptide and another of mol. wt. 67,000 in a variety of tissues. We regard all these proteins as forming a group of 'calelectrins'. Synexin (mol. wt. 47,000) described by Crentz et al. may also belong to this group. Brain is particularly rich in calelectrins. **R.J. Owens, answering A. Fotedar.**- We have not investigated whether our protein is expressed in reticulocytes. **Answering Jill Clayton.**- We have not examined whether, in the light of the generalized tissue location of receptors for $1,2S(OH)_2D_3$, it causes any stimulation of the $Ca^{2+}$-binding proteins.

*Comments on* #NC(D)-2, M.J. Crumpton - NATURE OF T3 ANTIGEN
        #NC(D)-3, M.M. Davis et al. - T-CELL ANTIGEN RECEPTOR

   **Remark by J. Kappler to M.J. Crumpton.**- By adjusting the ratio of protein to DOC, you should be able to have one T3 complex or less per micelle, and then distinguish between a 2-molecule and 3-molecule complex by immunoprecipitation.

   **P. Marrack, to M.M. Davis.**- Might the frequency of V regions in cDNA libraries reflect not the frequency with which individual T-cells express them, but rather the existence of very effective promoters for certain V's? (**Reply:** unlikely). Concerning the difference between human and mouse suppressor T-cells in respect of the rearrangement of β-chain genes, this might depend on the way in which the cells are defined in the two species. Mouse suppressors are usually defined by their secretion of antigen-binding factors, whereas in man assays usually depend on the suppressor activity of the intact cell which may not necessarily have receptors of the type that binds free antigen. **Remark by A. Fotedar.**- The high frequency of some T-cell genes (β) might be due to these recognizing antigens (TNP, DNP, CRBC) which some workers would say are 'background', primed to environmental antigens. **E.A. Kabat** wondered what the α chain does, if indeed antigen and MHC binding are both on the β chain.

*Some membrane-antigen refs. contributed by Senior Editor*

•Coombs, R.R.A. & Lachmann, P.J. (1968) *Br. Med. Bull. 24*, 113-117. - 'Immunological reactions at the cell surface'; a classical review.
•Fabre, J.W., Sunderland, C.A. & Williams, A.F. (1980) *Transplantation 30*, 167-173.- 'Immunosuppressive properties of rabbit Ab's directed against a major glycoprotein restricted to rat leukocyte membranes.'
•Kay, M.M.B., Sorensen, K., Wong, P. & Bolton, P. (1982) *Mol. Cell. Biochem. 49*, 65-85.- 'Antigenicity, storage and aging: physiologic auto-Ab's to cell membrane and serum proteins and the senescent cell antigen' (latter a glycopeptide, on erythrocytes); e.g. spectrin Ab's.
•Hauri, H.P. (1983) *Ciba Found. Symp. 95*, 132-149.- Intracellular transport & biosynthesis of brush-border proteins, investigated with MAb's.
•Park, T.S.......& Gelboin, H.V. (1984) *Biochem. Pharmacol. 33*, 2071-2081.- MAb to drug-induced cyt. P-450 will aid phenotyping & P-450 studies.

*Comments on* #D-4, P. Marrack et al. - T-CELL RECEPTOR

   **Reply to A. Fotedar.**- The expression of the T-cell allotypic determinant is probably not cell cycle-dependent. **To S.J. Kaufman.** - We have no information as to the fate of the T-cell receptor subsequent to binding by free or bead-bound Ab. **Remarks by C.M. Lewis.-**

Concerning the anti-T-cell receptor MAb's, in particular that directed against the hybridoma responding to I-A$^d$ and ovalbumin, your
studies suggest that this might recognize the binding site of the
receptor. Maybe anti-Id Ab's to the MAb bind to antigen in association with I-A on the antigen-presenting cell, which might help in
studying the interaction of APC and T cells. Also, I would like to
challenge the concept that Ab's bind antigen alone and T cells bind
antigen in association with the MHC molecule. I wonder whether this
might not be a concept which results from the techniques we use.
Suppressor cells have been shown to bind antigen directly, and one
wonders whether anyone has investigated whether some Ab's recognize
antigen in association with MHC.

**Comments by C. Bona.**- Your data suggest that MAb KJ16-133 probably recognizes two cross-reactive determinants shared by genes from
a $V_H$ family, e.g. $V_H J558$ for Ig. With respect to the allelism
recognized by this MAb, did you study the expression in H-2$^d$ IgC$_H$non
or on IgC$_H$$^a$ with different H-2's? The outcome might indicate
that the idiotype is not $C_H$-linked as was suggested by several
studies. **Comment by P. Alexander.**- It would be of interest to know
whether T-cell receptor modulates with anti-Id Ab.

**P. Marrack, answering the foregoing remarks.** #To C.M. Lewis.-
Yes, it is possible that Ab's directed at the idiotypes of anti-T
cell receptor idiotype Ab's might cross-react with the antigen/I-A
complex on antigen-presenting cells. We have given some thought to
the matter, but have been unable to persuade any of our students to
try the experiment. Concerning the difference in requirements for
MHC association with antigen for T cell rather than Ab recognition,
I don't think the difference has been observed for technological
reasons. All the helper and cytotoxic T cells we have studied have
this dual requirement for antigen and MHC recognition, and we have
never seen such a characteristic of an Ab, though Klinmen and his
associates have made some interesting observations on such Ab's. As
for suppressor T cells, they seem to be a completely different
matter. #To C. Bona.- Reactivity with KJ16-133 is not linked to Igh
since SJL (Igh$^b$) and SJA (Igh$^a$) mice are KJ16-133-negative, and
BALB/c (Igh$^a$) and CB20 (Igh$^b$) mice are KJ16-133-positive. # To P.
**Alexander.**- T-cell receptor does not modulate with anti-Id Ab.

#**NC(D)-3,** continued.- C. Bona wondered if the high frequency of
the 4 $V_T$'s, as mentioned by M.M. Davis for both thymocyte and spleen
hybridomas, might imply a high degree of cross-reaction (cross-
recognition) at the level of neo-antigens (self + X).

————————

*A cytochrome ref., supplementing that on preceding p.:-*

Boobis, A.R., et al. (1985) *Biochem. Pharmacol.* 34, 1671-1681.- Anti-
cyt. P-448 MAb that recognizes an epitope found in other cyt. P-450's.

Section #E

ANTIBODIES IN RELATION TO ONCOLOGY

#E-1

# REQUIREMENTS BY CANCER CELLS FOR POLYPEPTIDE GROWTH FACTORS SUGGEST NEW THERAPEUTIC STRATEGIES

Peter Alexander

Cancer Research Campaign Medical Oncology Unit
University of Southampton
Southampton General Hospital
Southampton SO9 4XY, U.K.

*Division of normal cells is evidently an adaptive response to a sequence of external signals. This often holds too for cancer cells except that they produce and release growth factors for which they bear the appropriate surface receptor, and this can result in autocrine stimulation.*

*Evidence is presented which indicates that autocrine stimulalation can be interrupted, and this suggests that the growth of tumours may be inhibited by interfering with the action of growth factors.*

Medical treatment of cancer which is directed at control mechanisms is inherently more attractive than the use of conventional cytotoxic agents, the action of which is aimed at the biosynthesis of nucleic acids and which therefore inevitably damage normal cells. It is the toxicity for normal cells that sets a limit to the amount of drug which can be administered, and frequently this amount is inadequate to eradicate the tumour. Such tumours respond dramatically to hormonal manipulations with minimal adverse effects, but inevitably within 1-5 years the biochemistry of the tumour changes so that they become independent of steroid hormones and the cancer re-emerges. Recent discoveries relating to the factors involved in cell division raise the possibility that cancer may be controlled by blockading the response to polypeptide growth factors on which cancer cells depend, and there is no indication that cancer cells can become independent of such growth factors.

## CELL DIVISION IS NOT THE NORM

Two mutually exclusive hypotheses have been advanced to explain the control of cellular proliferation *in vivo*. One states that cells enter the mitotic cycle as an adaptive response to controlling stimuli. The other states that cells comprising tissues capable of proliferating are normally programmed to divide and are prevented from doing so, by inhibitory substances which have been called chalones. The evidence is now overwhelming that the latter does not apply and that division of normal cells requires a sequence of signals mediated by polypeptides that bind to high-affinity membrane receptors. While positive mitogenic signals are essential, it is possible that there are some host factors which inhibit the growth of such triggered cells; this, however, is not the same as the chalone concept, which implies that cell division is the norm.

The principal characteristics of the mitogenic polypeptides can be summarized as follows.-

1. Act via high-affinity membrane receptors.
2. Biological activity resides in receptors.
3. Two general classes:
   *Category A:* Renders cells competent to respond to Category B factors.
   *Category B:* Initiates progression to mitosis (GO → GI).

*A type,* e.g. EGF, platelet-derived growth factor (PDGF), T(?B) lympho-cyte growth factor (IL-2), growth factor for myeloid cells (CSF; IL-3):
 - Restricted target specificity.
 - Act locally.
 - Produced by one cell for another (i.e. paracrine hormone).

*B type,* e.g. insulin-like; somatomedins, (transferrin?):
 - Broad target specificity.
 - Present in plasma at mitogenic concentrations (endocrine hormone).

## AUTOCRINE STIMULATION OF CANCER CELLS

In 1978 DeLarco & Todaro [1] observed that supernatants from tissue cultures of transformed cells, but not those from cultures of non-transformed cells, contained a potent mitogen, quickly shown to be polypeptide, which promotes the growth of normal cells. When this was found to be a general property of malignant cells, Sporn & Todaro [2] advanced the autocrine stimulation hypothesis for malig-nancy which states that malignant cells require mitogenic signals in order to divide, but that they have acquired the capacity to synthesize the necessary factors themselves.

The discoverers of the mitogenic polypeptides made by cancer cells considered that their biological properties were qualitatively distinct from those of the normal A-type growth factors as they caused normal cells to proliferate *in vitro* with a malignant phenotype expressed in morphology and growth pattern. For example, normal cells do not grow as anchorage-independent spheroids when suspended in semi-solid culture medium, whereas many types of malignant cell readily give rise to colonies in medium containing soft agar. Addition of polypeptides from culture supernatants from malignant cells enabled normal cells to grow as colonies in soft agar. Similarly, extracts from cancer cells - usually employing acidified ethanol - produced the same effect. On the other hand, typical A-type growth factors like EGF and PDGF were claimed to be incapable of causing normal cells to adopt a malignant phenotype *in vitro*. These properties led Todaro to designate the mitogenic polypeptides from cancer cells as *transforming growth factors* ('TGF's).

TGF activity is now known to be caused by the interaction of two polypeptides. α-TGF's show close structural homology with EGF and bind to EGF receptors. β-TGF is a larger molecule made up of two chains linked by S-S bonds and does not bind to EGF receptors. In some investigations both α- and β-TGF by themselves have been found to be mitogenic, but to be synergistic when given together, whereas other groups of investigators have concluded that β-TGF is not mitogenic alone, but increases the response of cells to EGF and α-TGF by inducing the synthesis of increased numbers of receptors for EGF.

Sporn & Todaro [2] postulated that an essential feature of the malignant transformation was the acquisition by cells of the capacity to synthesize TGF's which are then led by autocrine stimulation to autonomous growth.

## Simultaneous expression of a growth factor and its membrane receptor by cells exhibiting transformed phenotype

The hypothesis that production of TGF's is a correlate or consequence of malignancy cannot, however, be sustained because:
(1) at high concentrations normal growth factors such as EGF, PDGF and particularly mixtures of normal growth factors can, with normal cells, induce anchorage-independent growth;
(2) acid-ethanol extracts of normal tissues, in particular kidney and placenta, yield polypeptides which mimic in every respect the TGF activity of polypeptides from cancer cells;
(3) on degranulation, platelets release polypeptides showing potent TGF activity [3];
(4) cDNA coding for one of the TGF polypeptides hybridizes with an mRNA in normal cells although the amount of message made greatly increases after the cells have been virally transformed [4].

All this suggests that the essential difference between normal and malignant cells is not the synthesis by cancer cells of a special

type of growth factor that endows cells with a transformed phenotype
and that the TGF activity can be brought about by polypeptides
present in normal cells.   Indeed, there is no compelling evidence
that the TGF's in cancer cells differ qualitatively from polypeptides
made in normal cells.   This led Currie and myself [5] to suggest
that a key factor in malignancy is that the cancer cells constitutively synthesize both an A-type growth factor and the receptor for it.

According to our hypothesis, autocrine stimulation can arise in
one of two ways: either, on becoming transformed, a cell that normally makes a particular growth factor expresses the membrane receptor for it, or a cell that normally expresses the receptor acquires
the capacity to synthesize the growth factor.   Very recently, this
hypothesis has been validated for some T-cell lymphomas [6].   T-cells
synthesize their growth factor, IL-2, when activated by the specific
antigen to which the T-cells have been allergized.   Thereby clonal
expansion of antigen-specific T-cells occurs during an immune response.   The T-cell lymphomas constitutively synthesize both the
normal IL-2 and its receptor, and this results in uncontrolled proliferation.   Another example where rapid proliferation coincides
with the simultaneous expression of a growth factor and its receptor
is during a restricted period in foetal development when embryonic
fibroblasts acquire the capacity to synthesize PDGF's for which  all
fibroblasts have a receptor.   This results in rapid growth which is
however strictly controlled as the capacity of fibroblasts to make
PDGF is phase-specific. Interestingly, many sarcomas constitutively
make a growth factor which is indistinguishable from PDGF.

**Can autocrine stimulation be interrupted?**

Evidence that the interaction between the growth factor(s)
released by a malignant cell and its binding to the membrane receptor
is not tightly coupled, and that autocrine stimulation is an inevitable consequence when cells make growth factors amd have receptors
for them, comes from both *in vitro* and *in vivo* studies, which show
that the process of autocrine stimulation does not in general operate
at the level of an individual cell.   Malignant cells, *in vitro*, will
grow only in serum-free media if the initial cell concentration is
high, and growth in serum-free medium from low-cell inocula requires
the addition of growth factors such as TGF's, either extracted from
cancer cells or found in the supernatants from dense cultures of
tumour cells.   Apparently the process of autocrine stimulation can
be interrupted because the concentration of TGF around an isolated
cell is not sufficient to be mitogenic and the TGF diffuses away
from the cell environment before it has bound to the receptor.   A
concentration of TGF sufficient for proliferation *in vitro* in the
absence of serum is achieved only in cultures containing relatively
high cell concentrations. In serum, clonal growth can occur because
cooperation between a number of transformed cells is not required
since the need for  an adequate amount of TGF has been by-passed by
the release of TGF's from platelets.

The concept that cancer cells which have not undergone powerful selection by prolonged passage are incapable of giving rise to a tumour from a single cell, and that true autonomy of cancer requires a cluster of malignant cells which have to create a micro-environment adequate for proliferation, would at first sight appear to be in conflict with blood-borne metastasis. Single cells are capable of causing blood-borne metastasis, especially in organs other than the lung to which they must have gained access via the arterial circulation. However, one of the most striking aspects of the metastatic process is the peculiarity of the relative frequencies of metastases in different organs. In experimental animals the organ preference can be demonstrated by injecting cancer cells into the left ventricle (so as to avoid the filtering effect of the lung which arises if cells are given intravenously) whence they are distributed via the arterial circulation to all of the organs. Several investigators had found that following this procedure few, if any, metastases occurred in gut and muscle which received the greater part of the blood, but occurred instead in adrenal, bone, ovary and other organs that took only a small fraction of the cardiac output.

We [7] have made a detailed study of the initial distribution, trapping, cell death and eventual incidence of metastases for three histologically different rat tumours following intracardiac injection of their cells. In our studies the proportion of the cells arrested in different organs parallelled the blood flow to the organs (i.e. the cells went where the blood went), but the probability that a cancer cell deposited in an organ causes a macroscopic metastasis varied very widely between different organs. Thus one out of ten cells trapped in the adrenal resulted in a metastasis, whereas in skeletal muscle the figure was one in $10^5$. This organ preference does not have an immunological basis as the same distribution is seen in genetically athymic (nu/nu) rats and rats immunosuppressed with cyclosporin A. The data suggest that *in vivo* as *in vitro* an isolated cancer cell is not capable of autonomous growth unless it finds itself in a tissue capable of supplying it with growth factors which act like TGF's, or which potentiate TGF's. Once growth has started, it will be self-sustaining since a cluster of cancer cells will ensure the necessary concentration of TGF in the fluid around the metastasis.

The existence of dormant metastases in organs distant from the primary tumour could be explained similarly. In animal models [8] the presence in the lung of dormant cancer cells which stemmed via blood-borne spread from a distant primary tumour could be demonstrated by transplantation. In the lung the cells do not grow, but when a cell suspension from the lung taken from animals from which the 'primary' had been surgically removed a week previously was injected into the peritoneal cavity, then tumours indistinguishable from the 'primary' grew out. We have evidence that it is growth factors released by peritoneal macrophages that allow the outgrowth of iso-

lated cancer cells that in the environment of the lung (where macrophages are on the air side) remain dormant.

## TUMOUR CONTROL BY ANTAGONISTS TO GROWTH FACTORS

Since the binding of TGF's to the receptors in the membrane is
not instantaneous but appears to require a minimum concentration of
TGF in the extracellular fluid, it should in principle be possible
to block autocrine stimulation pharmacologically and thus arrest
tumour growth.   An Ab directed against the EGF receptor has been
found to be an agonist (i.e. mimic the action of EGF on the target
cells) while another behaved like an antagonist (i.e. prevent stimulation by added EGF) [9]. This raises the hope that Ab's against TGF
may be capable of arresting division of cancer cells.

For therapy *in vivo*  low mol. wt. antagonists would probably be
needed.  When the binding sites of the TGF's have been characterized
it may be possible to design polypeptides with antagonist properties,
but it may be difficult for relatively low mol. wt. peptides to have
an affinity for the receptor comparable to that of the growth factor.
The discovery of non-polypeptide antagonists may call for extensive
screening.   However, that the hope of finding such antagonists is
not pure 'pie in the sky' is shown by the serendipitous observation
that the anti-trypanosomal agent suramin interferes with the action
of PDGF on fibroblasts [10].

*References*

1.   DeLarco, J.E. & Todaro, G.J. (1978) *Proc. Nat. Acad. Sci. 75*,
     4001-4005.
2.   Sporn, M.B. & Todaro, G.J. (1980) *New Engl. J. Med. 303*, 878-880.
3.   Sporn, M.B., Anzano, M.A., Aasoian, R.K., DeLarco, J.E., Frolik,
     C.A., Meyers, C.A. & Roberts, A.M. (1984) *Cancer Cells 1*, 1.
4.   Lee, D.L., Rose, T.M., Webb, N.R. & Todaro, G.J. (1985) *Nature
     313*, 489-491.
5.   Alexander, P. & Currie, G.A. (1984) *Biochem. Pharmacol. 33*, 941-943.
6.   Urdal, D.L., March, C.L., Gillis, S., Larsen, A. & Dower, S.K.
     (1984) *Proc. Nat. Acad. Sci. 81*, 6481-6485.
7.   Murphy, P., Taylor, I. & Alexander, P. (1985) in *Treatment of
     Metastases - Problems and Prospects* (Hellman, K. & Eccles, A.,
     eds.), Taylor & Francis, Basingstoke, pp. 195-199.
8.   Alexander, P. (1983) *J. Path. 141*, 379-383.
9.   Schreiber, A.B., Lax, I., Yarden, Y., Eshar, Z. & Schlessinger,
     J. (1981) *Proc. Nat. Acad. Sci. 78*, 7535-7539.
10.  Garrett, J.S., Coughlin, S.R., Niman, H.L., Tremble, P.M.,
     Giels, G.M. & Williams, L.T. (1984) *Proc. Nat. Acad. Sci. 81*,
     7466-7470.

#E-2

# MONOCLONAL ANTIBODIES TO HUMAN TUMOURS FOR TARGETING CYTOTOXIC DRUGS

R.W. Baldwin

Cancer Research Campaign Laboratories
University of Nottingham
Nottingham NG7 2RD, U.K.

*Monoclonal antibodies (MAb's) reacting with human tumours provide novel approaches for targeting anti-cancer agents, as studied with MAb 791T/36 which localizes in human tumours including carcinoma of the colon and ovary and sarcomas. Localization has been evaluated initially in human tumour xenografts in immune-deprived mice and, with radiolabel in the imaging context, in patients. Ab-linked cytotoxic drugs investigated have included vindesine and methotrexate; already it appears that they significantly inhibit the growth of osteogenic sarcoma xenografts in immuno-deprived mice. The possible advantage of different Ab's or of Ab fragments is discussed.*

A fundamental objective in cancer chemotherapy is to destroy malignant cells whilst minimizing damage to normal cells. But the differential cytotoxicity of cytotoxic drugs between cancer cells and normal cells is not sufficiently great to permit curative doses of agents to be administered without producing widespread manifestations. Consequently methods are being sought for the selective delivery of chemotherapeutic agents to tumours, and this has led to investigations on the use of MAb's for targeting [1].

These basic developments are now considered with respect to the application of an anti-human tumour MAb designated 791T/36. This Mab, produced by a hybridoma obtained following fusion of spleen cells from a mouse immunized against cells of a human osteogenic sarcoma 791T and a mouse myeloma P3NS1 reacts with cells from a high proportion of human osteogenic sarcoma lines and, furthermore, the antigen is expressed in many human sarcomas. The 791/36 Ab-defined antigen is not restricted to sarcomas, however, and it has been found to localize in other tumours, including colorectal and ovarian carcinomas [2].

## TUMOUR LOCALIZATION OF MONOCLONAL ANTIBODIES

MAb targeting of agents to tumours requires that the selected Ab localizes in the tumour with only low levels accumulating in normal tissues.  Localization of MAb's in human tumours has been demonstrated using techniques designed to examine their potential for diagnostic imaging (immunoscintigraphy) following injection of preparations labelled with γ-emitting radioisotopes [3].  This approach can be illustrated by a series of clinical trials in which MAb 791T/36 has been used for immunoscintigraphy of human tumours.  In one trial [131]I-labelled 791T/36 Ab has been used to detect a primary osteogenic sarcoma in the right knee of a young girl.  This procedure has been used in a more extensive trial in colorectal cancer, and imaging has been reported in 56 patients with gastrointestinal cancer and also 4 patients with benign colorectal  tumours [2].  Positive images were recorded in 8 of 11 patients with primary colon cancer and in 5 of 12 with rectal tumours.  [131]I from catabolized Ab in urine rendered imaging difficult with the rectal tumours.  Of the 15 patients with metastatic/recurrent tumours, all but two showed positive *in vivo* imaging of 791T/36 Ab (Table 1).  None of the 4 patients with benign colon tumours gave positive Ab images.

Analysis of tissue levels of radioactivity in resected colon tumours permitted an estimate to be made of the amount of Ab which was localized.  Representative tests summarized in Table 2 indicate that this ranged from 0.008% to 0.003% of the injected dose over 1-4 days.  Since these patients received 200 µg of Ab the absolute amounts deposited in tumours ranged from 16 to 6 ng/g tissue.  It is interesting to compare these findings with more extensive trials on the localization of Ab 791T/36 in human  osteogenic sarcoma xenografts in immuno-deprived mice where, at optimal Ab dosing (100 mg/kg body wt.), tumour levels of ~80 µg Ab/g could be achieved, and maintained.  A general approximation from the tumour  xenograft studies suggests that a dose of 100 mg Ab 791T/36 in colorectal cancer patients could optimally deliver to tumours amounts of Ab of the order of 2 µg/g tissue.

## ANTIBODY TARGETING OF ANTI-TUMOUR AGENTS

### Vindesine conjugates

MAb 791T/36 has been conjugated to the vinca alkaloid analogue vindesine (VDS) by a procedure involving conversion of desacetyl-vinblastine hydrazide to the azide and protein interaction at pH 9.0. This procedure yielded VDS-791T/36 Ab conjugates containing 6 mol VDS/mol Ab with virtually no loss of Ab activity. VDS-791T/36 conjugates were tested for cytotoxicity for various target cells in comparison with free VDS using a post-labelling assay [4].  In this test, target cells are treated  for 15 min with VDS or conjugate, washed and then cultured for 24 h.  Target-cell survival is then measured by labelling with [75Se]selenomethionine.  Fig. 1 shows an

**Table 1.** Immunoscintigraphy of primary and secondary tumours with [131]I-labelled 791T/36 MAb: imaging detection rates, as no. seen.

| Primary tumour site | Primaries seen | Secondaries seen (& sites) |
|---|---|---|
| Bone | 7/9 | – |
| Colon | 8/11 ⎫ | 13/15 (liver, lung, brain) |
| Rectum | 5/11 ⎭ | |
| Breast | 3/5 | 5/5 (liver, lymph nodes, pleura) |
| Ovary | 10/11 | – |

**Table 2.** Localization of 791T/36 MAb in colorectal cancer. Patients received 200 µg (70 MBq) [131]I-labelled Ab, and tissue distribution of radioactivity was assayed on resected specimens.

| Patient | Day tumour resected | % injected dose/g tumour |
|---|---|---|
| 1 | 1 | 0.008 |
| 2 | 2 | 0.009 |
| 3 | 3 | 0.003 |
| 4 | 4 | 0.004 |
| 5 | 4 | 0.006 |

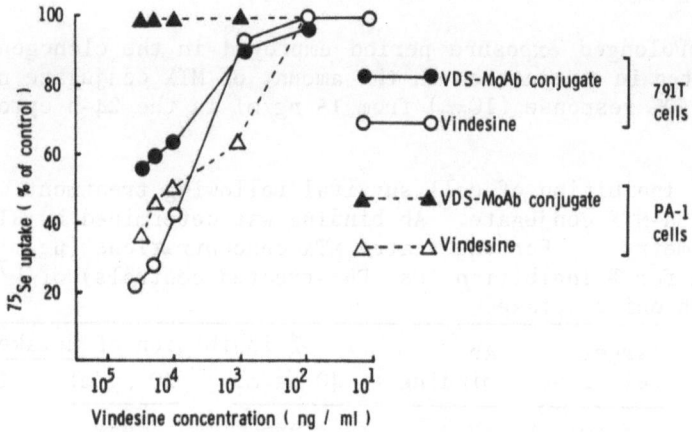

**Fig. 1.** Effect of vindesine (VDS) and VDS-791T/36 MAb on survival of tumour cells in culture. Cells were treated with the agent for 15 min, then washed and cultured for 24 h; they were then labelled for 16 h with [75Se]selenomethionine and washed 3 times. Controls were treated with PBS. 791T/36, osteogenic sarcoma; PA1, ovarian carcinoma.

example where incubation of osteogenic sarcoma 791T cells for 15 min
with VDS-791T/36 conjugate followed by extensive washing of the cells
produced marked inhibition of protein synthesis when assayed 24 h
later.   Similar treatment of ovarian carcinoma PA1 cells, which do
not express the 791T/36 Ab-defined antigen did not affect survival
of these cells.

## Methotrexate-791T/36 conjugates

Direct conjugation of agents to Ab is limited by the requirement to retain Ab structure in order to preserve immunological reactivity. It is possible, however, to increase the drug:Ab molar ratio
by using carrier-drug conjugates.   This approach is illustrated by
the synthesis of methotrexate-791T/36 Ab conjugates [5].   Methotrexate (MTX) was first reacted with human serum albumin (HSA) in the
presence of water-soluble carbodiimide to yield MTX-HSA products
with up to 32 mol MTX/mol HSA.   MTX-HSA conjugates were then conjugated to iodoacetyl-substituted 791T/36 Ab through free -SH groups
introduced through dithiopyridyl substitution to yield MTX-HSA-791T/
36 Ab conjugates.   These retained Ab activity and were specifically
cytotoxic for target cells expressing the Ab-defined antigen.   Thus,
Table 3 shows the cytotoxicity as assayed in the [$^{75}$Se]selenomethio-
nine uptake test of MTX-HSA-791T/36 Ab conjugates for 'antigen-
positive' osteogenic sarcomas 791T and T278 and 'antigen-deficient'
cells (bladder carcinoma T24 and melanoma RPMI 5966).   A clonogenic
assay was also used to demonstrate the tumour-inhibiting properties
of MTX-791T/36 Ab conjugates and, as illustrated in Fig. 2, 99% to
100% inhibition of 791T colony formation was achieved with continuous
exposure (5 days) to 500-100 ng MTX/ml [5].

The prolonged exposure period employed in the clonogenic assay
also resulted in a reduction in the amount of MTX conjugate needed to
produce a 50% response ($IC_{50}$) from 15 ng/ml in the 24-h cytotoxicity

**Table 3.**   Inhibition of cell survival following treatment with
791T/36-HSA-MTX conjugate.   Ab binding was determined by RIA and
flow cytometry.    For the stated MTX concentrations (µg/ml) values
are given for % inhibition (*vs.* PBS-treated controls) of [$^{75}$Se]-
selenomethionine uptake.

|  | Target cell line | Ab binding | % inhibition of uptake | | |
|---|---|---|---|---|---|
|  |  |  | 40 µg/ml | 20 µg/ml | 10 µg/ml |
| Osteogenic sarcoma | 791T | ++ | 79 | 82 | 79 |
|  | T278 | + | 73 | 79 | 81 |
| Bladder carcinoma | T24 | − | 7 | 0 | 0 |
| Melanoma | RPMI 5966 | − | −5 | 29 | 9 |

Fig. 2. Cytotoxicity of
methotrexate (MTX) and
MTX-HSA-791T/36 conjugate
against osteogenic sarcoma
791T cells as measured by
column inhibition assay
(clonogenic assay). The
treatment was for 5 days.

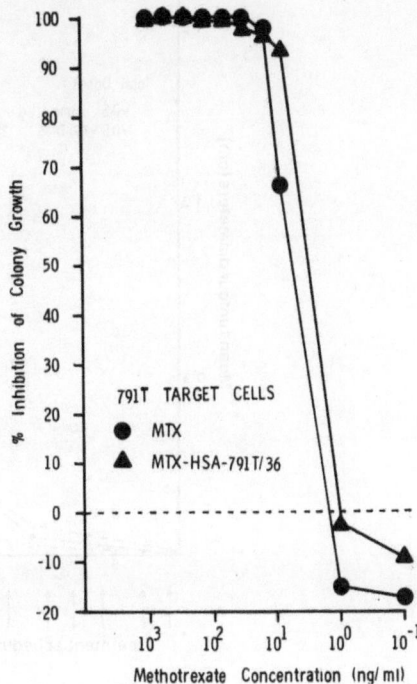

assay to 4 ng/ml in the colony inhibition tests. These observations
are relevant in relation to the amount of 791T/36 Ab that can be
deposited in tumours. This has been investigated in considerable
detail with respect to the localization of 125I-labelled 791T/36 in
osteogenic sarcoma 791T xenografts in immuno-deprived mice, where it
was found that there was a correlation between Ab dose injected (up
to 100 μg, i.e. 5 mg/kg body wt.) and the amount localized in tumours
[6]. The maximum level of 125I-labelled Ab under these conditions
was 70 μg/g tumour, and it was also found that Ab localization was
proportional to tumour size. The indication from these studies is
that doses of the order of 2 mg 791ZT/36 Ab injected at 3-4 day
intervals would maintain Ab levels at up to 70 μg/g, equivalent to
5 μg MTX with an MTX-HSA-791T/36 conjugate. This level of MTX
deposition approaches that which is known from pharmacokinetic
studies with other tumours to be tumour-inhibitory.

## THERAPY TRIALS WITH 791T/36 MONOCLONAL ANTIBODY CONJUGATES

Trials have been set up to determine whether conjugates of
791T/36 MAb with VDS and MTX suppress growth of osteogenic sarcoma
xenografts in immuno-deprived mice.

Although these investigations are at an early stage of development with respect to treatment protocols, it has been established
that treatment with both VDS-791T/36 and MTX-HSA-791T/36 conjugates

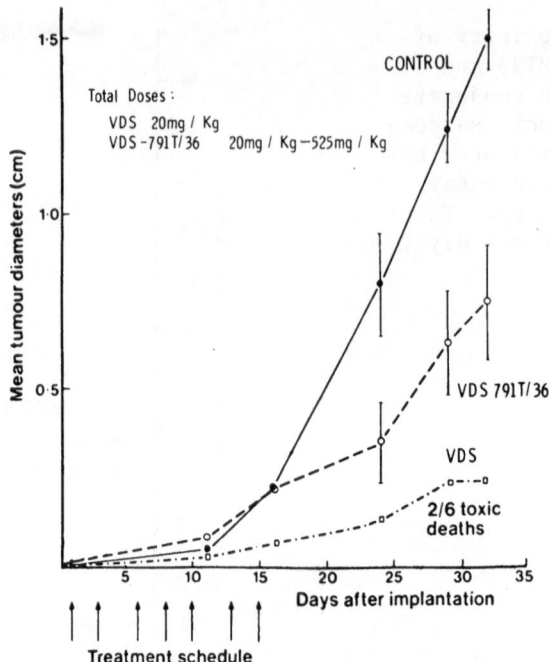

**Fig. 3.** Effect of VDS and VDS-791T/36 Ab conjugate on growth of xenografts of osteogenic sarcoma 791T. Each agent was given i.p. according to the treatment schedule shown. Total dose/mouse: VDS alone, 20 mg/kg; conjugate, 525 mg/kg.

produces a significant inhibition of tumour growth. This is illustrated by experiments with VDS in which mice received i.p. doses of VDS or VDS-791T/36 conjugate starting 1 day after s.c. injection of 791T cells (Fig. 3). The VDS-Ab conjugate inhibited tumour growth and a significant feature of this treatment was the lack of observable toxic effects. At the VDS dose selected (75 µg/kg body wt.) VDS was more tumour-inhibitory but produced more toxic effects, with only 50% of the treated mice surviving [7].

**CONCLUSIONS**

From the foregoing considerations it is concluded that Ab targeting of cytotoxic drugs is a valid concept, but there are many stages in the design of drug-Ab conjugates which require further development. Foremost is the need to generate a broader spectrum of anti-tumour MAb's. Ideally these should interact with a high proportion of tumour cells and show high affinity. In this respect, design of 'second generation Ab's' should include the production of human Ab's which will partly overcome the problem of patients generating Ab responses to infused murine MAb's.

Adequate Ab deposition in tumours is a prerequisite for drug targeting, and may be improved by selection of new Ab's as well as by the use of Ab fragments [Fab and $F(ab^1)_2$]. There are still insufficient data to generalize about the relative potential of Ab fragments, although a common experience is that fragments are excreted more rapidly than intact Ab so that any increase in specific localization as a result of removal of the Fc fragment may require higher doses of Fab or $F(ab^1)_2$ for drug localization.

Finally it is evident that the design of drug-Ab conjugates with respect to the use of degradable and non-degradable linkages needs to be developed. In addition, stoichiometric consideration indicates that direct linkage of drug to Ab molecules may not be sufficient. In this respect, therefore, drug carrier systems such as albumin [5] need to be defined more precisely.

## Acknowledgements

These studies were carried out with departmental grants from the Cancer Research Campaign, U.K. The author thanks his many colleagues in the Cancer Research Campaign Laboratories and the Departments of Surgery and Medical Physics at the University Hospital, Nottingham, for permission to include their data.

## References

1. Baldwin, R.W. (1983) *Pharmacy International 4*, 137–141.
2. Armitage, N.C., Perkins, A.C., Pimm, M.V., Farrands, P.A., Baldwin, R.W. & Hardcastle, J.D. (1984) *Br. J. Surgery 71*, 407–412.
3. Pimm, N.V. & Baldwin, R.W. (1984) *Clin. Immunol. Newsletter 5*, 150–152.
4. Embleton, M.J., Rowland, G.F., Simmonds, R.G., Jacobs, E., Marsden, C.H. & Baldwin, R.W. (1983) *Br. J. Cancer 47*, 43–49.
5. Garnett, M.C., Embleton, M.J., Jacobs, E. & Baldwin, R.W. (1983) *Int. J. Cancer 31*, 661–670.
6. Pimm, M.V. & Baldwin, R.W. (1984) *Eur. J. Cancer 20*, 515–524.
7. Rowland, G.F., Axton, C.A., Baldwin, R.W., Brown, J.P., Corvalan, J.R.F., Embleton, M.J., Gore, V.A., Hellstrom, I., Hellstrom, K.E., Jacobs, E., Marsden, C.H., Pimm, M.V., Simmonds, R.G. & Smith, W. (1985) *Cancer Immunol. Immunother.*, in press.

#E-3

# ANTIGENIC MODULATION AS A LIMITING FACTOR IN THE
# TREATMENT OF B-CELL LYMPHOMA WITH ANTI-IDIOTYPE ANTIBODY

M.J. Glennie and G.T. Stevenson

Lymphoma Research Unit
Tenovus Research Laboratory
Southampton General Hospital
Tremona Road, Southampton SO9 4XY, U.K.

*In exploring antibody (Ab) approaches to the treatment of malignant disease, one major problem is antigenic modulation, which allows tumour cells confronted by Ab to avoid damage from effector systems such as complement through redistribution and internalization of immune complexes. Working with anti-idiotype (anti-Id) Ab directed against neoplastic B lymphocytes we describe how Ab derivatives which are univalent but retain an intact Fc region can avoid this problem. Progress in this work has led to the construction of chimaeric molecules in which univalent Fab'γ from Ab is linked by thioether bonds to normal immunoglobulin (Ig) of the species to undergo immunotherapy. Two such derivatives, FabIgG and FabFc which use IgG and Fcγ respectively as effector partners of the Ab Fab'γ, appear superior to parent Ab in their ability to invoke complement and K cell killing of target lymphocytes. They also show promise of utilizing Fab'γ from any source including monoclonal reagents and, because they present homologous Ig, should prove minimally immunogenic and efficient recruiters of host effector systems.*

For some time we have been interested in idiotypic determinants of surface Ig found on neoplastic B lymphocytes as targets for Ab attack [1]. Our understanding of Ig structure and the restricted tissue distribution of idiotype has offered a unique opportunity to study the therapeutic potential of Ab. Throughout this work, antigenic modulation, during which tumour cells under attack by Ab exhibit diminished expression of antigen, has proved a major mechanism whereby cells escape destruction. Boyse & Old [2] were the first to describe this process for the TL antigen of murine thymic leukaemia. It has since been found that most surface antigens when cross-linked by Ab form complexes which are cleared from cell surfaces rendering them resistant to lysis by various effector systems [3]. Work in this and other laboratories has shown that tumour cells are generally only protected through modulation  when the Ab they encounter is bivalent [4].

One possible means of circumventing this escape mechanism is to use univalent Ab. It is well known that the univalent Ab fragment Fab'$\gamma$, derived by peptic digestion of Ab, will bind to cell surfaces without perturbing antigen expression. Unfortunately the absence of an Fc zone means that such molecules are ineffectual in activating effector systems such as complement. We have now developed a series of Ab derivatives which are univalent, but still retain an intact Fc region (Fig. 1). The simplest of these is the Fab/c molecule prepared by enzymatic removal of one Fab arm from rabbit IgG. This molecule, in failing to modulate, has proved superior to the parent Ab both *in vitro* and in immunotherapy of leukaemia in guinea pigs [5]. It has not been possible to evaluate Fab/c therapeutically in man because of logistic problems. However, with this in mind we have engineered chimaeric univalent derivatives. In these, Fab'$\gamma$ from Ab is linked to normal Ig of the Ab recipient. Fig. 1 shows two such molecules, FabFc and FabIgG, in which the Ab Fab'$\gamma$ is linked to Fc or IgG via a disulphide bond or a thioether bond respectively. *In vitro* these reagents performed in a similar manner to Fab/c, but they offer the advantage of being available in large quantities, being less immunogenic, and, because they present the recipient's Fc, should prove efficient recruiters of host effector systems [6].

We now describe the production and characterization of monoclonal Ab's (MAb's) which will be used to supply Fab'$\gamma$ fragments for the construction of chimaeric derivatives. These reagents are specific for the Id determinants found on the $L_2C$ leukaemia of strain 2 guinea pigs. They have behaved in a similar manner to their polyclonal counterparts, both in the clearing of surface Ig from tumour cells during culture and in immunotherapy. Their availability in large quantity has also presented an opportunity to investigate the maximum therapeutic effect which might be expected by giving regular daily treatment with unmodified Ab.

## MATERIALS AND METHODS

The $L_2C$ lymphoblastic leukaemia of strain 2 guinea pigs was used in the present study [7]. These cells express surface IgM and in culture secrete free $\lambda$ chains together with a small amount of idiotypic IgM. Idiotypic material to serve as antigen was prepared from the serum of guinea pigs in the terminal phases of the leukaemia by the method of Stevenson et al. [8].

MAb's specific for this idiotype were produced by the method of Köhler & Milstein [9] as described previously [10]. Briefly, spleen cells of BALB/c mice which had been immunized with idiotypic

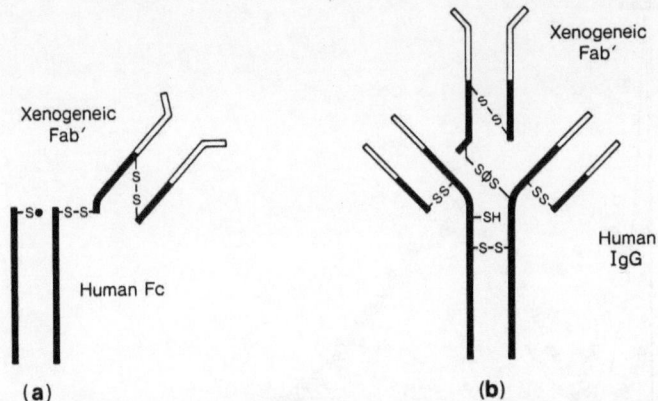

**Fig. 1.** The univalent Ab derivatives prepared in our laboratory. (a) FabFc hybrid derived by linking Fab'γ, from peptic digestion of Ab, with Fc from papain digestion of human IgG [11]. (b) FabIgG hybrid derived by linking Ab Fab'γ to human IgG via a thioether bond formed between hinge-region -SH groups of the Fab'γ and the IgG [6].

IgM were fused with the NS-1 (P3-NS-1/1-Ag 4.1) mouse myeloma line at a ratio of 2:1 using 50% (v/v) polyethylene glycol 4000 (PEG). Hybrid cells were suspended directly in supplemented Dulbecco's Minimal Essential Medium (MEM + HAT) containing 100 µM hypoxanthine, 0.4 µM aminopterin, 16 µM thymidine and distributed into 96-well micro-culture plates. Colonies were visible in 8-10 days, and culture supernatants were screened for Ab activity. Wells with a useful level were cloned by limiting dilution onto a feeder layer of mouse thymocytes. Larger amounts of Ab were prepared from ascitic fluid by precipitation with ammonium sulphate followed by chromatography on columns of DEAE-Trisacryl and Ultragel AcA34 (LKB Produkter).

Screening of Ab activity was carried out by the enzyme-linked immunosorbent assay (ELISA) [12]. Antigen in the form of purified guinea pig Ig was bound to 96-well microplates at 20 ng/ml in carbonate buffer. After washing, MAb's were added and the bound mouse Ig detected by exposure to horse-radish peroxidase-labelled rabbit anti-mouse Ig (Nordic Laboratories Ltd., Maidenhead, U.K.).

Estimates of the relative quantity of MAb bound to $L_2C$ cells were obtained indirectly with detection by [125]I-conjugated rabbit anti-mouse Ig, after exposing the cells ($2 \times 10^7$/ml) to MAb at 0° for 15 min and performing washing and various manipulations as below.

Immunotherapy tests were performed on matched groups of strain 2 guinea pigs. They received $10^5$ viable $L_2C$ cells i.p. and, 24 h later, a single treatment (unless stated otherwise) of anti-Id MAb or a non-reactive MAb, likewise i.p. Survival was monitored daily.

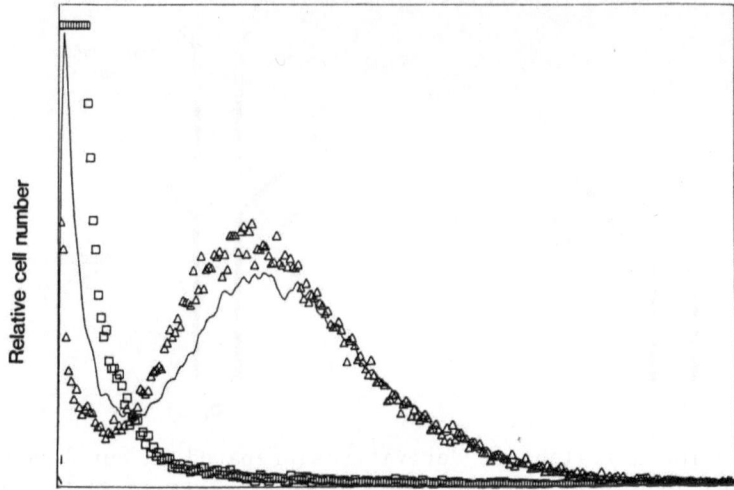

**Relative fluorescence intensity**

Fig. 2.   Binding of anti-Id Ab to $L_2C$  leukaemia cells.  Cells were exposed to MAb (5 µg/ml) which had been pre-incubated for 15 min at 37° with an equal vol. of phosphate-buffered saline (PBS, pH 7.2; ———), normal guinea pig serum (Δ) or serum obtained from guinea pigs in the terminal phase of the $L_2C$ leukaemia (□).

## RESULTS AND DISCUSSION

MAb's raised to $L_2C$ idiotypic IgM were assessed by the ELISA technique against a range of guinea pig Ig antigens coated onto microplates.  No reactivity was seen against normal pentameric IgM, normal κ or tumour-derived λ light chains or normal IgG at concentrations of Ab which gave absorbances of >1.0 in the assay using Id-IgM as antigen.  The absence of reactivity with light chains produced by the tumour and recovered from the urine of leukaemic guinea pigs indicates that the idiotypic determinants depend on either heavy chain or heavy-plus-light chain combination, as has been found for other idiotypes [13].

Ab specificity was confirmed by the reactivity with $L_2C$ cells in cytofluorimetry using the FACS III.  Strong fluorescence was seen (Fig. 2) using Ab at 5 µg/ml, and such staining was not affected by pre-incubating the Ab with normal guinea pig serum.  In contrast, serum obtained from guinea pigs in the terminal phase of $L_2C$ leukaemia, or purified Id-IgM at 4 µg/ml (data not shown), completely blocked reactivity with the cells.

Three specific anti-Id Ab's have been selected during this investigation;  but mostly in this article we will concentrate on just one, 'M6/3D10', of IgG1 isotype with a κ light chain.

Previous investigations with cultured $L_2C$ cells have shown that surface Ig is rapidly cleared after interaction with a polyclonal anti-Id Ab [4]. We can now show that an equivalent MAb can also perturb surface Ig, despite the fact that the monoclonal reagent must rely on a single antigenic determinant for cross-linking surface Ig. The clearance of Ab from the surface of $L_2C$ cells has been monitored at specific intervals by detecting mouse Ab with $^{125}I$-conjugated rabbit anti-mouse IgG. Two MAb's have been investigated, one reactive with the constant region of the light chain (anti-$C\lambda$) and one specific for the Ig idiotype. During the production of anti-Id Ab's, inevitably a number of anti-constant region Ab's are produced; the anti-$C\lambda$ used in this experiment is one such reagent.

Fig. 3 shows that at saturating levels the two Ab's differ in their ability to clear surface Ig and that both are less efficient in this respect than a polyclonal anti-Id serum. Interestingly, when both MAb's interact with the $L_2C$ cell at the same time then the rate and extent to which Ig is cleared increases and mimics closely that seen with the polyclonal reagent. This synergistic effect probably reflects the enhanced cross-linking activity of the two MAb's when recognizing different determinants on the same antigen.

Evidence that Ab is internalized during clearance rather than shed into the supernatant is presented in Fig. 4. Cells treated with $^{125}I$-conjugated anti-Id MAb retained radioactivity for several hours in culture with only a gradual depletion. In contrast, univalent Fab'γ fragments prepared by peptic digestion of this reagent showed significant shedding when first placed into culture. Such a loss probably occurs at this time, but is not manifest in the parent Ab due to the 'multiple binding bonus' provided by bivalent binding.

This result has important implications for the selection of MAb's destined to supply the Fab'γ fragments of chimaeric derivatives as discussed above. Previously we have always prepared Fab'γ fragments from high-affinity polyclonal Ab, thus ensuring a useful proportion of univalent derivatives which bind firmly to the target cells. If MAb's are to perform in the same way, then it is essential that we select only those with the highest affinity.

In immunotherapy experiments the present MAb's have performed in a similar manner to their polyclonal counterparts [14]: a single dose of Ab prolonged the median survival of $L_2C$ leukaemic guinea pigs by ~3 days. This therapeutic effect can only be described as modest although, based on a tumour-doubling time of ~19 h, it would suggest that ~90% of the inoculated cells were killed as a result of Ab treatment.

Fig. 5 shows an attempt to improve this survival curve through intensive treatment with unmodified MAb. Commencing 24 h after receiving $10^5$ $L_2C$ cells, guinea pigs were given a daily dose of

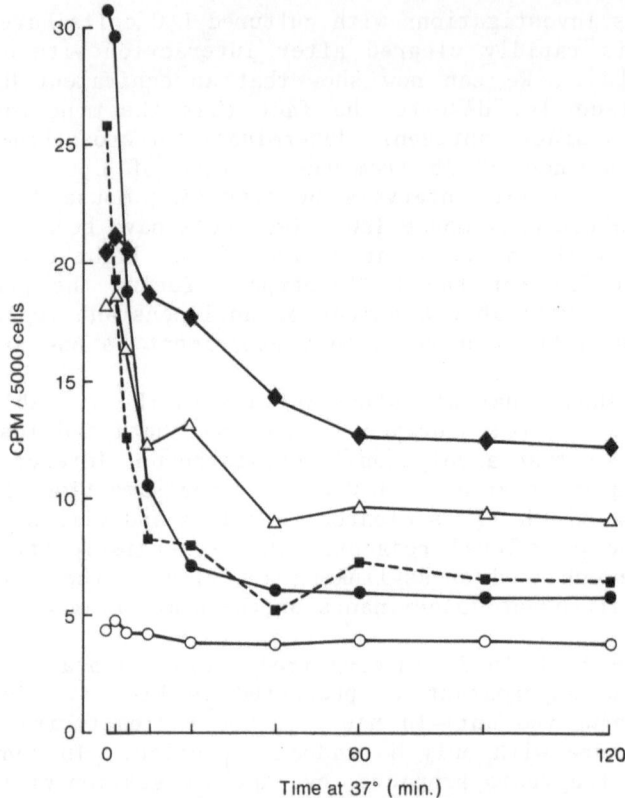

**Fig. 3.** Loss of surface Ig from $L_2C$ cells during incubation with anti-Id. Cells were exposed to normal mouse serum 1:100 dilution (○), mouse anti-Id IgM serum 1:100 dilution (■), monoclonal anti-Id IgM 20 μg/ml (△), monoclonal anti-Cλ 20 μg/ml (◆) or a mixture of monoclonal anti-Id 10 μg/ml plus anti-Cλ 10 μg/ml (●) for 15 min at 0°. After washing they were cultured ($5 \times 10^6$/ml) at 37° in supplemented Eagle's medium, and at intervals sampled to quantify the surface Ig by labelling with radiolabelled anti-Ig.

anti-Id for as long as they survived. Despite such treatment tumour development was only slightly suppressed compared with that seen after a single injection of Ab, and animals succumbed to the leukaemia while still receiving regular treatment. Apparently there is little to be gained from prolonged administration of Ab, at least in this system.

Circulating lymphocytes recovered from these animals during the terminal phases of the disease exhibit complete modulation. Fig. 6 shows that the surface Ig has been reduced to almost undetectable levels by prolonged exposure to Ab *in vivo*. This is in contrast to the situation *in vitro* (Fig. 3) where culture (37°) of $L_2C$

**Fig. 4.** Persistence of
monoclonal anti-Id and its
fragments on $L_2C$ cells
during culture.   Cells
were exposed to radio-
labelled IgG ($\Diamond$), F(ab')2
($\bullet$) or Fabγ ($\blacksquare$) at 20 μg/
ml for 15 min at 0°.
After washing, cells were
cultured in supplemented
Eagle's medium ($5 \times 10^6$/ml)
and sampled at intervals
to determine bound radio-
activity.

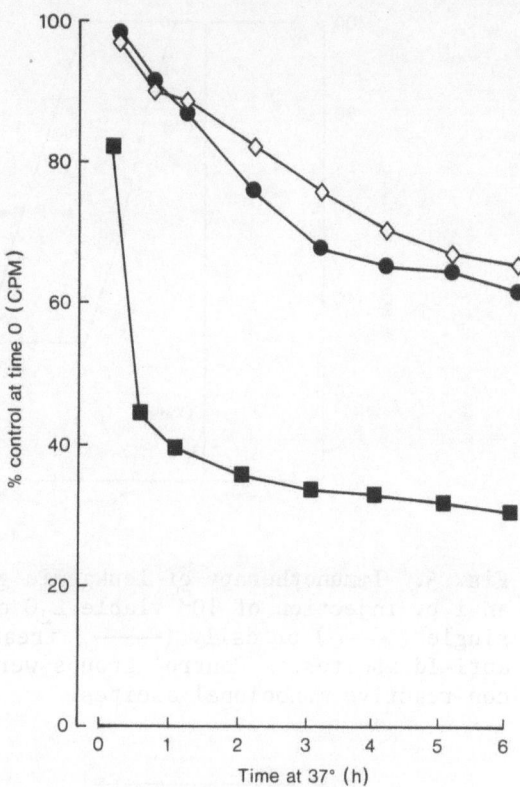

cells with a single MAb failed to remove all the surface Ig.  It is
not known why Ab can perturb a surface antigen more effectively *in
vivo* than *in vitro*.   This phenomenon has been described for several
systems using both polyclonal [2] and monoclonal [15] reagents in
animals and man.   Perhaps prolonged exposure to Ab in combination
with favourable metabolic conditions for the tumour cells are sufficient explanation.   However, it is always possible that exogenous
factors other than Ab are available *in vivo* to assist modulation [16].

The possibility that an Ig-negative variant of $L_2C$ had been
selected by Ab treatment was excluded by short-term culture of these
cells in the absence of Ab. During this time surface Ig was restored
(data not shown), thus confirming the role of Ab in maintaining their
negative phenotype during therapy.

The present work strongly confirms that the therapeutic potential of unmodified Ab is severely limited due to modulation and that
this limitation is unlikely to be solved simply by more intensive
treatment.  Circumventing this problem with univalent derivatives as
discussed above seems a logical solution, and we now look forward to
preparing chimaeric reagents from these MAb's for a direct comparison
of therapeutic performance.

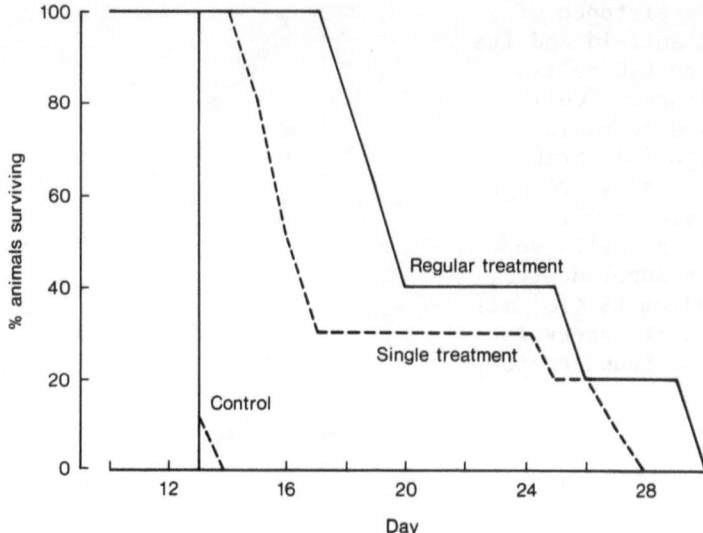

**Fig. 5.** Immunotherapy of leukaemic guinea pigs. Animals received an i.p. injection of $10^5$ viable $L_2C$ cells followed 24 h later by a single ( - - - - ) or daily ( ———— ) treatment with 0.5 ml of monoclonal anti-Id ascites. Control groups were treated similarly but with a non-reactive monoclonal ascites.

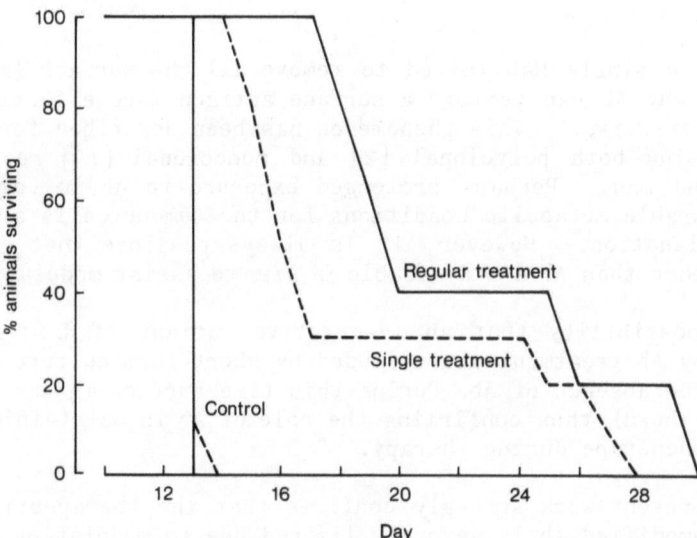

**Fig. 6.** Density of surface Ig on $L_2C$ cells during immunotherapy. Circulating lymphocytes recovered from leukaemic animals receiving daily treatment with a non-reactive Ab (Δ) or anti-Id (□) were stained with monoclonal anti-Id followed by fluorescent rabbit anti-mouse Ig. PBS-treated controls shown ————.

*References*

1.  Stevenson, G.T. & Stevenson, F.K. (1983) *Springer Semin. Immunopathol. 6*, 99–115.
2.  Boyse, E. A. & Old, L. J. (1969)   *Ann.  Rev.  Genet.  3*, 269–290.
3.  Chatenoud, L. & Bach, J.F. (1984) *Immunol. Today 5*, 20–25.
4.  Gordon, J., Anderson, V.A. & Stevenson, G.T. (1982) *J. Immunol. 192*, 2763–2769.
5.  Glennie, M.J. & Stevenson, G.T. (1982) *Nature 295*, 712–714.
6.  Stevenson, G.T., Cole, V.M., Summerton, J. & Watts, H.F. (1984) *Med. Oncol. Tumor Pharmacother. 1*, 275–278.
7.  Stevenson, G.T., Eady, R.P., Hough, D.W., Jurd, R.D. & Stevenson, F.K. (1975) *Immunology 28*, 807–820.
8.  Stevenson, F.K., Morris, D. & Stevenson, G.T. (1980) *Immunology, 41*, 313–321.
9.  Köhler, G. & Milstein, C. (1975) *Nature 256*, 495–497.
10. Stevenson, F.K., Glennie, M.J., Johnston, D.M.M., Tutt, A.L. & Stevenson, G.T. (1984) *Br. J. Cancer 50*, 407–413.
11. Stevenson, G.T., Glennie. M.J. & Gordon, J. (1982) *UCLA Symp. Mol. Cell Biol. 24*, 459–472.
12. Engvall, E. & Perlmann, P. (1972) *J. Immunol. 109*, 129–135.
13. Capra, J.D. (1977) *Fed. Proc. 36*, 204–206.
14. Stevenson, G.T., Elliott, E.V. & Stevenson, F.K. (1977) *Fed. Proc. 36*, 2268–2271.
15. Chatenoud, L., Baudrihaye, M.F., Kreis, H., Goldstein, G., Schindler, J. & Bach, J.F. (1982) *Eur. J. Immunol. 12*, 979–982.
16. Stackpole, C.W., Jacobson, J.B. & Galuska, S. (1978) *J. Immunol. 120*, 188–197.

#E-4

# TUMOUR-ASSOCIATED CARBOHYDRATE ANTIGENS OF GLYCOPROTEINS

E.F. Hounsell, H.C. Gooi and T. Feizi

Applied Immunochemistry Research Group
MRC Clinical Research Centre
Watford Road, Harrow, Middx. HA1 3UJ, U.K.

*Many antigens recognized by monoclonal antibodies (MAb's) as tumour markers are carbohydrate structures related to the major blood group system. They represent an inappropriate expression of saccharide antigens for the tissues in question (e.g. due to incomplete biosynthesis). Carbohydrate structures which behave as oncofoetal antigens, and the use of modern methods of oligosaccharide separation and structural analysis in the elucidation of antigenicity, are exemplified in this article.*

Many antibodies (Ab's) have been raised, with whole tumour cells, against tumour-associated and differentiation antigens using the hybridoma technique of Köhler & Milstein [1]. The antigenic determinants recognized by several of these MAb's are carbohydrate structures [2, 3]. The carbohydrate nature of cell-surface antigens is hardly surprising if one considers that cells, as represented in Fig. 1, are surrounded by carbohydrate chains on membrane-bound glycoproteins and glycolipids. For the characterization of carbohydrate antigens, we have relied on the fact that many oligosaccharide structures at cell surfaces are also present on the carbohydrate chains of secreted glycoproteins (particularly those of mucin glycoproteins) which can be obtained in relatively large amounts for structural analysis and for use as reference materials in RIA [4, 5].

## THE STRUCTURES AND ANTIGENICITIES OF CARBOHYDRATE CHAINS OF MUCINS

We now are well informed on the structures of the carbohydrate chains of mucin glycoproteins, and the blood group-related and tumour-associated changes that they express [4, 6]. The carbohydrate chains fall broadly into three domains (Fig. 1) – *core, backbone,* and *periphery,* each having a distinct set of antigens, some

Fig. 1. A schematic cell
membrane showing carbohydrate
chains of glycolipids and glycoproteins forming a 'glycocalyx' over the surface.  ⊙:
*Core region* monosaccharides,
viz. Glc in glycolipids; GalNAc,
Gal & GlcNAc in *O*-glycosidically
linked chains, and GlcNAc &
Man in *N*-glycosidically linked
chains.  o : *Backbone region*
monosaccharides, usually Gal &
GlcNAc.  ■ & ● : *Peripheral
region* monosaccharides, e.g.
sialic acid & Fuc respectively.

Conventional abbreviations used:
Fuc = fucose, Man = mannose, etc.

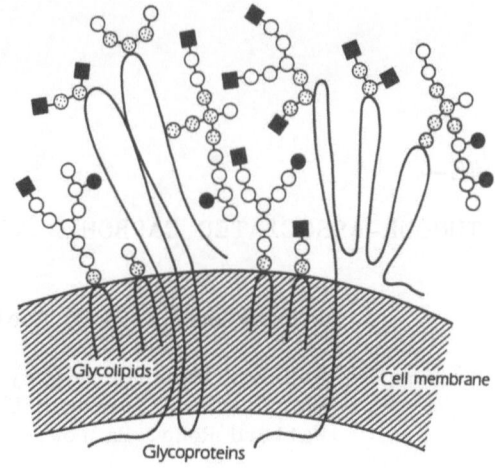

The carbohydrate chains of
examples of which are shown in Table 1.  The carbohydrate chains of
mucins are *O*-glycosidically linked via GalNAc to serine or threonine
residues of proteins.  Five types of *O*-linked core region have been
described (reviewed in [4]), viz.:

**Galβ1-3GalNAc**

**GlcNAcβ1-3GalNAc**

**GlcNAcβ1-6GalNAc**

**GlcNAcβ1**⟍
            ⁶⁄₃GalNAc
**Galβ1**⟋

**GlcNAcβ1**⟍
            ⁶⁄₃GalNAc
**GlcNAcβ1**⟋

The *backbone* regions of these carbohydrate chains are built up
by sequential addition of sugars to the core region.  The monosaccharides in the *backbones* are usually Gal and GlcNAc which are joined
by linkages of two alternative types [7], Galβ1-3GlcNAcβ1-3 (Type 1)or
Galβ1-4GlcNAcβ1-3 (Type 2); branches are formed by GlcNAcβ1-6Gal linkage
[8].  The chains are often terminated by peripheral monosaccharides
which have an α configuration, e.g. the fucose residues of blood group
H and the Lewis antigens, Le[a] and Le[b] (Fig. 1, Table 1).

The aberrant expression of carbohydrate antigens found in the
glycoproteins of tumours, e.g. those extracted from gastric adenocar-
cinomas (Table 1), is thought to be caused by incomplete [9, 10] or
aberrant [4] biosynthesis rather than degradation of the carbohydrate chains.  Thus, on the glycoproteins of secretors with gastric
cancer (Table 1) there is exposure of *backbone* and *core* region antigens FC 10.2 and IMa and the receptor for peanut lectin.

The assignment of several MAb-defined carbohydrate antigens, for
example IMa [8], FC 10.2 [11] and the stage-specific embryonic anti-

**Table 1.** Expression of the peanut lectin receptor (PNL) and of the carbohydrate antigens IMa, FC 10.2, Le[a], H and Le[b] in paired glycoprotein extracts from non-neoplastic mucosa (N) and tumours (T) of patients of known secretor status with carcinoma of the stomach. Symbols for monosaccharides in the three domains are as in Fig. 1.

| Structure | | Expression in Gastric Mucosa | | | |
| --- | --- | --- | --- | --- | --- |
| | | Secretors | | Non-Secretors | |
| | | N | T | N | T |
| **Core domain** | | | | | |
| Galβ1–3GalNAc – peptide | PNL | – | + | + | + |
| Galβ1–4GlcNAcβ1↘ ⁶GalNAc – peptide | IMa | – | + | + | + |
| **Backbone domain** | | | | | |
| Galβ1–4GlcNAcβ1↘ ⁶Galβ1– | IMa | – | + | + | + |
| Galβ1–3GlcNAcβ1–3Galβ1–4GlcNAcβ1– | FC10.2 | – | + | + | + |
| **Peripheral domain** | | | | | |
| Galβ1–3GlcNAcβ1– &#124;1.4 Fucα | Le[a] | – | + | + | + |
| Galβ1–3/4GlcNAcβ1– &#124;1.2 Fucα | H | + | ± | – | – |
| Galβ1–3GlcNAcβ1– &#124;1.2  &#124;1.4 Fucα  Fucα | Le[b] | + | ± | – | – |

gen of mouse, SSEA-1 [12, 13] was established by studies of the reaction patterns of the MAb's with mucins followed by binding-inhibition studies with structurally characterized oligosaccharides [8, 11–13].

Currently under investigation are an additional set of onco-foetal antigens recognized by natural Ab's of the mouse [14] which bind to mucin glycoproteins from human meconium [14, 15]. Characterization of these Ab's illustrates a second approach which involves the isolation, purification and structural/antigenic analysis of oligosaccharides of glycoproteins to which they bind. The natural Ab's of mouse were shown to bind strongly to a fraction of glycoproteins obtained from blood group H-active meconium by pronase digestion and ethanol precipitation. Reactivity of Ab with this glycoprotein preparation was increased on treatment with mild acid (20 mM $H_2SO_4$, $100°$, $1\frac{1}{2}$ h) to remove fucose residues. Structural analysis was done on the oligosaccharides obtained from these glycoproteins.

## ISOLATION AND PURIFICATION OF OLIGOSACCHARIDES OF MECONIUM

The glycoprotein preparation from blood group H-active meconium was treated with mild acid to remove fucose residues, then with mild alkali/borohydride (50 mM NaOH in 1 M NaBH$_4$, 50°, 16 h) in order to release intact, reduced $O$-glycosidically linked chains [16, 17]. Dialyzable oligosaccharides were lyophilized, evaporated with methanol/ 1% (v/v) acetic acid several times, and deionized on consecutive columns of Dowex AG50W×8 (hydrogen form) and Dowex 1×2 (acetate form). Size fractionation was carried out by Biogel P4 chromatography [18, 19]. The oligosaccharide isomers in each fraction were then separated by HPLC with detection at 208 nm [15, 20], initially using an ODS (RP) column eluted with water or with water/acetonitrile gradients for larger oligosaccharides. Further purification was performed by HPLC on an aminopropyl-silica column with isocratic elution by acetonitrile/water, 65:35 by vol.

## STRUCTURAL ANALYSIS OF OLIGOSACCHARIDES

The composition of the purified oligosaccharides was assayed by GC [21] and fast atom bombardment MS [22-24]. Structural analysis to deduce the sequence of monosaccharides, their linkage type and anomeric configuration (α or β) was carried out by NMR [15, 25, 26]. The NMR spectra were interpreted using a computer search program to compare signals in the spectra with those in a computerized library compiled from published oligosaccharide NMR data [26].

Oligosaccharides which were not identical with any of those included in the computerized library required additional structural information for complete assignment. This was obtained by electron impact (EI) MS of permethylated oligosaccharides [19, 24, 27] and by capillary GC-MS analysis of partially methylated alditol acetates [28-30]. For these analyses purified oligosaccharides were permethylated and isolated by LH20 column chromatography, according to Hakomori's method [30]. An aliquot of the permethylated material was analyzed by EI-MS which gave characteristic fragmentation patterns defining the sequence of the constituent monosaccharides and also some linkage information [15, 24]. The remainder was further derivatized to give partially methylated alditol acetates, these derivatives being characteristic for monosaccharide type and position of linkage.

The structures of the most abundant core-region oligosaccharides obtained from the meconium glycoproteins that react with the natural Ab's in mouse sera are as shown opposite. The structures of larger carbohydrate chains in this series and the antigenicities of their individual oligosaccharides will be the subject of future reports.

**MANY TUMOUR-ASSOCIATED ANTIGENS ARE CARBOHYDRATE STRUCTURES**

The carbohydrate antigens discussed in this article are only a few among tumour-associated cell-surface antigens recognized by MAb's [2, 3, 6, 31, 32]. Characterization of the structures recognized by these anti-tumour Ab's is important for rationalizing their use in tumour biology and clinical oncology. Without exception, each of the tumour-associated antigens occurs as a normal component in certain cell types. Thus, great caution should be exercised in considering the use of these MAb's for immunotherapy. On the other hand, the tumour-associated carbohydrate antigens may be released into the serum or body cavities in sufficiently large amounts to be of value in monitoring tumour size. Detailed studies will be required before the value of these Ab's in routine clinical oncology is ascertainable. In addition anti-carbohydrate MAb's are invaluable as immunological tools for studying the glycosylation changes in tumour cells.

*References*

1.  Köhler, G. & Milstein, C. (1975) *Nature 256*, 495-497.
2.  Feizi, T. (1984) *Biochem. Soc. Trans. 12*, 545-549.
3.  Hakomori, S. (1984) *Ann. Rev. Immunol. 2*, 103-126.
4.  Hounsell, E.F. & Feizi, T. (1982) *Med. Biol. 60*, 227-236.
5.  Wood, E., Lecomte, J., Childs, R.A. & Feizi, T. (1979) *Mol. Immunol. 16*, 813-819.
6.  Feizi, T., Gooi, H.C., Childs, R.A., Picard, J.K., Uemura, K., Loomes, L.M., Thorpe, S.J. & Hounsell, E.F. (1984) *Biochem. Soc. Trans. 12*, 590-595.
7.  Watkins, W.M. (1980) in *Advances in Human Genetics, Vol. 10* (Harris, H. & Hirschhorn, K., eds.), Plenum, New York, pp. 1-136 & 379-385.
8.  Feizi, T., Kabat, E.A., Vicari, G., Anderson, B. & Marsh, W. (1971) *J. Exp. Med. 133*, 39-52.
9.  Picard, J.K., Waldron-Edward, D. & Feizi, T. (1978) *J. Clin. Lab. Immunol. 1*, 119-128.
10. Hakomori, S. & Young, Jr., W.W. (1978) *Scand. J. Immunol. 7, Suppl. 6*, 97-117.

*[continued*

*Meconium glycoprotein oligosaccharides (SEE OPPOSITE)*

GalNAc-

Galβ1-3GalNAc-

GlcNAcβ1-3GalNAc-

Galβ1-3GlcNAcβ1-3GalNAc-

Galβ1-4GlcNAcβ1-3GalNAc-

Galβ1-4GlcNAcβ1-6GalNAc-

GlcNAcβ1
$\searrow$
§GalNAc-
Galβ1 $\nearrow$

Galβ1-4GlcNAcβ1
$\searrow$
§GalNAc-
Galβ1 $\nearrow$

11. Gooi, H.C., Williams, L.K., Uemura, K., Hounsell, E.F., McIlhinney, R.A.J. & Feizi, T. (1983) *Mol. Immunol. 20*, 607-615.
12. Gooi, H.C., Feizi, T., Kapadia, A., Knowles, B.B., Solter, D. & Evans, M.J. (1981) *Nature 292*, 156-158.
13. Hounsell, E.F., Gooi, H.C. & Feizi, T. (1981) *FEBS Lett. 131*, 279-282.
14. Gooi, H.C. & Feizi, T. (1982) *Biochem. Biophys. Res. Comm. 106*, 539-545.
15. Hounsell, E.F., Lawson, A.M., Feeney, J., Gooi, H.C., Pickering, J.M., Stoll, M.S., Lui, S.C. & Feizi, T. (1985) *Eur. J. Biochem.*, in press.
16. Iyer, R.N. & Carlson, D.M. (1971) *Arch. Biochem. Biophys. 142*, 101-105.
17. Hounsell, E.F., Pickering, N.J., Stoll, M.S., Lawson, A.M. & Feizi, T. (1984) *Biochem. Soc. Trans. 12*, 607-610.
18. Yamashita, K., Tachibana, Y. & Kobata, A. (1977) *J. Biol. Chem. 252*, 5408-5411.
19. Hounsell, E.F., Wood, E., Feizi, T., Fukuda, M., Powell, M.E. & Hakomori, S. (1981) *Carbohyd. Res. 90*, 282-307.
20. Hounsell, E.F., Rideout, J.M., Pickering, N.J. & Lim, C.K. (1984) *J. Liq. Chromatog. 7*, 661-674.
21. Bhatti, T., Chambers, R.E. & Clamp, J.R. (1970) *Biochim. Biophys. Acta 222*, 339-347.
22. Forsberg, L.S., Dell, A., Walton, D.J. & Ballou, C.E. (1982) *J. Biol. Chem. 257*, 3555-3563.
23. Kamerling, J.P., Heerma, W., Vilegenthart, J.F.G., Green, B.N., Lewis, I.A.S., Strecker, G. & Spik, G. (1983) *Biomed. Mass Spect. 10*, 420-425.
24. Hounsell, E.F., Madigan, M.J. & Lawson, A.M. (1984) *Biochem. J. 219*, 947-952.
25. Vilegenthart, J.F.G., Van Halbeek, H. & Dorland, L. (1981) *Pure & Appl. Chem. 53*, 45-77.
26. Hounsell, E.F., Wright, D.J., Donald, A.S.R. & Feeney, J. (1984) *Biochem. J. 223*, 129-143.
27. Karlsson, K.-A. (1976) in *Glycolipid Methodology* (Wittig, L.A., ed.), Am. Oil Chemists' Soc., Champaign, IL, pp. 96-122.
28. Björndal, H., Hellerqvist, C.G., Lindberg, B. & Svensson, S. (1970) *Angew Chem. Int. Ed. Enge 9*, 610-619.
29. Stellner, K., Saito, H. & Hakomori, S. (1973) *Arch. Biochem. Biophys. 155*, 464-472.
30. Lawson, A.M., Hounsell, E.F. & Feizi, T. (1983) *Int. J. Mass Spectrom. & Ion Phys. 48*, 149-152.
31. Feizi, T. (1985) *Nature 314*, 53-57.
32. Feizi, T. (1985) *Cancer Surveys 4*, in press.

#E-5

# GLYCOLIPID ANTIGENS: POTENTIAL IN CANCER DETECTION AND DIAGNOSIS

Gary R. Matyas, [1]Vivian P. Walter-Doelling, Christine Ferroli,
[2]K.R. Pennington, Diane Pikaard & D. James Morré

Purdue Cancer Center, Purdue University
West Lafayette, IN 47907, U.S.A.

*A number of tumour-specific monoclonal antibodies (MAb's) have been produced which are directed to unique carbohydrate structures associated with cellular membranes as glycolipids rather than glycoproteins (as reviewed by S.-I. Hakomori [1]). The potential of other more common glycolipid antigens for use in cancer detection and diagnosis is now addressed with especial emphasis on a specific role in early detection. The assays now reported concern glycolipids in sera from hepatocarcinogen-fed rats and cancer patients.*

Recent observations [1] revealed a number of 'tumour-specific' MAb's directed to carbohydrate moieties of glycolipid antigens. Some of these are directed to a novel structure produced by neosynthesis involving the activation of glycosyltransferases that are inactive in progenitor cells. These include sialosyl Le[a] MAb prepared by Koprowski and co-workers [2] and a MAb directed to polyfucosyl type 2 chains with fucosyl $\alpha1\rightarrow3$ linkages at the penultimate GlcNAc residue [3, 4]. Others are directed to the precursors of complex glycolipids that appear to accumulate as the result of blocked or incomplete biosynthesis. These include Ab's to the precursor ganglioside antigen $G_{D3}$ [5, 6] expressed by human melanomas and to the antigen identified as globotriaosyl ceramide [7], also known as blood group P[k] antigen, expressed on Burkitt tumour cells.

Evidently certain glycolipids are tumour-associated antigens with potential in tumour diagnosis and imaging and as targets for immunotherapy (see R.W. Baldwin's article, #E-2). Moreover, these antigens are relatively specific to one or a few tumour types and

[1]Present address: Medical Research Division, American Cyanamid Co., Pearl River, NY 10965
[2]Kenneth R. Pennington, M.D., Arnett Clinic, Lafayette, IN 47904

are largely restricted to the tumour *per se*. As such, they exhibit
only limited potential as aids in early cancer detection, e.g. for a
test based on serum analyses of circulating glycolipids.

This report summarizes findings from our laboratory that indicate a more general elevation of precursor glycolipids in the ganglioside series in a rat model. In addition, we give preliminary
findings on raised levels of these glycolipids in human carcinoma
patients and discuss the findings as the basis for developing an
immunodiagnostic blood test for early detection of cancer.

## MATERIALS AND METHODS

*Animals and tumours.*- Male Fischer-strain rats were used, some
bearing transplanted hepatomas derived initially from primary liver
tumours induced by 2-acetylaminofluorene [8]; others had primary
tumours induced by feeding the carcinogen ([9]; schedule as in [10]).

*Total lipid-associated sialic acid* (LASA; primarily ganglioside
[11]).- This was determined in sera by solvent extraction, precipitation with citric acid to remove glycoproteins, and estimation of
sialic acid by the thiobarbituric acid method [12] following acid
hydrolysis as detailed by Kloppel et al. [11].

*Gangliosides.*- Levels of gangliosides (primarily the monosialo
precursor ganglioside $G_{M3}$) in carcinoma patients were determined as
follows. A 1 ml serum sample was extracted overnight with 10 ml
chloroform/methanol (1:1, by vol.) at 4°. After centrifugation (10
min, 2500 g), the supernatant was evaporated to dryness by heating
at <35°. The pellet was re-extracted with 10 ml chloroform/methanol
(2:1) for 2 h at room temperature. The supernatant was added to the
residue from the first (1:1) extract. The residue obtained by re-
evaporation was dissolved in 0.2 ml chloroform/methanol (1:1) with
vigorous mixing. Then 0.3 ml chloroform was added, giving a 4:1
chloroform/methanol ratio.

To isolate gangliosides, the solution was passed through a 4 cm
Unisil column - a 15 cm disposable pipette plugged with glass wool,
slurry-packed in chloroform (activated silicic acid). The sample tube
was rinsed with 0.5 ml chloroform/methanol (4:1) and the wash also
applied to the column; a further 5 ml was applied to the column to
remove interfering neutral lipids and phospholipids. Gangliosides
were eluted with 6 ml of 1:1 chloroform/methanol, and the eluate
collected in 10 ml screw-cap tubes. The solvent was removed by evaporation, and the sialic acid content was determined by a periodate-
resorcinol procedure [13]; the standard was *N*-acetylneuraminic acid.

*Neutral glycosphingolipids.*- These ganglioside precursors were
extracted from the sera of animals bearing transplanted hepatomas or
from the livers of carcinogen-fed [9] rats. Animals were anaestheti-

**Table 1.**  Levels of ganglioside and citrate-procedure (LASA) sialic acid in sera of control and hepatoma-bearing rats. Tumour designations (histological classification and ability to form pulmonary metastases): PD/WD/HD, poorly/well/highly differentiated; M, metastatic, NM non-metastatic. Values represent serum sialic acid, nmol/ml and *exptl./control*. For extracted gangliosides, the values are uncorrected for losses during extraction; in controls the estimates of absolute amounts of circulating ganglioside (Fig. 1) varied from 3.5 to 6.5 nmol/ml serum. *Based on Table in ref. [8].*

| Animals from which sera derived | | Ganglioside extract'n | LASA procedure |
|---|---|---|---|
| Controls (normal animal) | | 1.7 | 34 |
| Bearing induced hepatomas | | 4.0, *2.4* | – |
| Bearing transplantable hepatomas | $RLT_1$ (PD-M) | 4.8, *2.8* | 78, *2.3* |
| | $RLT_2$ (PD-NM) | 3.8  *2.2* | 59, *1.5* |
| | $RLT_7$ (PD-NM) | 6.9, *4.1* | 52, *1.8* |
| | $RLT_8$ (WD-M) | 1.2, *0.7* | 54, *1.6* |
| | $H_2$ (PD-NM) | 7.0, *4.1* | – |
| | $H_7$ (PD-M) | 2.7, *1.6* | 46, *1.7* |
| Bearing transplantable jaw tumours, $JT_1$ (WD-NM) | | 3.3, *1.9* | – |

zed with $CO_2$ and sacrificed by cervical dislocation. The livers were perfused *in situ* with ice-cold 0.15 M KCl and excised. Chloroform/methanol lipid extracts were applied to a DEAE-Sephadex column, prepared in acetate form to ensure binding of acidic lipids [14]. Neutral lipids were eluted, applied to Unisil columns, and crude neutral glycosphingolipids eluted with acetone/methanol (9:1 by vol.). After several washing steps [9], neutral glycosphingolipids were analyzed by TLC (silica gel G, Uniplate; Analtech Inc., Newark, DE). Chromatograms were developed with chloroform/methanol/water (70:22: 3), sprayed with 50% (v/v) sulphuric acid to detect neutral glycosphingolipids, and quantitated with a densitometer (Photovolt Model 530, New York, NY). Homogenates of two livers were combined for each extraction.

*Protein.* – The Lowry procedure ([15]; BSA as standard) was used.

## ANIMAL STUDIES

Rats bearing transplanted tumours show an elevation in LASA and total ganglioside levels in either serum (Table 1) or plasma (not shown). The elevation is not proportional to tumour mass; in fact small tumours give a disproportionately raised LASA level which dissipates within 48 h after surgical removal of the tumour. This elevation in $G_{M3}$ levels likewise arises with autochthonously grow-

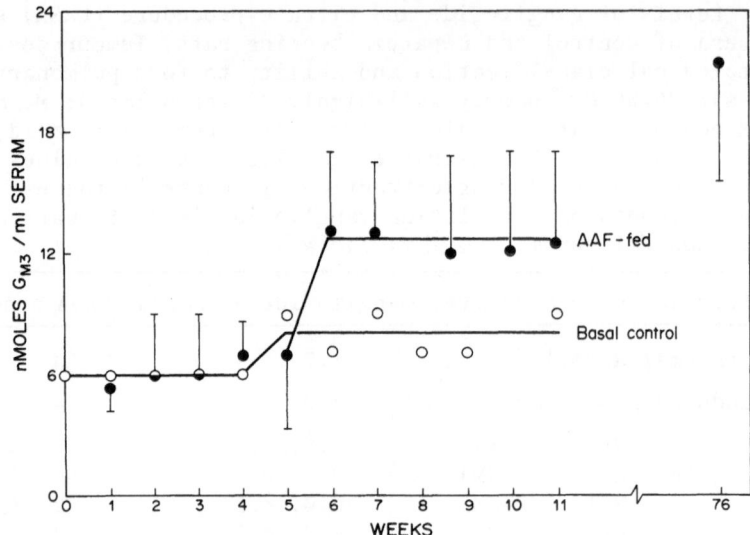

**Fig. 1.** Serum $G_{M3}$ in rats during feeding with hepatocarcinogen (AAF). Levels were elevated by week 6 when liver lesions first appear, and showed a 2- to 3-fold rise by week 76 when hepatocellular carcinomas had developed. Bars denote S.D.; 3 rats (and 2 controls) at each time point up to 76 weeks, then 20 rats.

ing tumours, as demonstrated during a carcinogen-feeding study where cardiac-puncture samples were analyzed for LASA [9] and $G_{M3}$ (Fig. 1). The mean $G_{M3}$ level was evidently elevated between weeks 6 and 11, an increase being typically found in 2 of the 3 rats killed at each time point; small (<1 mm) hepatocellular carcinomas were in evidence at week 6. The levels further increased between weeks 11 and 76 ($P$ 0.05, n = 20), notably if carcinomas were present (from 3 months).

An even earlier change was an increase in neutral glycosphingolipids (NGSL) which were elevated in sera of rats fed carcinogen or bearing tumours but not of animals with regenerating livers (Table 2). The dominant NGCL was ceramide monohexoside (galactosyl- or glucosylceramide) [16]. In livers of animals fed carcinogen, NGSL reached a maximum after 6 weeks and then declined [9]. Again, the increase was largely in material matching ceramide monohexoside (Fig. 2).

## PATIENT STUDIES

While patient data are more limited, both LASA [11] and lactosylceramide [17] have been reported to be elevated in human sera. Though the basis for the rather large elevations reported earlier by Kloppel et al. [11] remains problematic [18], gangliosides undoubtedly contribute to the increases in LASA. TLC reveals the major ganglioside of serum to be $G_{M3}$ (Fig. 3). The sialic acid is present as the *N*-acetyl derivative in both normal and cancer patients (Fig. 4).

**Table 2.**  Neutral glycosphingolipid levels in rat sera, as affected by transplanted hepatomas or 16 weeks of hepatocarcinogen (AAF) feeding [9, 10] that gave both hepatic nodules and small hepatomas. For transplantable hepatoma designations, see heading to Table 1. To get regenerating liver, 1 or 2 liver lobes were excised with Na pentobarbital anaesthesia and with ligation by suture to prevent haemorrhaging; the hyperplastic tissue was taken 1 week later. The NGSL levels are expressed as nmol/mg protein and as *experimental/control*. For the former, the S.D. was ±0.03.    *Based on a Table in ref. [16].*

| Animals from which sera derived | Serum NGSL |
|---|---|
| Control | 0.09 |
| With regenerating liver | 0.07, *0.8* |
| Fed hepatocarcinogen | 0.13, *1.4* |
| Bearing transplantable hepatomas: $RLT_1$ (PD–M) | 0.45, *5.0* |
| $RLT_2$ (PD–NM) | 0.38, *4.2* |
| $RLT_3$ (WD–M) | 0.82, *9.2* |

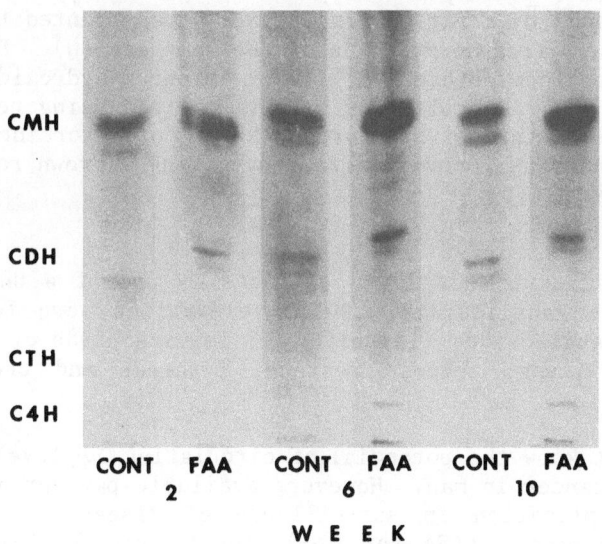

**Fig. 2.**  TLC analysis of neutral glycosphingolipids from livers of rats fed control basal diet or ('FAA') basal diet with 0.025% AAF. Plates were developed and sprayed as described in the text. CMH, ceramide monohexoside; CDH, ceramide dihexoside; CTH, ceramide trihexoside; C4H, ceramide tetrahexoside.

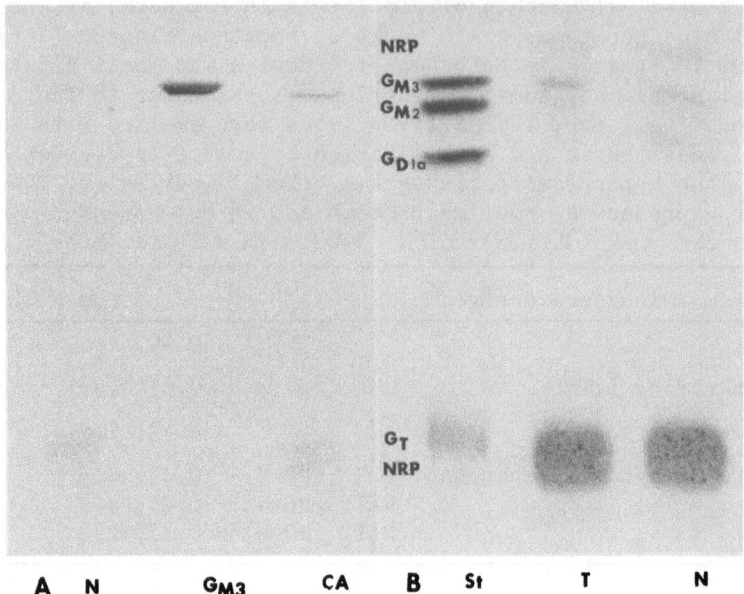

Fig. 3.   Elevation in $G_{M3}$ (sialyl-galactosyl-glucosylceramide)
shown by TLC in sera from (A) a 72-year old patient with stomach
adenocarcinoma (CA) compared to pooled sera from apparently normal
adults (N), and (B) a rat bearing an $RLT_1$ transplanted tumour (T)
compared to sera from normal rats of the same age (N).   Development
was with chloroform/methanol/28% (w/w) ammonium hydroxide/water,
60:37:7:3 by vol.   Gangliosides were visualized using resorcinol
reagent. St = mixture of standards.   NRP = not resorcinol-positive;
needs acid charring for visualization, giving a brown rather than
violet colour.


With purification of lipid extracts by use of a Unisil column
to concentrate gangliosides, 20-100% elevations were found in the
circulating gangliosides (largely $G_{M3}$) in two-thirds of the sera of
patients with lung, colon, prostate, pancreas and breast cancer
(Fig. 5).

We do not know the potential of circulating $G_{M3}$ levels for early
detection of cancer in man.   However, available patient data suggest
a clinical application in surveillance of disease progression in
diagnosed patients.   LASA (or $G_{M3}$) levels swiftly fell following
surgery (DH, Fig. 6) or successful chemotherapy (DV, Fig. 6) and rose
quickly with disease recurrence (EA, JS, OP & PP, Fig. 6).   Interestingly, several patients with active disease tended to show a decli-
ning titre of $G_{M3}$ in the last weeks preceding death (JM, BP & DO,
Fig. 6).   Stable disease (e.g. AB, Fig. 6) resulted in stable LASA
or $G_{M3}$ levels.

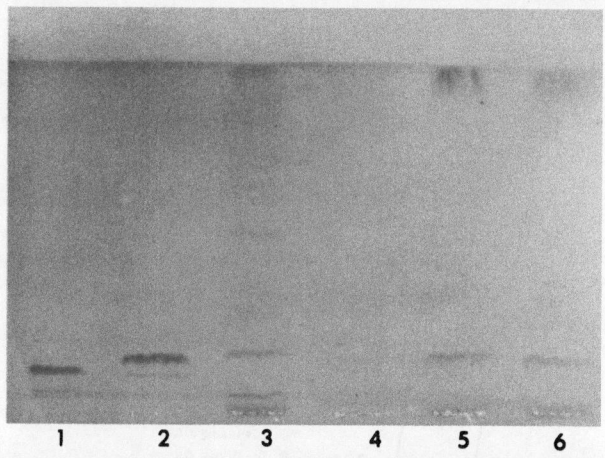

Fig. 4. TLC of sialic acids isolated from normal and cancer-patient sera. The plates were pre-run with 0.1 M HCl and developed in n-butanol/n-propanol/water (1:2:1 by vol.). Sialic acids were visualized by orcinol reagent diluted 3:1 with water [19]. Lanes: 1, *N*-glycolyl- and 2, *N*-acetyl-neuraminic acid (sialic acids as standards); 3 & 4, pooled patient sera; 5 & 6, pooled carcinoma (prostate, ovary, breast & colon) patient sera.

Fig. 5. Tumour sera from the patient population of Arnett Clinic, Lafayette, Indiana. A = normal (n = 40); B, lung carcinoma (n = 20); C, colon carcinoma (n = 10); D, prostate carcinoma (n = 9); E, pancreas carcinoma (n = 16; * = well-differentiated slow-growing); F, breast carcinoma (n = 15). NANA denotes *N*-acetylneuraminic acid.

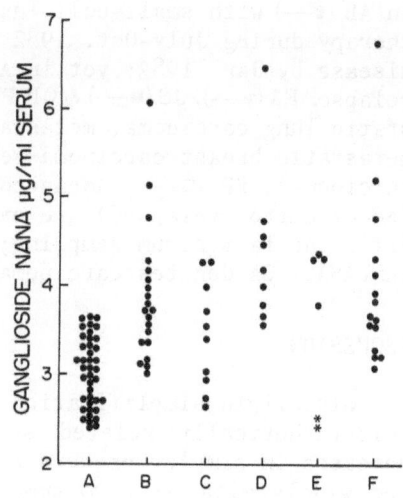

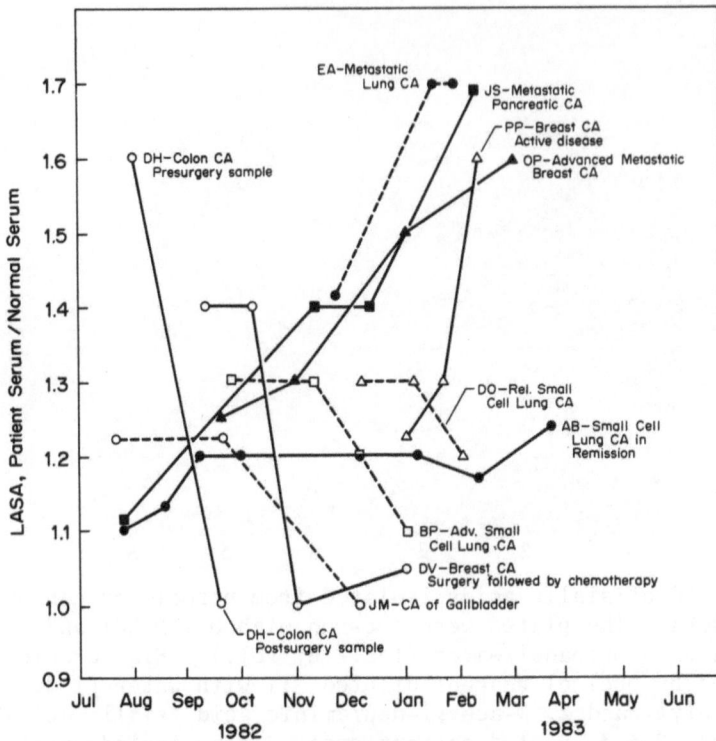

Fig. 6. Serial LASA assays for total ganglioside, compared in 10 carcinoma patients at the Arnett Clinic. Two (o—) show the response to surgery (DH) or surgery followed by successful chemotherapy (DV). In AB (●—) with small-cell lung carcinoma, after control by chemo-therapy during July–Oct. 1982 there was no clinical evidence of disease by Jan. 1983; yet in Aug. 1983 (not shown) there was a CNS relapse. EA(●---), JS(■—) & OP, PP (▲—, △—) exemplify active meta-static lung carcinoma, metastatic pancreatic carcinoma and advanced metastatic breast carcinoma respectively. JM (o---; gall-bladder carcinoma), BP (□---; advanced small-cell lung carcinoma) and DO (△---; ditto, relapsed) are examples where death occurred shortly after the last serum sampling; death was preceded by a marked fall in LASA. CA denotes carcinoma; Rel. denotes relapsed.

## DISCUSSION

Glycolipid simplification (depletion of more complex members of a biosynthetically related series of glycolipids with concomitant increases in simpler precursor gangliosides) is a phenomenon associ-ated widely with *in vivo* tumorigenesis [20–22]. In rats fed AAF to induce hepatomas, precursors also increase in early tumorigenesis (because of activation of the glycosyltransferases concerned) [22]. These increases appear also in normal liver around the neoplastic lesions [23] and in blood [8].

Comparing several animal models, but mainly the canine [24], elevations of serum LASA were described for a variety of naturally occurring solid tumours. No elevations were seen in most non-neoplastic disorders including infectious and reparative diseases [24]. In many carcinoma patients, the increases are primarily in the simple monosialoganglioside $G_{M3}$ (unpublished; cf. Fig. 3). Recently Bremer et al. [25] have shown that the level of $G_{M3}$ in membranes of Swiss 3T3 cells may modulate platelet-derived growth factor (PDGF) receptor function and subsequent growth behaviour by affecting the degree of tyrosine phosphorylation and receptor affinity for PDGF.

Our findings clearly suggest a potential for precursor gangliosides in cancer diagnosis, especially for early detection, as the basis of a simple blood test to monitor disease progression in patients receiving therapy (Fig. 6). Present methodologies could be supplanted by sensitive and rapid immunoassay protocols. Progress, especially for the common membrane gangliosides such as $G_{M3}$, has been hampered by numerous technical difficulties. While some carbohydrate chains of gangliosides appear to be strongly immunogenic, the property does not appear to be shared by all members of even the same series. We and others [26] have attempted to obtain MAb's to $G_{M3}$ gangliosides in mice, but with no success so far despite the systematic application of different methodologies to the problem.

Possibly the carbohydrate chains of $G_{M3}$ are only weakly immunogenic or maybe not immunogenic at all [26]. Alternatively, we find early non-cloned hybridoma cultures that have strong apparent anti-$G_{M3}$ activity only to lose it as the cultures are cloned. Anti-$G_{M3}$-producing clones, in our experience, are always slow-growing and prone to die out as if the anti-$G_{M3}$ produced, by combining with $G_{M3}$ molecules exposed at the surface of the producing cells, was preventing clonal expansion. Next we shall attempt production of polyclonal Ab's to $G_{M3}$ including species such as the chicken where production of anti-haematoside ($G_{M3}$) sera has been reported [27].

Other difficulties associated with quantitative immuno-detection of simple glycolipid antigens relate to their monovalency and solubility in water primarily as micelles. Best results to date have been obtained with an assay procedure where antigen is combined with phospholipid and cholesterol on the wells of plastic titre dishes [28]. Direct precipitation is precluded by the monovalent nature of glycolipid antigens; but quantitation by a standard radioimmunoassay as used, for example, for cholesterol should be feasible.

*Acknowledgement*

The work was supported in part by an award from the Trask Foundation, Purdue University, W. Lafayette, IN.

## References

1. Hakomori, S.-I. (1984) *Ann. Rev. Immunol. 2*, 103-126.
2. Koprowski, H., Steplewski, Z., Mitchell, K., Herly, M., Herly, D. & Fuhrer, P. (1979) *Soma. Cell Genet. 5*, 957-962.
3. Kannagi, R., Nudelman, E, Levery, S.B. & Hakomori, S. (1982) *J. Biol. Chem. 257*, 14865-14874.
4. Hakomori, S., Nudelman, E., Kannagi, R. & Levery, S.B. (1982) *Biochem. Biophys. Res. Comm. 109*, 36-44.
5. Pukel, C.A., Lloyd, K.O., Trabassos, L.R., Dippold, W.G., Oettgen, H.F. & Old, L.J. (1982) *J. Exp. Med. 155*, 1133-1147.
6. Nudelman, E., Hakomori, S.-I., Kannagi, R., Levery, S., Yeh, M.-Y., Hellstrom, K.E. & Hellstrom, I. (1982) *J. Biol. Chem. 257*, 12752-12756.
7. Nudelman, E., Kannagi, R., Hakomori, S.-I., Parsons, M., Lipinski, M., Wiels, J., Fellons, M. & Tursz, T. (1983) *Science 220*, 509-511.
8. Kloppel, T.M. & Morré, D.J. (1980) *J. Nat. Cancer Inst. 64*, 1401-1411.
9. Creek, K.E., Walter, V.P., Evers, D., Yeo, E., Elliott, W.L., Heinstein, P.F., Morré, D.M. & Morré, D.J. (1984) *Biochim. Biophys. Acta 793*, 133-140.
10. Elliott, W.L., Sawick, D.P., DeFrees, S.A., Heinstein, P.F., Cassady, J.M. & Morré, D.J. (1984) *Biochim. Biophys. Acta 800*, 194-201.
11. Kloppel, T.M., Keenan, T.W., Freeman, M.J. & Morré, D.J. (1977) *Proc. Nat. Acad. Sci. 74*, 3011-3013.
12. Warren, L. (1959) *J. Biol. Chem. 234*, 1971-1975.
13. Jourdian, G.W., Dean, L. & Roseman, S. (1971) *J. Biol. Chem. 246*, 430-435.
14. Ledeen, R.W., Yu, R.K. & Eng, L.F. (1973) *J. Neurochem. 21*, 829-839.
15. Lowry, O.H., Rosenbrough, N.J., Farr, A.L. & Randall, R.J. (1951) *J. Biol. Chem. 193*, 265-275.
16. Walter, V.P., Kloppel, T.M., Deimling, I.G. & Morré, D.J. (1980) *Cancer Biochem. Biophys. 4*, 145-151.
17. Kościelak, J., Jóźwiak, W., Pacuszka, T. & Miller-Podraza, H. (1979) in *Glycoconjugates* (Schauer, R., Boer, P., Buddecke, E., et al., eds.), Georg Thieme Verlag, Stuttgart, pp. 619-620.
18. Sallay, S.I., Prosise, W.N., Morré, D.J. & Matyas, G.R. (1985) *Cancer Res. 45*, in press.
19. Ginsburg, V. (1978) *Meths. Enzymol. 50*, 64-84.
20. Hakomori, S.-I. (1973) *Adv. Cancer Res. 18*, 265-315.
21. Brady, R.O. & Fishman, P. (1974) *Biochim. Biophys. Acta 335*, 121-148.
22. Richardson, C.L., Baker. S.R., Morré, D.J., Morré, D.M. & Keenan, T.W. (1975) *Biochim. Biophys. Acta 417*, 175-186.
23. Merritt, W.D., Richardson, C.L., Keenan, T.W. & Morré, D.J. (1978) *J. Nat. Cancer Inst. 60*, 1313-1328; *also* 1329-1337.
24. Kloppel, T.M., Franz, C.L., Morré, D.J. & Richardson, R.C. (1978) *Am. J. Vet. Res. 39*, 1377-1380.
25. Bremer, E.G., Hakomori, S.-I., Bowen-Pope, D.F., Raines, E. & Ross, R. (1984) *J. Biol. Chem. 259*, 6818-6825.
26. Hakomori, S.-I. (1984) in *Monoclonal Antibodies and Functional Cell Lines* (Kenneth, R.H., et al., eds.), Plenum, New York, pp. 67-100.
27. Higashi, H., Ikuta, K., Ueda, S., Kato, S., Hirabayashi, Y., Matsumoto, M. & Naiki, M. (1984) *J. Biochem. 95*, 785-794 (*see also 94*, 327-330).
28. Kannagi, R., Levery, S.B., Ishigami, F., Hakomori, S.-I., Shevinsky, L.H., Knowles, B.B. & Solter, D. (1983) *J. Biol. Chem. 258*, 8934-8942.

#NC(E)

**NOTES** and COMMENTS related to the foregoing topics

Comments related to particular contributions:

#E-1 & #E-2, p. 341
#E-4, p. 342
Additional discussion based on #E-4 (& #E-5), p. 337
#E-3, p. 339

#NC(E)-1

*A Note on*

IMMUNE SURVEILLANCE

N.A. Mitchison

Tumour Immunology Unit, Zoology Department
University College London, London WC1E 6BT, U.K.

The best evidence bearing on immune surveillance comes from studies on the incidence of tumours in human transplant recipients [1]. The most recent U.K./Australasian figures for relative risk (and no. of cases) are as follows: non-Hodgkins lymphoma, 49 (42); skin, 9 (24); melanoma, 9 (2); all others, 1.4 (94). Some and perhaps all of this is indicative of viral oncogenesis. The best animal data concern tumour-specific transplantation antigens (TSTA) in chemically induced [2] and UV-induced [3] cancer.

The major effector mechanisms in host control of tumour growth are T cells and NK cells.

The major groups of tumour antigens are (i) differentiation serum markers such as carcinoembryonic antigen (CEA), $\alpha$-foetoprotein and placental alkaline phosphatase; (ii) differentiation surface markers such as common acute lymphatic leukaemia antigen (CALLA) and transferrin receptor; (iii) viral gene products; and (iv) oncogene products.

The major prospects for immunological intervention relate to (i) diagnosis: identification/classification of tumour cells *in vitro*, and tumour localization *in vivo*; and (ii) therapy: bone-marrow cleansing, graft-*versus*-host disease prevention *in vitro*; immunotoxins and drug-Ab conjugates *in vivo*. Mention should also be made of forthcoming viral vaccines, such as Epstein Barr virus and Hepatitis B virus.

*References*

1. Mitchison, N.A. & Kinlen, L. (1980) in *Immunology '80 [4th Int. Congr. Immunol., Paris, July 1980]* (Fougereau, M. & Dausset, J.), Academic Press, London, pp. 641-650.
2. Boon, T. (1983) *Adv. Cancer Res. 39*, 121-151.
3. Urban, J.L., Holland, J.M., Kripke, M.L. & Schreiber, H. (1982) *J. Exp. Med. 157*, 642-656.

#NC(E)-2

*A Note on*

WORKSHOP DISCUSSION focused on #E-4, #E-5 & #E-3

Senior Editor's adaptation of a summary compiled by
**J.W. BUCKIE & G.M.W. COOK** (Pharmacology Dept., Univ. of Cambridge)

with help from Sect. #E contributors
especially **E.F. HOUNSELL** (MRC Clin. Res. Centre, Harrow)

*Following this Workshop Report there are some 'Comments' that relate to an earlier Discussion period, in the Forum itself. For some aspects there is no overlap because of absence from the present Report, as in the case of R.W. Baldwin's subject-matter (art. #E-2).*

**CARBOHYDRATE TUMOUR ANTIGENS** – discussion points centered on #E-4

The discussion, opened by **P. Alexander** (Workshop Chairman) and initially reminiscent of a published discussion [1], was concerned with the desire to find markers of tumour cells that distinguish them from their normal counterparts. **E.F. Hounsell** reiterated what she had stressed in her talk, that among the monoclonal antibodies (MAb's) raised against tumour cells by many laboratories in the hope of identifying specific cell markers, the majority of those that have been characterized in detail are recognizing carbohydrate (CHO) structures. So far none of the tumour-associated antigens that have been identified are unique to tumour cells, and therefore the Ab's recognizing them are not tumour-specific; they merely detect an aberrant glycosylation. That all these anti-CHO Ab's may be of use in diagnosis, and that they help understand some of the cell-surface changes that occur in neoplasia, also featured in the discussion (cf. concluding portion of #E-4, E.F. Hounsell et al.).

Arising from the characterization of possible anti-tumour Ab's in normal mouse sera (discussed by Hounsell, #E-4), he was especially concerned to assess the potential of anti-CHO Ab's as natural anti-tumour Ab's and with respect to the variety of CHO and [cf. #E-5, Matyas et al.] glycolipid antigens identified on tumour cells. He wondered how they relate to the recognition that in some experimental cancers the cell surface expresses certain macromolecules that can induce a weak rejection reaction, i.e. the tumour-specific transplantation-type antigens (TSTA's). However –

(a) after 25 years we still have no idea of their nature;
(b) after 25 years it is still not known whether they are a trivial or a general feature of tumours;
(c) as a biological property they induce graft rejection in the host, yet by-and-large do not evoke much in the way of Ab's.

**Hounsell** was asked whether (1) she had any idea what the mouse serum antigen is that evokes the response to produce Ab's against these determinants found in a tumour-bearing host, and (2) any experiments have been carried out to determine whether tumours grow better, or worse, in animals that have the natural Ab's? Reply.- Both types of study are being carried out, and current structural work should determine what these Ab's are directed towards.

In response to points (a)-(c) above and to the question whether gastric tumour patients or animals produce auto-Ab's against antigens such as Lewis[a] or FC10.2 (cf. #E-4), Hounsell pointed out that, in relation to the $I$ and $i$ antigens which are normally associated with adult and foetal erythrocytes respectively, there is no increase in circulating Ab's to these antigens in patients with gastric tumours which express them. However, in cold-agglutinin syndrome, a condition giving rise to some cases of haemolytic anaemia, patients form Ab's against $I$ antigens. There is also a transient rise in anti-$I$ titre coincident with *Mycoplasma pneumoniae* infection and a rise in anti-$i$ following Epstein-Barr virus infection. The question arises in all three cases why are these anti-$I/i$ autoimmune responses suddenly elicited.

One recent finding, mentioned by Hounsell with respect to this question, is that *M. pneumoniae* binds to sialylated $I$ antigen structures on the host red cell [2]. It is therefore suggested that the interaction of the mycoplasma with glycoproteins or glycolipids bearing the sialylated $I$ antigen somehow triggers the autoimmune response [3], perhaps by redistribution or clustering of the molecules bearing the determinants within the membrane. In summary (**Alexander**), quantitative factors can confer immunogenicity, as agreed by **Hounsell** and **Kabat** who pointed out the special monoclonal nature of auto-anti-$I$ responses.

Response to a question by Hounsell: the fact that some TSTA's of virally induced tumours are viral products is not in doubt. However, for tumours that are not virally induced, we do not know the nature of the TSTA's. Maybe they are of CHO or glycolipid nature, whereas for the past few years investigators have tried to isolate them on the basis that they are protein in nature, with inevitable failure.

**J. Wyke** then pointed out that from what is known of the mechanisms of oncogenes, variation in cell-surface structures are probably an effect rather than a cause of altered growth regulation, etc., in neoplasia.  In reply, **Hounsell** mentioned that there is a distinct possibility that surface carbohydrates may be involved in growth regulation (ref. [31] in #E-4).

## LYMPHOID MALIGNANCY TREATMENT WITH ANTIBODY DERIVATIVES
- some of the discussion that arose from #E-3

Questions (**P. Alexander**) and **M.J. Glennie's** replies.-
(1) How effective is the use of the univalent anti-idiotypic (anti-Id) Ab in treating neoplastic disease? - Anti-Id Ab specific for neoplastic B cells appears an ideal model system to investigate the therapeutic potential of Ab.  In this system univalent derivatives have proved superior to conventional Ab due to their ability to avoid antigenic modulation, viz. the specific clearance of immune complexes from  cell surfaces making them inaccessible to attack by natural effectors such as complement.
(2) Are neoplastic cells very sensitive to modulation? - They are extremely sensitive under normal circumstances, and it is for this reason that we have constructed univalent derivatives which can attack target cells without perturbing the surface idiotype (surface Ig).
(3) Do all surface antigens modulate? - Most will modulate  when cross-linked by Ab; however, one notable exception to this is the major histocompatibility antigens which seem very resistant to modulation.   Furthermore there is good evidence from several centres that  modulation *in vivo* is far  more  efficient than  *in vitro*, suggesting that other factors besides cross-linking Ab may be involved.

Question by **D.J. Morré**.- Do we really understand modulation in mechanistic terms, and does the Ab cover the whole cell surface or does it bind to distinct regions?  Reply.- It would appear that Ab will initially cross-react with antigen throughout the cell surface, causing cross-linking and build-up of immune complexes. However, on metabolically active cells the complexes are rapidly redistributed to one pole of the cell, giving the typical 'patching' and 'capping' often seen even when cells are labelled with fluorescent Ab.   The capped immune complexes can now be internalized or shed from the cell. This redistribution of immune complexes is known to  require  metabolic  energy  and  is  inhibited  by  sodium azide or low temperatures; it also involves the cytoskeleton of  the cell, in particular  the  actin  and  myosin  filaments.   Removal of the immune complexes obviously leaves the cell inaccessible to attack  by  effectors;  but  we  have  shown  for  complement-mediated cytotoxicity that  resistance  to  lysis  can  occur  while  Ab  still persists on the cell surface and simply requires the Ab to become redistributed.

Further reply by **Glennie**, to a question on what evidence exists about the fate of cells on *in vivo* modulation: since it is known that cells are removed from the blood, and there is no evidence that they are actually killed upon modulation, could it be that they are merely temporarily sequestered? - We have some evidence that cells may be killed following Ab infusion. When polyclonal anti-Id is infused into chronic lymphocytic leukaemia (CLL) patients, then circulating dead cells (as shown by trypan blue uptake) may result, probably following complement activation. However, as pointed out in the question, in most cases neoplastic cells are simply removed from the circulation, and it is not known whether they are destroyed or simply reappear at a later date. We know in the guinea pig leukaemia model that animals which have been actively immunized against the tumour Id or received daily treatment with anti-Id still succumb to the tumour and that emerging cells have completely modulated their surface Ig. From this evidence and from human studies it seems clear that modulation is a major limiting factor in tumour therapy with Ab. Whether univalent Ab will overcome this problem remains to be seen.

*References*

1. Evered, D. & Whelan, J., eds. (1983) *Foetal Antigens and Cancer [Ciba Foundation Symp.]*, Pitman, London: see pp. 243-247.
2. Loomes, L.M., ......... & Feizi, T. (1984) *Nature 307*, 560-563.
3. Loomes, L.M., et al. (1985) *Infection & Immunity 47*, 15-20.

## Comments on material in #E

*besides the foregoing Workshop Discussion comments*

*Comments on* #E-1, P.A. Alexander - GROWTH FACTORS

**Comments by J. Wyke.-** Concerning autocrine stimulation as a
mechanism of neoplasia, the *sis* oncogene product ($\equiv$ PDGF) is of
cytoplasmic location, not secreted, so may not precisely mimic PDGF
action; moreover, in HTLV-induced T-cell tumours, autocrine stimulation by tumour-secreted IL-2 with receptor on tumour cell cannot
explain all cases, since some lack one component of this stimulation.
**Comment by S.M. Hobbs.-** For the autocrine hypothesis to explain dormant metastasis through an insufficient local concentration of
growth factor, it is implicit that the growth factor and the
receptor must be intracellularly synthesized separately and transported to the membrane in different vesicles; otherwise the receptor
would arrive at the membrane already carrying the bound growth
factor.

*A 'landmark' ref. cited by Senior Editor (cf.* [6] *in* #E-1*).-*
Ullrich, A., Coussens, L., Hayflick, J.S., Dull, T.J., Gray, A., Tam,
A.W., Lee, J., Yarden, Y., Libermann, T.A., Schlesinger, J., Downward,
J., Mayes, E.L.V., Whittle, N., Waterfield, M.D. & Seeburg, P.H.
(1984) *Nature 309*, 418-425.- 'Human epidermal growth factor receptor
cDNA sequence and aberrant expression of the amplified gene in A431
epidermoid carcinoma cells'. A methodological implication is noted.-
"The structural studies described here provide a basis from which to
probe other cell lines and primary tumour tissue for alterations in
the EGF receptor, its gene, or its transcription pattern."

*Comments on* #E-2, R.W. Baldwin - MAb TUMOUR DETECTION AND THERAPY

**R.W. Baldwin, in response to questions.-** There is insufficient
information to say whether the therapeutic index of the drugs is
affected by Ab conjugation. We have not tried the stratagem of
using MAb derived from a surgically removed primary to either treat
or image secondary tumours in the same patient. **G.M.W. Cook asked:**
concerning the problem of getting sufficient drug onto Ab, do you
see any advantage of immunotoxins (i.e. A chain-Ab complex) over
drug-Ab complexes? **Reply.-** We have tended to keep away from immunotoxins because other laboratories are investigating them. One would
need not only A chain but some B chain to get the A chain internalized. Our preference is for drug-Ab complexes.

*Ref. noted by Senior Editor.*- Krolick, K., Uhr, J. & Vitetta, L. (1982) *Nature 295*,  604-605.- In the context of destroying cancer cells by drugs or irradiation, where the  vulnerability  of  normal cells calls for excision and ultimate return of marrow,  the  authors sought to purge the marrow of tumour cells before returning it. They had encouraging results in leukaemic mice through treating the marrow with the toxic moiety of the ricin molecule coupled to an Ab direc-ted against the leukaemic cells. Also of interest: Manabe, Y., et al. (1985) *Bioch. Pharmacol. 34*, 289-291.- Cytotoxicity of mitomycin-anti-HLA MAb. *In Vol. 13, this series:* several 'Notes' in Sect. #C deal with approaches that entail conjugating cytotoxic agents to Ab's, notably through use of liposomes for the targeting.

*Comments on #E-4*, E.F. Hounsell et al. - CARBOHYDRATE ANTIGENS

**D.J. Morré** asked: have you treated the membrane with trypsin or other proteases to determine whether there is an effect on antigeni-city?  **Reply.**- The ganglioside extracts are trypsin-resistant, but the experiment has not been done with membranes.  **A. Fotedar**: it is an interesting possibility that carbohydrate antigens might induce suppression in animals or cancer patients as an immune-surveillance escape mechanism.

*Editorial thoughts put to* E.F. Hounsell *re #E-4, & author's response*

**D.J. Morré** queried the basis on which domain assay was made, in relation to Fig. 1 where sialic acid is regarded as 'peripheral' although others would regard it as 'terminal' to the backbone; also (cf. Legend) might not one have  Gal, besides, Glc, in the core of glycolipids?  **Response.**- The terms 'core', 'backbone' and 'periphe-ral' are operational terms which have now been accepted in common usage by several authors and which have been extensively discussed in our cited publications. By definition, therefore, sialic acid is peripheral, not terminal, and the core region of glycolipids is Glc alone.

**G.M.W. Cook** commented that the term 'glycocalyx' seems inapp-ropriate. The term ('sweet husk') was coined by H.S. Bennett at a meeting of the Mexican Anatomical Society and discussed further by him in a subsequent paper [(1963) *J. Histochem. Cytochem. 11*, 14-23]. Bennett felt it desirable to assign a single, general inclusive term to describe what he called the extracellular, sugary, coating of cells; from his article it is clear that he regarded the polysaccha-ride-rich coating as being outside the p.m.  While contemporary e.m. studies were in accord with cells being  surrounded by a carbohyd-rate-rich layer, the carbohydrate moieties detected are now known to be attached to integral membrane components (e.g. glycoproteins) and therefore if the term 'glycocalyx' is used it should not be under-stood to be describing molecules which are quite separate from the p.m.  **E.F. Hounsell, responding by letter.**- We feel it *is* appropriate

to refer to the carbohydrate coat of the cell surface as a 'glycocalyx', for this carbohydrate coat visualized by e.m. is the major
carrier of the carbohydrate antigens in question.  Our present knowledge on the 3-D structures of the oligosaccharides of glycoproteins
gleaned from NMR studies also suggests  that such a coat is formed
by carbohydrate chains attached to integral membrane components.  We
therefore disagree with this comment.

*Supplementary contribution (a Forum Abstract):-*

MOLECULAR ANALYSIS OF THE PATHOGENICITY OF MYC GENE MUTANT

- PAULA J. ENRIETTO, Viral Leukemogenesis Lab., ICRF, London WC2A 3PX

     The oncogene myc  is found in an avian retrovirus, MC29, which
induces myelocytomas and endotheliomas in chickens.  The cellular
homologue of the myc gene has also been implicated in avian leukosis
virus-induced B-cell lymphomas,  where it is activated by insertion
of a viral promoter nearby; in mouse plasmacytomas and Burkitt's
lymphomas, where it is 'activated' by translocation;  and in the
human promyelocytic cell line HL60 where it has been shown to be
amplified.

     In an effort to study the mechanism by which this gene acts,
several deletion mutants of the myc gene in MC29 have been generated
and tested *in vivo* and *in vitro*.   These mutants, which can trans-
form  fibroblasts *in vitro*, are no longer tumorigenic.  One of these
mutants recovered myc sequences from the cellular homologue on passage *in vitro*.   When the pathogenicity of this virus was examined,
it was found to induce lymphomas - both B- and T-cell.  The molecular basis of this change in pathogenicity has been explored.

*Some refs. contributed by Senior Editor:-*

Kannagi, R., ........ & Hakomori, S.-J. (1983) *Cancer Res. 43*, 4997-5005.
- Glycolipid tumour antigens: influence of ceramide composition and
co-existing glycolipids on the antigenicity of gangliotriaosyl ceramide in murine lymphoma cells.

Banga, J.P.,...... & Roitt, I.M. (1984) *Scand. J. Immunol. 19*, 11-21.- A
leucocytes common epitope associated with a group of high mol. wt.
glycopeptides.

Downward, J., Yarden, Y., ...... & Waterfield, M.W. (1985) *Nature 307*,
521-527.- 'Close similarity of epidermal growth factor and v-*erb-B*-
oncogene protein sequences.'  Implications of this study, based on
protein sequencing, are outlined by T. Prentice in *The Times* (9 Feb.
1984).- "It suggests that a defective 'lock' is part of the mechanism
whereby an animal virus can produce leukaemia in chickens" (the 'lock'
being a cell-surface receptor onto which a 'key' such as a growth
factor can fit;  the oncogene contains the information to make the
defective 'lock').

# GUIDANCE FOR 'UNINITIATED' READERS

*The following compilation, by the Senior Editor, does not purport to be comprehensive. It merely aims to help recall some background points and guide 'further reading', for readers who are not immersed in the field and, in perusing this book, may welcome an aide-memoire.*

Quite wide coverage of principles and approaches in the molecular biology field is to be found in a techniques book [1] that can be afforded as a paperback desk companion. It explains, for example, plasmid mystique and a conventional nomenclature: 'pAZ123' signifies plasmid no. 123 developed by the investigator or group 'AZ'. There is good attention to sequencing, and to electrophoretic techniques including blotting although 'Immunoblotting' is not indexed - nor is 'Hybridoma' to be found, even in the valuable Glossary. 'Blind spots' such as the latter call for recourse to an account - suitably an elementary book [e.g. 2, 3] - or sketch such as a pair of diagnostically slanted reviews [4], the first of which has the title *Hybridoma - An Immunochemical Laser*. For general back-up the book *Antibody as a Tool* [5] may be helpful, or an immunology textbook notwithstanding shortcomings such as, in [6], non-mention of 'hybridoma' and 'monoclonal'. The routine use of MAb's is exemplified by articles on diagnosis in bacteriology and virology [7] and on freeze-drying in immunocytochemistry [8].

Complementing this bibliographic guidance, an outline of some basic points may be welcomed by some readers. Especially pertinent is the structure and chain nomenclature of the immunoglobulin (Ig) molecule, as depicted in Fig. 1 (and well presented in some textbooks, e.g. [6, 9]). Whereas the intact IgG molecule contains two binding sites, each Fab contains only one, i.e. is univalent and does not form a precipitate with antigen. (Precipitation is well known to be maximal when Ab and antigen are present in equivalent amounts.) Ig classes differ in the heavy chain: it is $\gamma$ in IgG's ($\gamma$1, $\gamma$2a, etc.) but $\mu$ in IgM, $\alpha$ in IgA, and $\delta$ and $\varepsilon$ respectively in the trace components IgD (surface receptor on immature B lymphocytes) and IgE. The 'secretory' form of IgA is a dimer linked to a $\beta$-globulin 'transport (T) piece' through the Fc portion. IgM is a pentamer; like IgD, it has a cell-surface role and contains >10% carbohydrate.

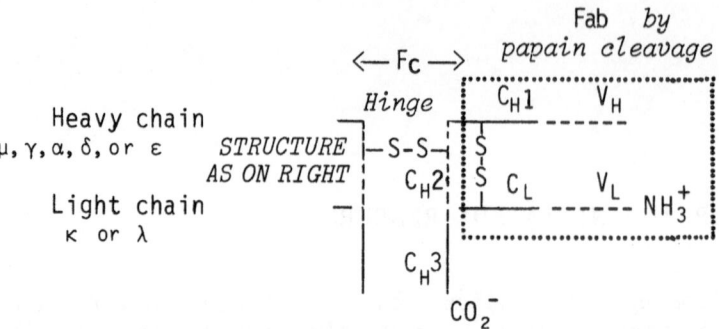

**Fig. 1.** Features of an Ig (schematically shown as T-shaped, but actually Y-shaped). Proteolytic cleavage gives two Fab's (each having an antigen-binding site) and one Fc (having a complement-binding site). IgG is a monomer, $L_2H_2$ ($H = \gamma$), from which chains are obtainable by reductive cleavage. Fd is the portion of the H chain contained in Fab. Bence-Jones protein is a dimer of the L chains. The constant regions (C; Greek-letter types) - one on the L chain and three 'domains' on the H chain - are homologous. *Editorial policy has been flexible in respect of subscripts, e.g.* $F_C$ *vs.* Fc. L *denotes* 'light' - *not (cf. p. 15)* 'low copy no.'!

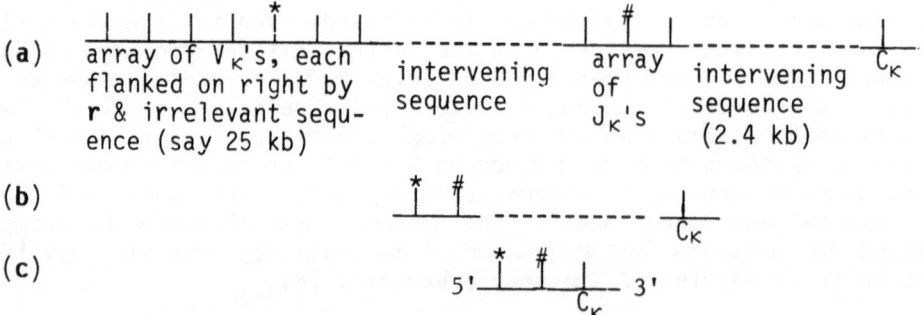

**Fig. 2.** Formation of mRNA for a light κ chain, exemplifying progression from (**a**) germ-line DNA in lymphocyte development. (**b**) The selected $V_K$ coding sequence (exon; *) is excised, with deletion of its recombination site **r** and the adjoining non-coding sequence (intron), and linked to the selected J (joining segment, #; codes for V residues 96-108) which, although excised, remains linked to the C gene via the intervening sequence. The latter remains when a pre-mRNA as in (**b**) is formed by transcription, but excision and splicing occur when the mRNA (**c**) is formed. *Not shown:* S = 'signal segment'. The progression refers only to mouse Lλ chain; one variation is that in heavy chains there is a D segment between J and C, and provision for Ig 'class switching'. (*Senior Editor's interpretation of the informative outlines in [10].*)
An abbreviation often used without definition: **kb**, **Kb** = kilobase.

The pathways by which genes mediate Ig synthesis entail complex steps, notably translocations and excisions of DNA stretches. This is exemplified in Fig. 2 for a particular type of light chain, the tailoring of which is completed at the mRNA stage.   Fig. 2 and its legend serve mainly to give guidance on nomenclature.   The nature of the different pathways for the various Ig chain types may be gleaned from terse reviews [10] of Ig gene rearrangement and of aberrations affecting light chains.   There is classical Mendelian inheritance of an allotypic difference in residue 191 of both types of L chain (constant region): this residue may be either lysine or arginine in the $\lambda$ type, and leucine or valine in the $\kappa$ type. In either type the variable region represents about half of the 214 residues in the chain.   Further consideration of sequences and genes is beyond the scope of this sketch.

*Cell-surface Immunology* warrants some guidance, given effectively in a survey by M.C. Raff in 1976 with this title [11]. It is worth consultation if only to gain perspective on advances during the subsequent decade.   Its historical starting-point is 1959 when Macfarlane Burnet attributed acquired immunity to clonal selection.   "The immune system functions as a warehouse with a vast store of ready-made antibodies and sensitized T cells" [11]. The antigen thy-1 (formerly termed theta) proved to be a valuable surface marker for T lymphocytes in mice (Mitchison & Raff), and anti-thy 1 could kill T cells in the presence of complement, the first of whose components binds to antigen-Ab complexes;   the final components insert into the membrane, causing leakiness and cell death (dye exclusion test applied). A second approach to detecting surface antigens is the immunofluorescence (IF) method: Ab-dye conjugate is taken up, as revealed by fluorescence, by cells that bear antigen.   Thirdly, Ab may be coupled to ferritin, manifest as black specks on antigen-bearing cells by e.m. examination.

In Ab-producing (B) lymphocytes, which originate from bone-marrow stem cells and are found in peripheral lymphoid tissues, the surface receptors for antigens are Ig molecules.  For T-lymphocytes, which are activated by antigen to produce cell-mediated immune responses, advances in understanding the surface receptors are evident from the present book.   (The 'Lymphocytes' index entry in Vol. 11, this series, can help in various respects, e.g. cell isolation and cancer relevance.)   The Ig on B cells is evenly distributed over the surface, as revealed by fluorescent labelling with an Fab; but divalent Ab induces antigen cross-linking and clustering into patches which, in metabolically active cells, migrate to one pole to form a 'cap' (with ensuing pinocytosis).   The surface relationship between two antigens is ascertainable by a 'co-capping' approach: exposure of the cell, already capped with dye-conjugated Ab, to the second Ab conjugated to a different dye gives a coinciding cap (illustrated in [11]) only if the two antigens have the same molecular location.   Thereby Raff and collaborators showed that "all the Ig receptors on an individual B cell have the same antigen specificity, a central prediction of the clonal-selection hypothesis" [11].   The capping phenomenon testifies to the 'fluidity' of membranes.      Fibrous cellular components such as microfilaments and the larger microtubules "are thought to play a role in controlling the distribution and movement of some membrane proteins" [11].

In respect of immunological concepts and terminology such as 'idiotype', the reader may find standard books [e.g. 6] frustrating and have to delve.* An idiotype is an epitype characteristic of an individual animal, antigen-specific.   As is explained in the present book (e.g. Preface and #A-1), an anti-idiotypic Ab can attach to the receptor site, which can be investigated with its help.

### References

1.  Walker, J.M. & Gaastra, W., Eds. (1983) *Techniques in Molecular Biology*, Croom Helm, London, 333 pp.
2.  Sikora, K. & Smedley, H.M. (1984) *Monoclonal Antibodies*, Blackwell, Oxford, 146 pp.
3.  Pritchard, R.H. & Holland, I.B., eds. (1984) *Basic Cloning Techniques*,  Blackwell, Oxford, 300 pp. approx.
4.  (a) David, G.S., et al., & [same lab.] (b) Sevier, E.D., et al. (1981)  *Clin. Chem. 27*, (a) 1580-1585, (b) 1797-1806.
5.  Marchalonis, J.J. & Warr, G.W., eds. (1982) *Antibody as a Tool*, Wiley, Chichester, 578 pp.
6.  Cooper, E.L. (1982) *General Immunology*, Pergamon, Oxford, 343 pp.
7.  Fleck, D.G. (1984) *Lab. Pract. 33 (12)*, 25-27.
8.  Gatter, K.C., et al. (1984) *Lab. Pract. 33 (11)*, 9-10.
9.  Stryer, L. (1981) *Biochemistry*, 2nd edn., Freeman, San Francisco, 949 pp.
10. (a) Gough, N., & (b) Kuehl, W.M. (1981) *Trends Biochem. Sci. 6*, (a) 203-205, (b) 206-208.
11. Raff, M.C. (1976) *Scientific American 234 (5)*, 30-39.

**Another techniques book (cf. 1.), favourably reviewed:** Glover, D.M., *Gene Cloning: the Mechanics of DNA Manipulation*, Chapman & Hall (AVAILABLE IN PAPERBACK).

---

* Ref. [98] in #A-1 may be helpful.

# INDEX

*This Index deals especially with phenomena and investigative approaches, indexed similarly to previous 'Biochemistry' volumes so as to facilitate back-consultation. For applications of a particular technique such as immunochromatography, and for observations concerning, for example, sequences, comprehensive listing is clearly precluded, and indexing has been selective as indicated by 'e.g.'.*

*Page entries such as 25- signify that the ensuing pages are also relevant, i.e. the '-' denotes a major entry.*

**Corrections to Vol. 13,** *Investigation of Membrane-located Receptors*

p. vi (Preface), line 8:   p. 540 **should read**  p. 546

p. 308, line 6:   [23, 23] **should read**  [22, 23]

p. 425:  **insert** missing Art. No.,  #NC(D)-3

(**Vol. 14:**  no significant errors noticed)